AUTOMOTIVE

MANUAL TRANSMISSIONS AND POWER TRAINS

OTHER BOOKS AND INSTRUCTIONAL MATERIALS
BY WILLIAM H. CROUSE AND *DONALD L. ANGLIN

The Auto Book*
Auto Shop Workbook*
Auto Study Guide*
Auto Test Book*
Automotive Air Conditioning*
Workbook for Automotive Air Conditioning*
Automotive Automatic Transmissions*
Workbook for Automotive Automatic Transmissions*
Automotive Body Repair and Refinishing*
Workbook for Automotive Body Repair and Refinishing*
Automotive Brakes, Suspension, and Steering*
Workbook for Automotive Brakes, Suspension, and Steering*
Automotive Dictionary*
Automotive Electronics and Electrical Equipment
Workbook for Automotive Electronics and Electrical Equipment*
Automotive Emission Control*
Workbook for Automotive Emission Control*
Automotive Engine Design
Automotive Engines*

Automotive Engines Sound Filmstrip Program
Workbook for Automotive Engines*
Automotive Fuel, Lubricating, and Cooling Systems*
Workbook for Automotive Fuel, Lubricating, and Cooling Systems*
Automotive Mechanics
Study Guide for Automotive Mechanics*
Testbook for Automotive Mechanics*
Workbook for Automotive Mechanics*
Automotive Service Business
Automotive Technician's Handbook*
Automotive Tools, Fasteners, and Measurements*
Automotive Tuneup*
Workbook for Automotive Tuneup*
General Power Mechanics* (With Robert Worthington and Morton Margules)
Motor Vehicle Inspection*
Workbook for Motor Vehicle Inspection*
Motorcycle Mechanics*
Workbook for Motorcycle Mechanics*
Small Engine Mechanics*
Workbook for Small Engine Mechanics*

AUTOMOTIVE ROOM CHART SERIES

Automotive Brake Charts
Automotive Electrical Equipment Charts
Automotive Emission Controls Charts
Automotive Engines Charts
Automotive Engine Cooling Systems, Heating, and Air Conditioning Charts

Automotive Fuel Systems Charts
Automotive Suspension, Steering, and Tires Charts
Automotive Transmissions and Power Trains Charts

AUTOMOTIVE TRANSPARENCIES
BY WILLIAM H. CROUSE AND JAY D. HELSEL

Automotive Air Conditioning
Automotive Brakes
Automotive Electrical Systems
Automotive Emission Control
Automotive Engine Systems

Automotive Steering Systems
Automotive Suspension Systems
Automotive Transmissions and Power Trains
Engines and Fuel Systems

SIXTH EDITION

AUTOMOTIVE
MANUAL TRANSMISSIONS AND POWER TRAINS

WILLIAM H. CROUSE
DONALD L. ANGLIN

GREGG DIVISION/MCGRAW-HILL BOOK COMPANY

New York ○ Atlanta ○ Dallas ○ St. Louis ○ San Francisco ○ Auckland ○ Bogotá ○ Guatemala
Hamburg ○ Lisbon ○ London ○ Madrid ○ Mexico ○ Montreal ○ New Delhi
Panama ○ Paris ○ San Juan ○ São Paulo ○ Singapore ○ Sydney ○ Tokyo ○ Toronto

ABOUT THE AUTHORS

William H. Crouse

Behind William H. Crouse's clear technical writing is a background of sound mechanical engineering training as well as a variety of practical industrial experience. After finishing high school, he spent a year working in a tinplate mill. Summers, while still in school, he worked in General Motors plants, and for three years he worked in the Delco-Remy division shops. Later he became director of field education in the Delco-Remy Division of General Motors Corporation, which gave him an opportunity to develop and use his writing talent in the preparation of service bulletins and educational literature.

During the war years, he wrote a number of technical manuals for the Armed Forces. After the war, he became editor of technical education books for the McGraw-Hill Book Company. He has contributed numerous articles to automotive and engineering magazines and has written many outstanding books. He was the first editor-in-chief of McGraw-Hill's *Encyclopedia of Science and Technology.*

William H. Crouse's outstanding work in the automotive field has earned for him membership in the Society of Automotive Engineers and in the American Society of Engineering Education.

Donald L. Anglin

Trained in the automotive and diesel service field, Donald L. Anglin has served both as a mechanic and as a service manager. He has taught automotive courses and has also worked as curriculum supervisor and school administrator. He is interested in all types of vehicles and vehicle performance, and he has served as a racing-car mechanic and as a consultant to truck fleets on maintenance problems.

Currently he devotes full time to technical writing, teaching, and visiting automotive instructors and service shops. Together with William H. Crouse he has co-authored magazine articles on automotive education and a number of automotive books published by McGraw-Hill.

Donald L. Anglin is a Certified General Automotive Mechanic, a Certified General Truck Mechanic, and he holds many other licenses and certificates in automotive education, service, and related areas. His work in the automotive service field has earned for him membership in the American Society of Mechanical Engineers and the Society of Automotive Engineers. In addition, he is a member of the Board of Trustees of the National Automotive History Collection.

Library of Congress Cataloging in Publication Data

Crouse, William Harry, (date)
  Automotive manual transmissions and power trains.

  (McGraw-Hill automotive technology series)
  "Expanded and updated version of half of
Automotive transmissions and power trains, fifth
edition"—Pref.
  Includes index.
  1. Automobiles—Transmission devices.
I. Anglin, Donald L. II. Crouse, William Harry, (date). Automotive transmissions and power trains. 5th ed. III. Title. IV. Series
TL262.C75   1982   629.2'44   81-17206
ISBN 0-07-014776-0   AACR2

AUTOMOTIVE MANUAL TRANSMISSIONS AND POWER TRAINS, Sixth Edition

          9 0   SMSM   8 9 8

Sponsoring Editor: D. Eugene Gilmore
Editing Supervisor: Evelyn Belov
Design and Art Supervisor: Caryl Valerie Spinka
Production Supervisor: Kathleen Morrissey

Text Designer: Linda Conway
Cover Designer: David Thurston
Technical Studio: J&R Services
Cover Illustration: Fine Line Inc.

ISBN 0-07-014776-0

CONTENTS

Preface   **vii**

Acknowledgments   **ix**

Chapter   **1.** Power-Train Components   1
              2. Clutch   15
              3. Clutch Service   32
              4. Manual Transmissions, Transaxles, and Transfer
                     Cases   48
              5. Manual-Transmission Trouble Diagnosis   82
              6. Manual-Transmission Removal and Installation   88
              7. Three-Speed Manual-Transmission Service   93
              8. Four-Speed Manual-Transmission Service   102
              9. Four-Speed Overdrive Manual-Transmission
                     Service   108
             10. Five-Speed Overdrive Manual-Transmission
                     Service   116
             11. Manual-Transaxle Trouble Diagnosis   126
             12. Manual-Transaxle Removal and Installation   131
             13 Manual-Transaxle Service   139
             14. Dual-Range Manual-Transaxle Service   152
             15. Transfer-Case Trouble Diagnosis, Removal, and
                     Installation   173
             16. Transfer-Case Service   178
             17. Drive-Line Construction and Operation   189
             18. Drive-Line and Universal-Joint Service   206
             19. Differentials and Drive Axles   223
             20. Differential and Drive-Axle Service   238
Glossary   **271**
Index   **285**
Answers to Review Questions   **291**

PREFACE

Automotive Manual Transmissions and Power Trains, Sixth Edition, is the new title of an expanded and updated version of half of *Automotive Transmissions and Power Trains,* Fifth Edition. When the time came to revise the old and publish a new edition of *Automotive Transmissions and Power Trains,* the authors were faced with the task of evaluating recent developments in the field. Automotive manufacturers had brought out many new manual and automatic transmissions. There were now numerous four- and five-speed manual transmissions with overdrive. Transaxles, both standard and dual-range, had become more common. A variety of new automatic transmissions had appeared on the scene, some with overdrive and automatic torque converter lockout during the cruising range. Transfer cases had become important.

All these new developments demanded space in the new edition of *Automotive Transmissions and Power Trains.* However, if all these new developments had been included in a single book, the result would have been a large, expensive, and unwieldy volume. It was decided, therefore, to divide the book into two volumes: *Automotive Manual Transmissions and Power Trains* and *Automotive Automatic Transmissions.*

This book is one of nine books in the McGraw-Hill Automotive Technology Series that cover in detail the construction, operation, and maintenance of automotive vehicles. They are designed to give you the information needed to become successful in the automotive service business. The books satisfy the recommendations of the Motor Vehicle Manufacturers Association-American Vocational Association Industry Planning Council.

The books also meet the requirements for automotive mechanics certification and state vocational educational programs, as well as the recommendations for automotive trade apprenticeship training. Furthermore, the comprehensive coverage of the subject matter in these books makes them valuable additions to the library of anyone interested in any aspect of automotive engineering, manufacturing, sales, service, and operation.

Meeting the Standards

The nine books in the McGraw-Hill Automotive Technology Series meet the standards set by the Motor Vehicle Manufacturers Association (MVMA) for an associate degree in automotive servicing and in automotive service management. The books also cover the subjects recommended by the American National Standards Institute in their detailed standard D18.1-1980, "American National Standard for Training of Automotive Mechanics for Passenger Cars, Recreational Vehicles, and Light Trucks."

Finally, the books provide in-depth coverage of the subject matter tested by the National Institute for Automotive Service Excellence (NIASE). The tests given by NIASE are used for certifying general automotive mechanics and technicians working in areas of specialization under the NIASE voluntary mechanic testing and certification program.

Getting Practical Experience

While you study, you should also get practical experience in the shop. You should handle automotive parts, tools, and servicing equipment; you should also perform actual servicing jobs. Each book in the Automotive Technology Series has a corresponding workbook which should be helpful to you in your shop work. For example, the *Workbook for Automotive Manual Transmissions and Power Trains* includes jobs which cover the procedures for servicing transmissions and power trains. If you do every job in the workbook, you will have hands-on experience with all types of automotive-transmission and power-train work.

While you are studying *Automotive Manual Transmissions and Power Trains,* and demonstrating your ability in the shop, you and your instructor should be constantly aware of the need for congruence between your school curriculum and your future needs as you face and cope with real life in the automotive field. Everything you learn about the automobile should be competency-based. Find out what the minimum standard of competence is in your chosen field. Then, with the aid of your instructor, develop your skills by mastering the necessary competencies and performance indicators outlined by the text.

If you are taking an automotive mechanics course in school, your instructor will guide you in your classroom and shop activities. If you are not taking a course, the workbook can act as your instructor and tell you, step by step, how to do the various servicing jobs. Perhaps you can meet people who are taking a school course in automotive mechanics. It would be helpful if you could talk over problems with them. Try to get to know the automotive mechanics of a local garage or new-car dealership. You will find them a great source of practical information. Watch them at work if you can. Make notes of important points for filing in a notebook.

Service Publications

While you are in the service shop, study the various publications received there. Automobile manufacturers and suppliers of parts, accessories, and tools publish shop manuals, service bulletins, and parts catalogs. All these publications help service personnel do a better job. In addition, there are numerous automotive magazines which deal with problems and methods of automotive service. All these publications will be of value to you.

These related activities—taking an automotive mechanics course, using the workbook as your instructor, observing and participating in shop work, reading service publications—will help you get practical experience in automotive mechanics. In time, this experience, plus the knowledge that you have gained from studying the books in the McGraw-Hill Automotive Technology Series, will permit you to step into the automotive shop on a full-time basis. Or, if you are already in the shop, you will be equipped to step up to a more responsible job.

NIASE-Type Multiple-Choice Tests

Anyone in automotive servicing should work toward certification. Certified mechanics have an edge over mechanics who are not certified, having been tested by the NIASE (National Institute for Automotive Service Excellence) and found to possess the necessary knowledge and skills to be qualified mechanics.

To help you pass the NIASE tests, NIASE-type tests are placed at the end of each chapter. These tests are a series of incomplete statements, each followed by a choice of words or phrases labeled *a, b, c,* and *d.* You are to pick the word or phrase that correctly completes each statement. The answers to the tests are at the end of the book.

These multiple-choice tests do two things. First, they test your knowledge of the chapter you have completed. (If you miss any questions, review the chapter.) Second, they prepare you for the NIASE tests. After passing the tests and gaining practical experience, you can become a NIASE Certified General Automotive Mechanic or a certified specialist.

William H. Crouse and Donald L. Anglin

ACKNOWLEDGMENTS

During the preparation of this new edition of *Automotive Manual Transmissions and Power Trains,* the authors were given invaluable aid and inspiration by many people in the automotive industry and in the field of education. The authors gratefully acknowledge their indebtedness and offer their sincere thanks to these people. All cooperated with the aim of providing accurate and complete information that would be useful in training automotive mechanics.

Special thanks are owed to the following organizations for information and illustrations that they supplied: American Motors Corporation; ATW; Axle Division, Eaton Corporation; Buick Motor Division of General Motors Corporation; Cadillac Motor Car Division of General Motors Corporation; Chevrolet Motor Division of General Motors Corporation; Chrysler Corporation; Ford Motor Company; General Motors Corporation; Nissan Motor Company, Ltd.; Oldsmobile Division of General Motors Corporation; Pontiac Motor Division of General Motors Corporation; Service Parts Division of Dana Corporation; Society of Automotive Engineers; Sun Electric Corporation; Toyo Kogyo, Ltd.; Toyota Motor Sales Co., Ltd.; and Volkswagen of America, Inc. To all these organizations and the people who represent them, sincere thanks.

We would also like to acknowledge our debt to Paul H. Smith, Vale Technical Institute, for his help over the many years he has been a user and reviewer of this text. His helpful suggestions and technical expertise have been of tremendous value in the preparation of this edition.

William H. Crouse
Donald L. Anglin

AUTOMOTIVE
MANUAL TRANSMISSIONS AND POWER TRAINS

POWER-TRAIN COMPONENTS

After studying this chapter, you should be able to:

1. List the components in the power train and explain the basic operation of each.
2. Explain why a clutch is needed.
3. Define the term *gear ratio* and explain the relationship between gear ratio and torque.
4. Describe the types of gears used in the automotive power train.
5. Discuss the purpose of the differential.
6. Define *transaxle.*

 1-1 Components of the automobile The automobile has five basic components, or parts. These are:

1. The power plant, or engine, which is the source of power.
2. The chassis, which supports the engine and body and includes the brake, steering, and suspension systems.
3. The power train, or *drive train,* which carries the power from the engine to the drive wheels. It consists of the clutch (on vehicles with manual transmissions), transmission, driveshaft, differential, and wheel axles.
4. The car body.
5. The car-body accessories, which include the heater and air conditioner, lights, radio and tape player, windshield wiper and washer, and electric windows and seat adjusters.

Many of these components can be seen in the phantom view of the automobile shown in Fig. 1-1. The engine (Fig. 1-2) produces power by burning a mixture of fuel and air in the engine combustion chambers. Combustion causes high pressure, which forces the engine pistons downward. The downward pushes on the pistons are carried through connecting rods to cranks on the engine crankshaft. These pushes on the cranks cause the crankshaft (Fig. 1-3) to rotate.

The engine flywheel is attached to the rear end of the crankshaft. The rear face of the flywheel is flat and smooth and serves as the driving member of the clutch (on vehicles with manual transmissions). When the clutch is engaged, the rotary motion of the engine crankshaft is carried from the flywheel, through the clutch, to the transmission. With the transmission "in gear," the rotary motion is delivered by the transmis-

sion through a driveshaft to the differential. From there, the rotary motion is carried by the axle shafts and wheels to the tires, which push against the ground.

1-2 Power train In a vehicle, the power train is the mechanism that transmits power from the engine crankshaft to the drive wheels (Figs. 1-2, 1-4, and 1-5). The power train can provide several gear ratios (explained in 1-5) between the engine crankshaft and the wheels. For example, with a three-speed manual transmission, the gear ratio can be changed so that the engine crankshaft will rotate approximately 4, 8, or 12 times to cause the wheels to rotate once. On cars equipped with automatic transmissions, these ratios may be different. Automatic transmissions with torque converters provide a wide range of gear ratios.

Automatic transmissions are described in other books. One is *Automotive Automatic Transmissions,* which is a book in the McGraw-Hill Automotive Technology Series. The book you are now studying covers automotive power trains using *manual* transmissions. These are transmissions that the driver must shift by hand, or *manually.* Power trains with manual transmissions usually have a clutch, transmission, driveshaft, differential, and wheel axles. Later chapters in this book cover the construction, operation, trouble diagnosis, and servicing of each of these power-train components.

1-3 Clutch A clutch is placed in the power train between the engine crankshaft and the transmission (Fig. 1-5). In a vehicle with a manual transmission, the clutch is operated by the driver's left foot (Fig. 1-6). By depressing and releasing the clutch pedal, the driver can connect and disconnect the engine crankshaft to and from the transmission.

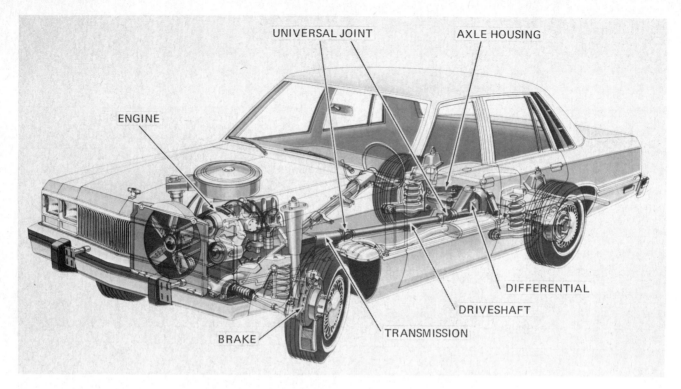

Fig. 1-1 Phantom view of an automobile showing the source of power, or engine; the chassis, which supports the engine and body; the power train, which carries the engine power to the drive wheels; and the car body and accessories. (*Ford Motor Company*)

The clutch does four jobs:

1. It permits the engine to run freely without delivering power to the power train.
2. It permits the operation of the transmission shift lever to select and engage various gear ratios between the engine crankshaft and drive wheels.
3. It slips while transmitting torque from the engine to the transmission, to allow acceleration and getting the vehicle into motion.
4. When engaged, it transmits power from the engine to the transmission.

To start an automobile engine, the clutch must be used to remove the load. The engine cannot deliver power while it is being started. For the engine to deliver usable power, the crankshaft must be rotating faster than the engine's minimum idle speed [for example, 600 revolutions per minute (rpm) or higher]. An engine may start at speeds below 100 rpm, but it will stall as soon as a load is applied. Use of the clutch removes the load and then permits it to be applied gradually. This allows starting the engine and bringing it up to normal speed, which prevents stalling as the load is applied.

Chapter 2 discusses clutches in detail.

✿ 1-4 Transmission A manual transmission (Fig. 1-4) is an assembly of gears and shafts that transmits power from the engine to the final drive, or drive axle. In an automobile, the transmission provides at least three forward-gear ratios (a "three-speed" transmis-

sion) and reverse. In addition, there is a neutral position, which permits disengaging the gears inside the transmission so that no power can flow through. The operation of a basic three-speed transmission is explained in ✿ 1-6.

Today most manual transmissions in cars have four forward speeds. These positions, or "gears," are identified as first, second, third, and fourth. However, *first* sometimes is called "low," and *third* (in a three-speed transmission) or *fourth* (in a four-speed transmission) may be referred to as "high" by various manufacturers.

Different gear ratios are necessary because the internal combustion engine develops relatively little power at low engine speeds. The engine must be turning at a fairly high speed before it can deliver enough power to get the car moving. To help overcome this problem, the transmission has several different forward-gear ratios. Through selection of the proper gear ratio, the engine torque or twisting force is increased. This permits putting the car into motion without stalling the engine or slipping the clutch excessively.

In an automatic transmission, the various ratios between the engine crankshaft and wheels are selected and changed automatically. The driver does not manually shift gears. The automatic controls inside the automatic transmission supply the proper ratio for the driving conditions.

✿ 1-5 Gears and torque Gears (Fig. 1-7) are wheels with teeth that transmit power between shafts (Fig. 1-8). The teeth may be on the edge, inside or outside,

MANUAL TRANSAXLE

FRONT

Fig. 1-2 Cutaway view of an engine and power train for a small front-wheel-drive car. (*Chevrolet Motor Division of General Motors Corporation*)

or at an angle. Usually a gear is attached to a shaft and is "meshed" with another gear fastened to a different shaft. Gears are meshed when the teeth of the two gears are interlaced (Fig. 1-8). In any set of gears, the smaller gear may be called the *pinion gear*.

When one gear of a two-gear external-tooth set is turned, the teeth force the other gear to turn in the opposite direction. The relative speed of the two meshing gears is determined by the number of teeth in the gears. This is called the *gear ratio*.

For example, when two meshing gears have the same number of teeth, both gears turn at the same speed (Fig. 1-8). However, when the driven gear has more teeth than the driving gear, the smaller driving gear will turn faster than the larger driven gear. This means that a driving gear with 12 teeth (as shown in Fig. 1-9) will turn twice as fast as a driven gear with 24 teeth. The gear ratio between the two gears is two to one, which is usually written 2:1.

If a 12-tooth gear is meshed with a 36-tooth gear, the 12-tooth gear turns three times for every revolution of the larger gear. The gear ratio between these gears is 3:1.

Gear ratio may also be determined by comparing the diameters of the two gears.

1. Torque   The gear ratio changes as the number of teeth on each meshing gear changes. At the same time, *torque* also changes. Torque is twisting or turning effort. When you try to turn a shaft with your hand (Fig. 1-9), you are applying a twisting force, or torque, to the gear. Torque is measured in pound-feet (abbreviated lb-ft), or in newton-meters (N-m) in the metric system of measurement. Torque should not be confused with work, which is measured in foot-pounds (ft-lb) or joules (J).

To calculate torque, multiply the force (in pounds or newtons) times the distance (in feet or meters) from the center of the shaft to the point where the force is exerted. For example, suppose you have a wrench 1 foot [0.31 meter (m)] long and you use it to tighten a nut (Fig. 1-10). When you apply a force of 10 pounds [44.5 newtons (N)] to the wrench (as shown in Fig. 1-10), you are applying 10 lb-ft [13.6 N-m] torque to the nut. If you apply a force of 20 pounds [89 N] on the wrench, you are applying 20 lb-ft [27.1 N-m] torque. If the

3

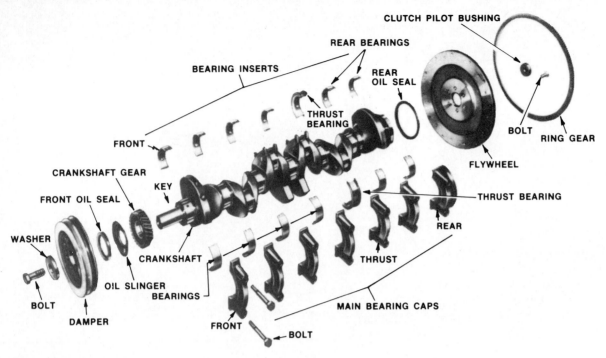

Fig. 1-3 Crankshaft and related parts for a six-cylinder in-line engine. (*Ford Motor Company*)

wrench is 2 feet [0.62 m] long and you apply a force of 10 pounds [44.5 N], the torque on the nut is 20 lb-ft [27.1 N-m].

Any shaft or gear that turns has torque applied to it. The engine pistons and connecting rods push on the cranks on the crankshaft. This applies torque to the crankshaft and causes it to turn. The crankshaft applies torque through the flywheel and clutch to the gears in the transmission, and so the gears turn. This turning effort, or torque, is carried through the power train to the drive wheels.

2. **Torque in gears** Torque on shafts and gears is measured as a straight-line force at a distance from the center of the shaft or gear. For example, suppose we want to measure the torque in the gears shown in Fig. 1-9. If we could hook a spring scale to the gear teeth and get a measurement of the pull on the scale, we could determine the torque. However, a spring scale is not actually used because the gear teeth are moving. Other devices are used to measure the torque of rotating parts. Torque available at the engine crankshaft to do work is measured with an engine dynamometer.

In a circle, the distance from the center to the outside edge is the *radius*. On a gear, the radius is the distance from the center to the point on the tooth when the force is applied. Now, suppose that a tooth on the driving gear is pushing against a tooth on the driven gear with a force of 25 pounds [111 N] (Fig. 1-11). When the force is applied at a distance of 1 foot [0.31 m], which is the radius of the driving gear, a torque of 25 lb-ft [33.9 N-m] is applied to the driven gear.

The 25 pounds [111 N] force from the teeth of the smaller (driving) gear is applied to the teeth of the larger (driven) gear (Fig. 1-11). But the force is applied at a distance of 2 feet [0.61 m] from the center. Therefore the torque on the shaft at the center of the driven gear is 50 lb-ft (25 × 2) [67.8 N-m (33.9 × 2)]. The same force is acting at twice the distance from the shaft center.

3. **Torque and gear ratio** When the smaller gear in Fig. 1-9 is driving the larger gear, the gear ratio is 2:1. However, the *torque ratio* is 1:2. The larger gear turns at half the speed of the smaller gear. As a result,

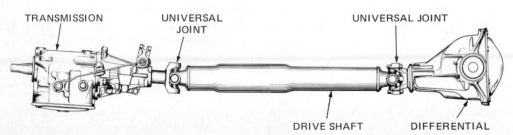

Fig. 1-4 Components of the power train for a car with a front-mounted engine and rear-wheel drive. (*Chevrolet Motor Division of General Motors Corporation*)

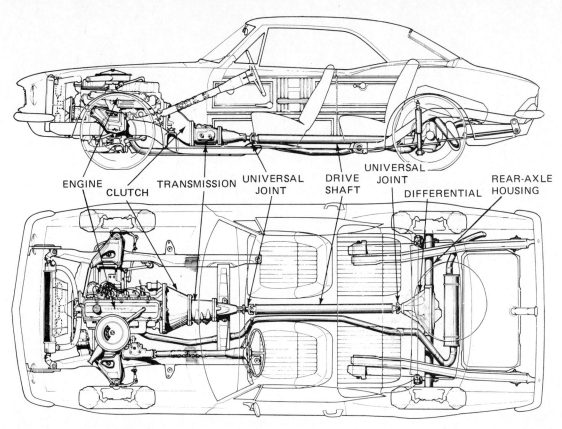

ENGINE CLUTCH TRANSMISSION UNIVERSAL JOINT DRIVE SHAFT UNIVERSAL JOINT DIFFERENTIAL REAR-AXLE HOUSING

Fig. 1-5 Top and side views of a passenger car showing the power train. (*Chevrolet Motor Division of General Motors Corporation*)

the larger gear will have twice the torque of the smaller gear.

In gear systems, *speed reduction means torque increase.* For example, when a typical three-speed transmission is in first gear, there is a speed reduction (or gear reduction) of 12:1 from the engine to the drive wheels (✿ 1-4). The crankshaft turns 12 times to turn the wheels once. Ignoring losses resulting from friction, this means that the torque increases 12 times. If the engine produces a torque of 100 lb-ft [135.6 N-m], then a torque of 1200 lb-ft [1627 N-m] is applied to the drive wheels.

To see how this torque produces the forward thrust or push on the car, refer to Fig. 1-12. In the example shown, the engine is delivering a torque of 100 lb-ft [135.6 N-m]. The gear reduction from the engine to the drive wheels is 12:1 with a torque increase of 1:12. The wheel radius is assumed to be 1 foot [0.31 m], for ease of figuring.

With the torque acting on the ground at a distance of 1 foot [0.31 m] (the radius of the wheel), the force of the tire pushing against the ground is 1200 pounds [5338 N]. Consequently, the push on the wheel axle and therefore on the car is 1200 pounds [5338 N].

NOTE: Actually, the torque is split between the two drive wheels. Each tire pushes against the ground with a force of 600 pounds [2669 N]. Both tires together push with a force of 1200 pounds [5338 N], giving the car a forward thrust of 1200 pounds [5338 N].

4. **Types of gears** Many types of gears are used in the automobile (Fig. 1-7). The most basic type is the *spur gear.* On the spur gear, the teeth are parallel to and align with the center of the gear. Other types of gears differ from the spur gear mainly in the shape and alignment of the gear teeth.

For example, *helical gears* are like spur gears except that, in effect, the teeth have been twisted at an angle to the gear center line. *Bevel gears* (Fig. 1-8) are shaped like imaginary cones with the tops cut off. The teeth point inward toward the apex, or peak, of the cone. Bevel gears are used to transmit motion through angles.

Some gears, called *internal gears,* have their teeth pointing inward. These gears form part of the planetary-gear set which is used in automatic transmissions and older *overdrive* transmissions. Internal gears, a planetary-gear set, and other gears are shown in Fig. 1-7.

✿ **1-6 Operation of the transmission** Passenger-car transmissions have three to five forward speeds, with most cars using a four-speed transmission. Trucks and buses use bigger transmissions with 4 to 16 forward speeds. Regardless of type, all manually shifted transmissions are similar in operation, although they may be different in size and construction.

Now, let's look at a simple three-speed transmission and find out how it works. This transmission, shown in Figs. 1-13 to 1-17, has three shafts and eight gears of various sizes. In the illustrations, only the moving parts

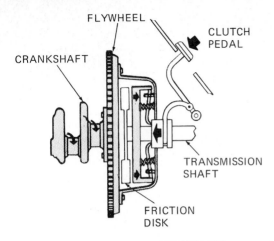

FLYWHEEL
CLUTCH PEDAL
CRANKSHAFT
TRANSMISSION SHAFT
FRICTION DISK

PEDAL DOWN, CLUTCH DISENGAGED.

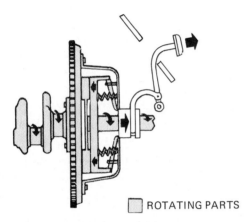

☐ ROTATING PARTS

PEDAL UP, CLUTCH ENGAGED.

Fig. 1-6 (*Top*) When the clutch pedal is pushed down, the clutch disengages so no power flows through to the transmission. (*Bottom*) When the clutch pedal is released, the clutch engages, transmitting power from the flywheel on the crankshaft to the transmission. (*General Motors Corporation*)

are shown. The transmission housing and bearings are not shown.

Four of the gears are rigidly connected to the countershaft (Fig. 1-13). These are the countershaft driven gear, second gear, low gear, and reverse gear. When the clutch is engaged and the engine is running, the clutch gear drives the countershaft drive gear. This turns the countershaft and the other gears on the countershaft. The countershaft rotates in a direction opposite, or *counter to,* the rotation direction of the clutch-shaft gear.

With the gears in neutral (Fig. 1-13) and the car stationary, the transmission main shaft is not turning. The transmission main shaft is mechanically connected by shafts and gears to the car wheels. The two gears on the transmission main shaft may be shifted back and forth along the splines on the shaft by operation of the gearshift lever in the driving compartment. The splines are mating internal and external teeth that permit the gears to slide on the shaft as the gears and shaft rotate together.

Figures 1-13 to 1-17 show a floor-shift lever. This type is shown because it illustrates more clearly the lever action in shifting gears. The transmission action is the same regardless of whether a floor-shift lever or a steering-column-shift lever is used.

1. **Low gear** Before the shift lever is moved from neutral to any gear position, the clutch must first be disengaged. This allows the clutch shaft and gear to stop rotating. Now, when the shift lever is operated to place the gears in low, or first, gear (Fig. 1-14), the large low-and-reverse gear on the transmission main shaft is moved along the shaft until the larger gear meshes with the smaller low gear on the *countershaft gear.* In a transmission, the countershaft gear (or *cluster gear*) is the cluster of gears that is driven by the clutch-shaft gear and turns ''counter'' to it.

When the clutch is again engaged, the transmission main shaft rotates. It is driven through the countershaft gear by the driving gear on the clutch shaft. The path that power takes through a transmission is called the *power flow.* The operation of a transmission is usually described in terms of its power flow from the clutch shaft (or *input shaft*) to the transmission main shaft (or *output shaft*).

In Fig. 1-14, the power flow can be traced by following the arrows. Power flows into the transmission through the gear attached to the clutch shaft. The clutch-shaft gear turns the countershaft driven gear, which is always in mesh. Power then flows through the countershaft to the countershaft low gear, and through the meshed low-and-reverse gear to the transmission main shaft, which is attached to the drive line.

Compare the neutral position shown in Fig. 1-13 with the low-gear position shown in Fig. 1-14. Notice how the low-and-reverse gear can turn free of the countershaft (Fig. 1-13), and then how it is in mesh with the gear on the countershaft (Fig. 1-14) to provide low-gear operation.

Since the countershaft gear turns more slowly than the clutch-shaft gear and the small countershaft gear is engaged with the larger transmission main-shaft gear (with the shift lever in low), a gear reduction of about 3:1 takes place. The clutch shaft turns three times for each revolution of the transmission main shaft. Further gear reduction in the drive axle or differential produces an even higher gear ratio of about 12:1 between the engine crankshaft and the drive wheels.

2. **Second gear** When the clutch is operated and the gearshift lever is moved to second (Fig. 1-15), the low-and-reverse gear on the transmission main shaft de-meshes from the countershaft low gear. The second-and-high-speed gear on the transmission main shaft slides into mesh with the second gear on the countershaft gear. This action provides a gear ratio for second gear that is lower than the ratio for low gear. Now the engine crankshaft turns only about twice while the transmission main shaft turns once, giving a gear ratio of about 2:1 through the transmission. Reduction through the drive axle increases this gear ratio to about 8:1.

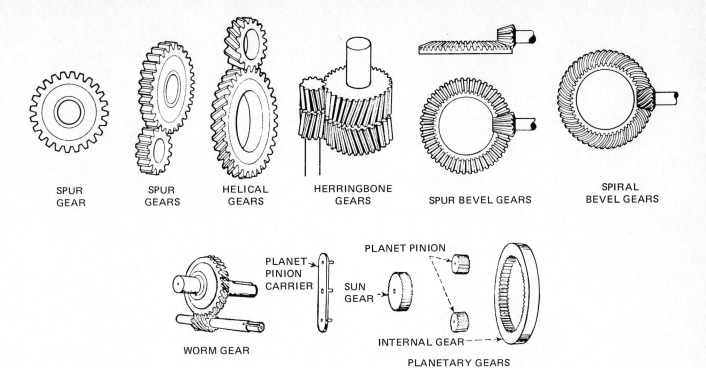

SPUR GEAR   SPUR GEARS   HELICAL GEARS   HERRINGBONE GEARS   SPUR BEVEL GEARS   SPIRAL BEVEL GEARS

PLANET PINION CARRIER   PLANET PINION   SUN GEAR   INTERNAL GEAR

WORM GEAR

PLANETARY GEARS

Fig. 1-7 Various types of gears. (*General Motors Corporation*)

3. **High gear** When the gears are shifted into third, or high, gear (Fig. 1-16), the two sliding gears on the transmission main shaft are de-meshed from the countershaft gears. The second-and-high-speed gear is slid along the transmission main shaft (moving to the left in Fig. 1-16) until the front of the gear contacts the back of the clutch gear.

Teeth on the front and back sides of the two gears mesh so that the clutch shaft and the transmission main

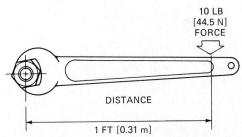

10 LB [44.5 N] FORCE

DISTANCE

1 FT [0.31 m]

Fig. 1-10 Torque is calculated by multiplying the applied force times the distance through which the force acts.

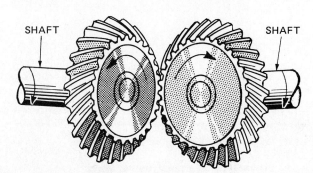

SHAFT   SHAFT

Fig. 1-8 Meshed spiral-bevel gears.

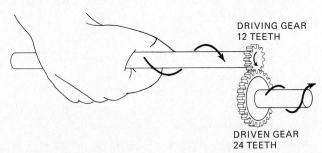

DRIVING GEAR 12 TEETH

DRIVEN GEAR 24 TEETH

Fig. 1-9 Two revolutions of the small gear are required to turn the large gear once. This is a gear ratio of 2:1.

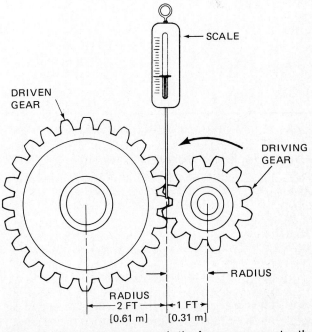

SCALE

DRIVEN GEAR

DRIVING GEAR

RADIUS

RADIUS 2 FT [0.61 m]

1 FT [0.31 m]

Fig. 1-11 The torque on a gear is the force on a gear tooth times the distance from the center of the shaft to the point on the tooth where the force is applied.

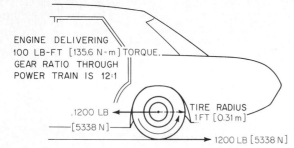

ENGINE DELIVERING 100 LB-FT [135.6 N-m] TORQUE. GEAR RATIO THROUGH POWER TRAIN IS 12:1

.1200 LB [5338 N]

TIRE RADIUS 1 FT [0.31 m]

1200 LB [5338 N]

Fig. 1-12 How torque at the drive wheels is translated into a forward push on the car. The tire is turned with a torque of 1200 lb-ft [1627 N-m]. Since the tire radius is 1 foot [0.31 m], the push of the tire on the ground will be 1200 lb-ft [1627 N-m]. As a result, the car is pushed forward with a force of 1200 pounds [5338 N].

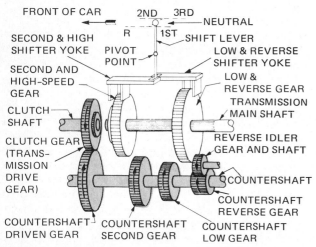

FRONT OF CAR — NEUTRAL — SHIFT LEVER — SECOND & HIGH SHIFTER YOKE — PIVOT POINT — SECOND AND HIGH-SPEED GEAR — LOW & REVERSE SHIFTER YOKE — CLUTCH SHAFT — LOW & REVERSE GEAR — TRANSMISSION MAIN SHAFT — CLUTCH GEAR (TRANSMISSION DRIVE GEAR) — REVERSE IDLER GEAR AND SHAFT — COUNTERSHAFT — COUNTERSHAFT REVERSE GEAR — COUNTERSHAFT DRIVEN GEAR — COUNTERSHAFT SECOND GEAR — COUNTERSHAFT LOW GEAR

Fig. 1-13 A sliding-gear transmission showing the gears in neutral.

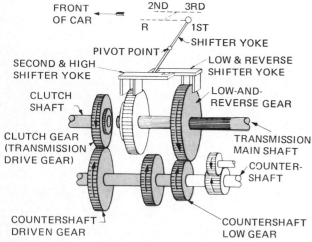

FRONT OF CAR — SHIFTER YOKE — PIVOT POINT — LOW & REVERSE SHIFTER YOKE — SECOND & HIGH SHIFTER YOKE — LOW-AND-REVERSE GEAR — CLUTCH SHAFT — TRANSMISSION MAIN SHAFT — CLUTCH GEAR (TRANSMISSION DRIVE GEAR) — COUNTERSHAFT — COUNTERSHAFT DRIVEN GEAR — COUNTERSHAFT LOW GEAR

Fig. 1-14 Transmission with gears in first, or low.

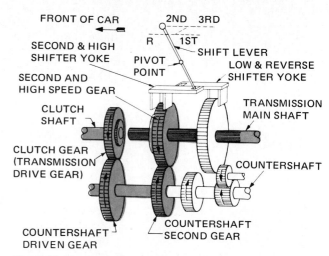

FRONT OF CAR — SECOND & HIGH SHIFTER YOKE — PIVOT POINT — SHIFT LEVER — LOW & REVERSE SHIFTER YOKE — SECOND AND HIGH SPEED GEAR — TRANSMISSION MAIN SHAFT — CLUTCH SHAFT — CLUTCH GEAR (TRANSMISSION DRIVE GEAR) — COUNTERSHAFT — COUNTERSHAFT DRIVEN GEAR — COUNTERSHAFT SECOND GEAR

Fig. 1-15 Transmission with gears in second.

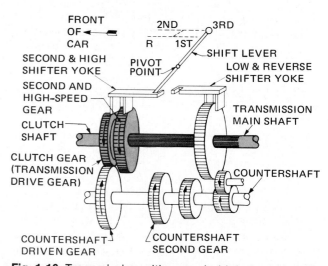

FRONT OF CAR — SECOND & HIGH SHIFTER YOKE — PIVOT POINT — SHIFT LEVER — LOW & REVERSE SHIFTER YOKE — SECOND AND HIGH-SPEED GEAR — CLUTCH SHAFT — TRANSMISSION MAIN SHAFT — CLUTCH GEAR (TRANSMISSION DRIVE GEAR) — COUNTERSHAFT — COUNTERSHAFT DRIVEN GEAR — COUNTERSHAFT SECOND GEAR

Fig. 1-16 Transmission with gears in high, or third (also called *direct*).

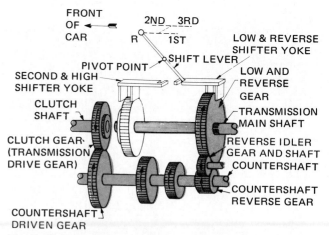

FRONT OF CAR — 2ND 3RD — R 1ST — PIVOT POINT — SHIFT LEVER — SECOND & HIGH SHIFTER YOKE — LOW & REVERSE SHIFTER YOKE — LOW AND REVERSE GEAR — CLUTCH SHAFT — TRANSMISSION MAIN SHAFT — CLUTCH GEAR (TRANSMISSION DRIVE GEAR) — REVERSE IDLER GEAR AND SHAFT — COUNTERSHAFT — COUNTERSHAFT REVERSE GEAR — COUNTERSHAFT DRIVEN GEAR

Fig. 1-17 Transmission with gears in reverse.

shaft are locked together. These teeth are shown in Fig. 1-15. Now the transmission is said to be in "direct drive," since the two shafts turn at the same speed with a gear ratio of 1:1. The final gear reduction through the

drive axle produces a gear ratio of about 4:1 between the engine crankshaft and the drive wheels.

4. **Reverse gear**   When the gears are placed in reverse (Fig. 1-17), the low-and-reverse gear on the trans-

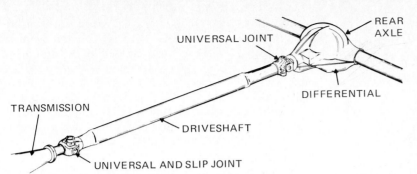

Fig. 1-18 The driveshaft connects the transmission with the differential. This is a one-piece driveshaft with two universal joints and one slip joint.

mission main shaft is meshed with the reverse idler gear. The reverse idler gear is always in mesh with the countershaft reverse gear. Placing the idler gear between the countershaft reverse gear and the low-and-reverse gear causes the main shaft to rotate in the opposite direction. This reverses the rotation of the drive line and wheels so that the car backs up.

The type of transmission discussed above and illustrated in Figs. 1-13 to 1-17 is a *sliding gear* transmission. Its operation is easy to illustrate and to understand. However, in cars, this type of transmission is noisy and difficult to operate. Some transmissions use sliding gears for low and reverse, but no recent automotive transmission shifts gears by actually moving the gears. In most transmissions, the gears are in constant mesh. Shifting is accomplished by locking the gears to each other or to shafts.

Modern automotive transmissions have helical gears with the teeth twisted at an angle to the main shaft (Fig. 1-7). Gears with helical teeth run more smoothly and make less noise than spur gears with straight-cut teeth. Automotive transmissions also include devices that cause the teeth of gears about to mesh to move at the same speed. This allows the teeth or locking devices to mesh without clashing. These devices are called *synchronizers*. A transmission with synchronizers on the gears is called a *synchromesh* transmission. Some transmissions have synchronizers on only second gear and high gear. Other transmissions have synchronizers on all gears.

✿1-7 Rear-wheel drive Many cars have front-mounted engines that drive the rear wheels through the power train (Figs. 1-1 and 1-5). Other cars have front-mounted engines that drive the front wheels of the car

(Fig. 1-2). Front-wheel drive (✿ 1-8) has become more common in recent years and is widely used in small cars. Rear-wheel drive uses a long driveshaft that connects the transmission to the differential in the rear axle (Fig. 1-18).

The driveshaft is a steel tube that usually incorporates two universal joints (Fig. 1-19) and a slip joint (Fig. 1-20). The driveshaft transfers power from the transmission main shaft, or output shaft, to the differential in the rear axle. The differential, in turn, transmits the power through the rear-wheel axles to the rear wheels, which then move the car forward or backward. Driveshafts differ in construction, lining, length, diameter, and type of slip joint. Typically, the driveshaft is connected at one end to the rigidly mounted transmission and at the other end to the rear axle, which moves up and down with wheel and spring movement.

Two separate effects are produced by the movement of the rear axle. First, the distance between the transmission and the rear axle changes as the springs compress and expand and the axle housing moves toward and away from the frame or underbody of the car. Second, the driving angle between the transmission and the differential in the rear axle changes with the spring movement.

1. **Universal joints** To take care of the differences in the angle of the drive as the axle moves up and down, the driveshaft has two or more *universal joints* (called *U joints*). A universal joint (Fig. 1-19) is essentially a double-hinged joint. The driving shaft can transmit power to the driven shaft through the U joint even though the two shafts are several degrees out of line with each other.

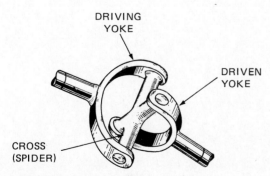

Fig. 1-19 A universal joint.

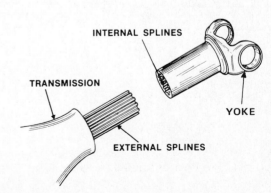

Fig. 1-20 A slip joint.

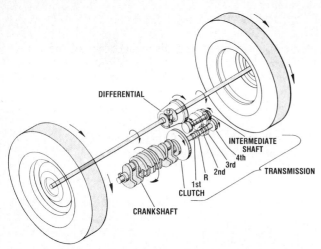

Fig. 1-21 Power flow from the engine crankshaft to the wheels of a car with front-wheel drive. (*Chrysler Corporation*)

Each of the two shafts has a Y-shaped yoke on the end. Between these yokes there is a center member that is shaped like a cross (or "spider"). The four arms of the center member are assembled in bearings in the ends of the shaft yokes. The bearings can turn on the crossarms to take care of any angularity between the shafts as the shafts rotate. The driving shaft causes the center member to rotate by its pressure on two of the crossarms. The other two crossarms cause the driven shaft to rotate. Many types of universal joints have been used on automobiles. Chapter 17 covers the types found on cars today.

2. **Slip joint** The driveshaft tends to shorten and lengthen with the rear-axle movement. Therefore the driveshaft must be able to change length to permit this action. The device used for this is the *slip joint*, usually located at the front end of the driveshaft (Fig. 1-20). The slip joint is an externally splined transmission output shaft and a matching internally splined universal-joint yoke. The two can slide back and forth with respect to each other while transmitting driving power.

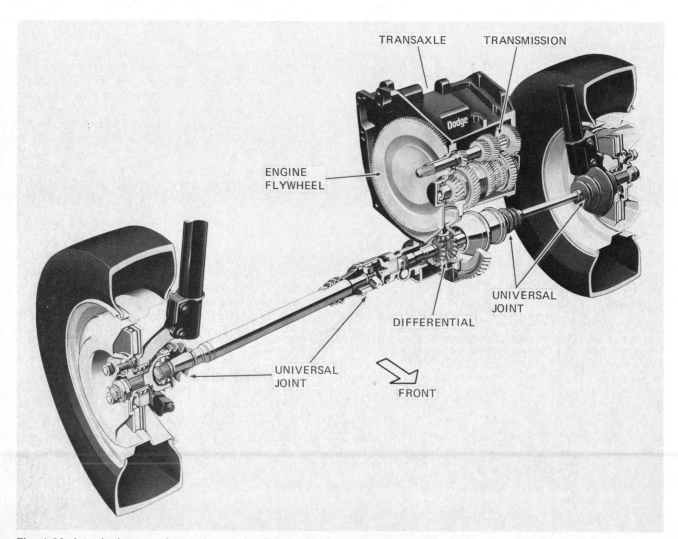

Fig. 1-22 A typical transaxle used on a small car with front-wheel drive when the engine is mounted transversely. (*Chrysler Corporation*)

⚙ 1-8 Front-wheel drive When the engine is mounted in the front of the car, the car may have either rear-wheel drive (⚙ 1-7) or front-wheel drive. With front-wheel drive, the power train is very short because no driveshaft is needed (Fig. 1-21). Engine power flows directly from the transmission into the differential or final drive. There are two basic arrangements. In one, the engine is mounted longitudinally, with its front end pointing to the front of the car (Fig. 1-5). In the other, the engine is mounted sideways, or transversely (Figs. 1-2 and 1-21). The engine in most small cars built today is transversely mounted.

Most front-wheel-drive cars use a *transaxle* instead of a separate transmission and differential. A transaxle is a transmission and a drive axle or differential combined in one unit. Figure 1-22 shows a typical transaxle used in small cars that have a transversely mounted engine in the front of the car. Differentials are described in ⚙ 1-11.

⚙ 1-9 Rear drive with rear-mounted engine Some vehicles have the engine mounted in the rear (Fig. 1-23). In these vehicles, a transaxle and short driveshafts carry the engine power to the rear wheels. Each driveshaft has universal joints and a slip joint. Various models of vehicles built by Volkswagen (and others, such as the Chevrolet Corvair) use this type of drive. The engine usually is a flat "pancake," or opposed, type to lower the height of the engine compartment.

⚙ 1-10 Four-wheel drive Some vehicles, especially those that are used off the road, can deliver power to all four wheels (Fig. 1-24). The driver can select two-wheel or four-wheel drive. In some four-wheel-drive vehicles, engagement and disengagement of the front axle is automatic. This is called *full-time* four-wheel drive. Each driveshaft to the front and rear axles has universal joints and a slip joint.

In these vehicles the power from the engine, after passing through the transmission, enters a *transfer case* (Fig. 1-24). The transfer case has gears that can be engaged to send power to only the rear wheels, or to both the front and the rear wheels. Construction and operation of the transfer case is covered in Chap. 4.

NOTE: Four-wheel-drive vehicles are often referred to as 4WD vehicles. Usually, front-wheel-drive cars are identified as FWD cars. However, when you see the abbreviation FWD, make sure that it means front-wheel drive and not four-wheel drive.

⚙ 1-11 Differential Power from the engine flows through the driveshaft or transaxle to the differential (Figs. 1-4 and 1-22). When the power arrives at the differential, the power is split and sent to the two driving wheels. If the car is moving in a straight line, both driving wheels travel at the same speed. But if the car is making a turn, the outer wheel must travel farther and faster than the inner wheel. The differential makes this possible.

A *differential* is a gear assembly located between the axle shafts of the drive wheels (Fig. 1-25). The action of the differential allows one wheel to turn at a different speed than the other while both are transmitting power from the driveshaft or transaxle to the wheel axles.

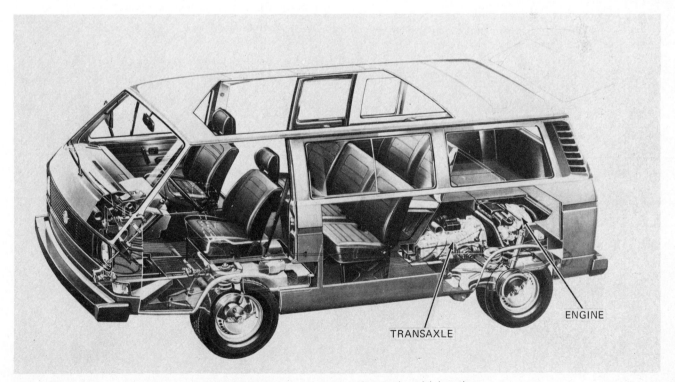

Fig. 1-23 Phantom view of a van that has a rear-mounted pancake engine driving the rear wheels through a transaxle. (*Volkswagen of America, Inc.*)

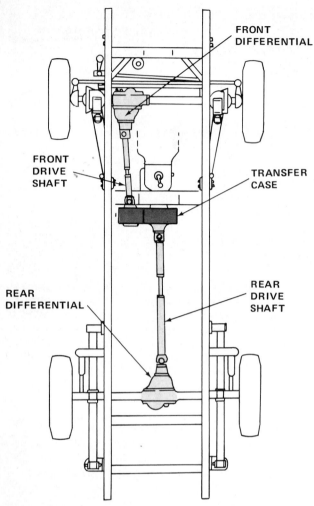

FRONT
DIFFERENTIAL

FRONT
DRIVE
SHAFT

TRANSFER
CASE

REAR
DIFFERENTIAL

REAR
DRIVE
SHAFT

Fig. 1-24 Suspension and drive-train parts for a typical four wheel-drive vehicle. (This is a Ford Bronco.) The transfer case allows the driver to select rear-wheel drive or four-wheel drive. (*Ford Motor Company*)

When a car is steered into a turn, the outer wheel on the drive axle must travel farther than the inner wheel. This action is shown for a rear-wheel-drive car in Fig. 1-26. If a 90° turn to the left is made with the inner wheel turning on a 20-foot [6.1-m] radius, this wheel travels about 31 feet [9.5 m]. The outer wheel, being nearly 5 feet [1.5 m] from the inner wheel, turns on a 24⅔ foot [7.5-m] radius and travels nearly 39 feet [11.9 m].

If the driveshaft is rigidly geared to both rear wheels so that they must rotate together, then each wheel would skid an average of 4 feet [1.2 m] in making the turn. As a result, tires would wear quickly. Also, the car would be difficult to control during turns. The differential eliminates these troubles because it allows the outer wheel to rotate faster as turns are made.

Differential construction and operation are covered in Chap. 19.

⚙ **1-12 Dynamometer tests of the power train** The chassis dynamometer can test engine power output and power-train operation under various operating conditions. Some dynamometers can duplicate almost any type of road test at any load or speed desired by the dynamometer operator. The part of the dynamometer that you can see consists of two rollers mounted at about floor level. The car is driven onto these rollers so that the car's drive wheels can spin the rollers (Fig. 1-27). Next, the engine is started and the transmission is put into gear. Then the car is operated as though it were out on an actual road test.

Located under the floor and connected to the rollers is a device called a *power absorber,* which can place various loads on the rollers. This allows the technician to operate the car under various conditions. The tech-

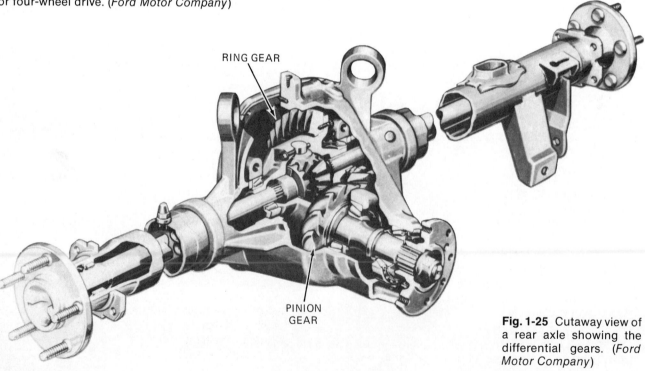

RING GEAR

PINION
GEAR

Fig. 1-25 Cutaway view of a rear axle showing the differential gears. (*Ford Motor Company*)

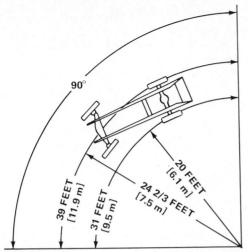

Fig. 1-26 The difference in wheel travel distance as the car makes a 90° turn with the inner wheel turning on a 20-foot [6.1-m] radius.

Fig. 1-27 Running a car on a chassis dynamometer. The rear wheels drive the dynamometer rollers. At the same time, instruments on the test panel can be connected to the vehicle to make many measurements, including engine speed, car speed, engine power output, and torque. From these and other readings, additional information can be calculated. (*Sun Electric Corporation*)

nician can find out how the car performs during acceleration, cruising, idling, and deceleration. Test instruments, such as the oscilloscope, tachometer, and vacuum gauge, are hooked into the engine. These instruments then show the actual state of the engine or other components during various operating conditions.

On the dynamometer, checks can be made of the clutch, transmission, and differential. For example, with a tachometer connected to the engine, the dynamometer can easily indicate a slipping clutch, even if the condition has not yet been noticed by the driver.

Special diagnostic dynamometers are used in some shops. These have many instruments attached to them and also have motored rollers that permit testing of wheel alignment, suspension, brakes, and steering.

Chapter 1 review questions

Select the *one* correct, best, or most probable answer to each question. Then check your answers against the correct answers given at the end of the book.

1. The power train includes the clutch, driveshaft, differential, and:
   a. suspension
   b. transmission
   c. brakes
   d. steering
2. The clutch allows the engine to:
   a. be started without load
   b. idle faster
   c. produce more power
   d. produce more torque
3. Two meshed gears have a gear ratio of 3:1. Every time the larger gear turns once, the smaller gear will turn:
   a. one-third time
   b. once
   c. three times
   d. six times
4. If two meshing gears have a 4:1 gear ratio and the smaller gear has 12 teeth, the larger gear will have:
   a. 12 teeth
   b. 24 teeth
   c. 36 teeth
   d. 48 teeth
5. The device that provides several different forward-gear ratios is the:
   a. differential
   b. transmission
   c. speed changer
   d. driveshaft
6. In gear systems, speed reduction means torque:
   a. reduction
   b. split
   c. increase
   d. none of the above
7. When shifting a sliding-gear transmission into low, a gear on the transmission main shaft is moved into mesh with the:
   a. countershaft low gear
   b. countershaft idler gear

    c. clutch gear
    d. output gear
8. In the transmission, the countershaft driven gear is meshed with a gear on the:
    a. output shaft
    b. main shaft
    c. clutch shaft
    d. driveshaft
9. In high gear, the transmission main shaft turns at the same speed as the:
    a. countershaft
    b. clutch shaft
    c. idler shaft
    d. wheel shaft
10. To take care of the difference in driving angle as the rear axle moves up and down, the driveshaft has two:
    a. slip joints
    b. bearings
    c. clutches
    d. universal joints
11. To take care of the lengthening and shortening of the driveshaft as the rear axle moves up and down, the driveshaft has a:
    a. slip joint
    b. flexible rubber cushion

    c. section of flexible tubing
    d. universal joint
12. The differential allows the outside wheel in a turn to rotate:
    a. faster than the inside wheel
    b. at the same speed as the inside wheel
    c. slower than the inside wheel
    d. none of the above
13. In front-engine cars with rear-wheel drive, the driveshaft transmits power from the transmission to the:
    a. transaxle
    b. transfer case
    c. front wheels
    d. differential
14. The transaxle combines the transmission and the:
    a. universal joints
    b. driveshafts
    c. differential
    d. slip joints
15. Vehicles with four-wheel drive engage and disengage the front axle through a:
    a. transmission
    b. transaxle
    c. transfer case
    d. differential

After studying this chapter, you should be able to:

1. Explain the purpose of the clutch.
2. Describe the construction and operation of three basic types of clutches.
3. Define *double-disk clutch.*
4. Discuss how clutches are mounted in transaxles.
5. List three types of clutch linkage.
6. Describe the operation of the hydraulic clutch linkage.

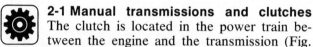 **2-1 Manual transmissions and clutches** The clutch is located in the power train between the engine and the transmission (Fig. 2-1) or between the engine and the transaxle (Fig. 1-2). At one time, all cars had a clutch. Today, most cars are equipped with an automatic transmission, which does not require a driver-operated clutch. For example, during one recent year, about 82 percent of all cars built in the United States were equipped with an automatic transmission. Only 18 percent had a manual transmission and clutch. However, that 18 percent, plus the cars and trucks now on the road with clutches, add up to many millions of clutch-equipped vehicles that require service.

Automotive clutches depend on friction for their operation. Although cars have been built with *fluid couplings* and *torque converters,* which take the place of the clutch, these devices depend on fluid friction for their operation. Clutches which operate on fluid friction are covered in *Automotive Automatic Transmissions,* another book in the McGraw-Hill Automotive Technology Series. The clutches discussed in this chapter operate by mechanical friction.

2-2 Purpose of clutch A clutch transmits rotary motion from one shaft to another while permitting engagement or disengagement of the shafts during rotation of one or both. In the car, these two shafts are the engine crankshaft and the transmission input (or clutch) shaft. Normally, these two shafts are in-line and rotate around the same axis. With the clutch engaged, power is delivered from the engine crankshaft to the clutch shaft. With the clutch disengaged (by the driver's depressing the clutch pedal), the crankshaft is uncoupled from the clutch shaft and no power is delivered to the transmission.

The clutch temporarily uncouples the engine and the transmission so that the transmission gears can be shifted. Without temporarily interrupting the flow of power between the two, it would be difficult to de-mesh and mesh the transmission gears. The force between gear teeth in a set of gears through which power is flowing makes it hard to shift the gears out of mesh. Also, without a clutch, forcing gears into mesh would damage the transmission. The driving gears and the driven gears would probably be running at different speeds. Broken gear teeth would result as meshing was attempted. The clutch, when operated, interrupts the flow of power so gear-tooth force is relieved for de-meshing.

With the gears de-meshed and the engine uncoupled by operation of the clutch, the transmission drive gear runs free. Now it can attain the same, or synchronous, speed with other transmission gears. This is accomplished by synchronizing devices in the transmission. Therefore, meshing can be accomplished without any clashing of gears.

At times during car operation, the clutch is operated to permit shifting of the transmission gears into neutral. In neutral, even when the clutch is engaged, engine power cannot be transmitted through the clutch to the transmission. The engine can be started and brought to speed without delivering power through the transmission and power train to the car wheels.

The amount of power that an engine can deliver during starting and at speeds below idle is small—too small to put the car into motion or keep it in motion. The engine must be rotating at several hundred revolutions per minute before it is able to deliver any usable power. After the engine has been started, the clutch is operated to permit shifting of gears through the various speed ratios (Chap. 4) so that the car can be set into motion and its speed increased.

Figures 2-2 and 2-3 show various clutch parts and

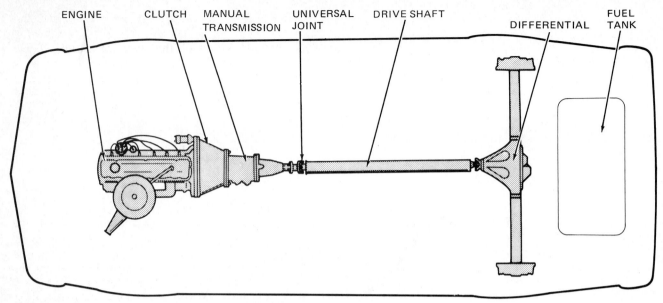

ENGINE   CLUTCH   MANUAL TRANSMISSION   UNIVERSAL JOINT   DRIVE SHAFT   DIFFERENTIAL   FUEL TANK

Fig. 2-1 Power train for a car with a manual transmission.

their relationship to the clutch pedal in the passenger compartment of the car. Figure 2-2 shows the clutch partly cut away. Figure 2-3 shows a sectional view of the clutch, with the linkage to the clutch pedal. Several linkage systems are used. The arrangements of the linkages may differ, but they all cause disengaging, or declutching, when the clutch pedal is pushed down. Clutch linkages are discussed in detail in ✿ 2-11 and 2-12.

✿ **2-3 Types of clutches** Automotive clutches may be classified in several ways. One classification is by the manufacturer's basic design, for example, Borg and Beck, Long, or Belleville spring (diaphragm) clutch. Each of these types is described in a later section. The major difference is in how spring force is applied against the pressure plate.

The size and type of clutch selected by the designer for use in a vehicle depends on several factors. These

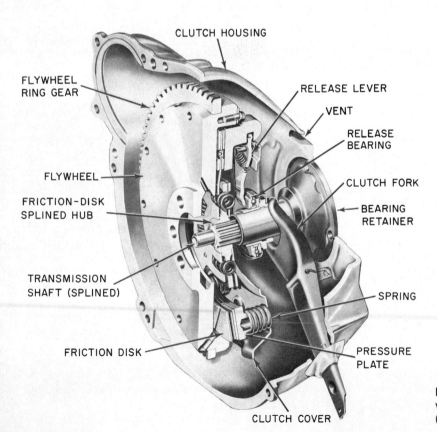

CLUTCH HOUSING

FLYWHEEL RING GEAR

RELEASE LEVER

VENT

RELEASE BEARING

FLYWHEEL

FRICTION-DISK SPLINED HUB

CLUTCH FORK

BEARING RETAINER

TRANSMISSION SHAFT (SPLINED)

SPRING

FRICTION DISK

PRESSURE PLATE

CLUTCH COVER

Fig. 2-2 Partial cutaway view of a typical clutch. (*Ford Motor Company*)

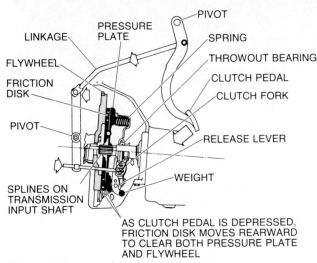

Fig. 2-3 Sectional view of a clutch, with the linkage to the clutch pedal. (*Buick Motor Division of General Motors Corporation*)

include the maximum engine torque developed, the type of transmission, and the nature of the service in which the clutch will operate. Single-plate, multiple-plate, and cone clutches have been installed in automobiles.

Today the most widely used automotive clutch is the dry single-plate spring-force clutch. Figures 2-2 to 2-5 show various views of this type of clutch. These clutches are known as *dry clutches* because they spin in air and not in oil. A clutch that operates in oil is a *wet clutch*. These are used in motorcycles, heavy trucks, and industrial equipment, but in very few automobiles.

Both coil springs and diaphragm springs are found in automotive clutches. They both use spring force against the pressure plate to clamp the friction disk to the flywheel.

✿ 2-4 Clutch construction Between the rear of the engine and the front of the transmission, there is a cast housing called the *clutch housing* (Fig. 2-6). The clutch housing shields the other parts of the clutch from dirt and water. It may have a vent to allow heat to escape. In addition, the clutch housing supports the clutch release lever and other parts of the clutch operating mechanism. On some cars, to make clutch inspection and service easier, the casting for the clutch housing is open across the bottom. Then a thin plate or pan is attached to complete the enclosing of the clutch. This pan or inspection plate can be removed so you can visually check the condition of the clutch. Some clutch

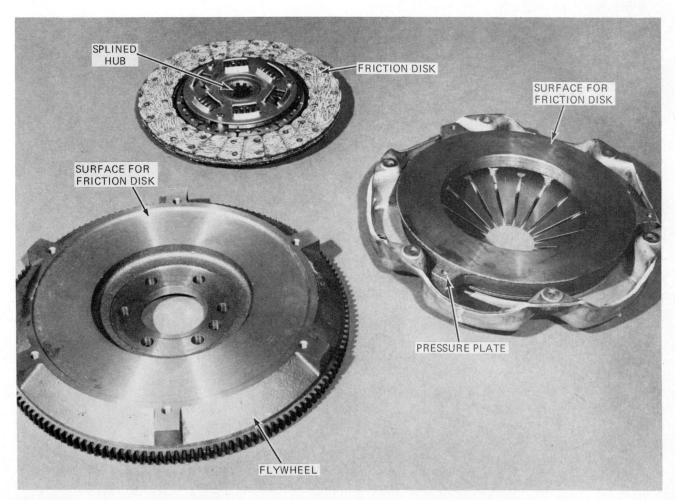

Fig. 2-4 The basic elements of the single-plate clutch used in automobiles. (*Chevrolet Motor Division of General Motors Corporation*)

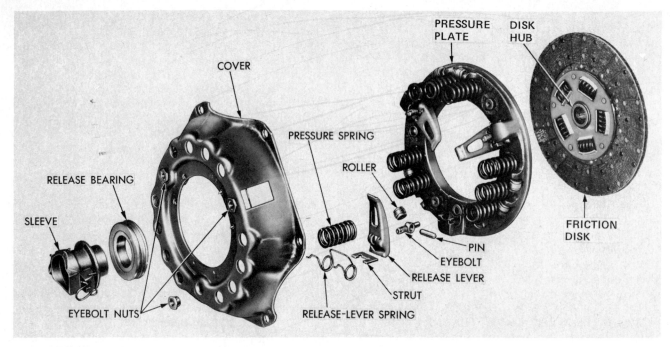

Fig. 2-5 Disassembled view of a single-plate clutch. (*Chrysler Corporation*)

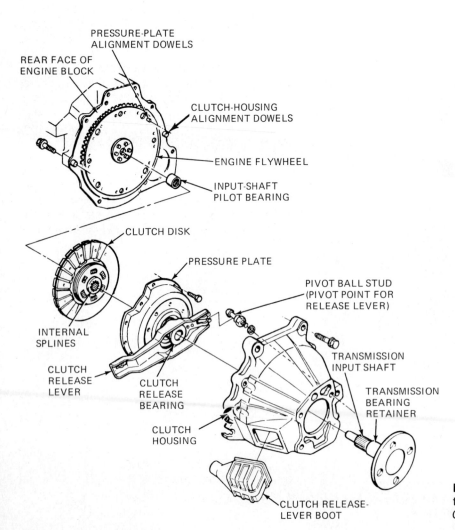

Fig. 2-6 Exploded view of the clutch. (*Ford Motor Company*)

housings also have the mounting pad for the starting motor cast in.

Inside the clutch housing, there usually is a single-plate clutch that has three basic elements. These are the crankshaft flywheel, the friction disk (or *clutch disk*), and the pressure-plate assembly (Fig. 2-6). When the friction disk is held against the rotating engine flywheel by the pressure plate, the disk will rotate with the flywheel. Since the disk is splined to the clutch shaft, power flows into the transmission.

⚙ **2-5 Friction disk** The clutch friction disk (Fig. 2-7) is a driven plate with a splined hub that slides along the splines of the clutch shaft. When the clutch is engaged, the disk drives the clutch shaft through these same splines. Grooves on both sides of the disk lining prevent any vacuum between the disk and the flywheel or the disk and the pressure plate. A vacuum would cause the members of the clutch to stick together instead of disengaging cleanly.

The friction disk (or *driven plate*) is about 12 inches [30.5 mm] in diameter. It is faced on both sides with a heat-resistant friction material. There are external splines on the forward part of the clutch shaft and matching internal splines in the hub of the friction disk (Fig. 2-6). The splines allow the disk to move slightly back and forth along the shaft. Also, when the disk is clamped to the flywheel, the splines force the clutch shaft to rotate at the same speed as the flywheel.

When the clutch is engaged, the disk is clamped tightly between the machined metal surfaces of the flywheel and the pressure plate (Fig. 2-8). These parts are usually made of cast iron, or nodular iron, ground to a smooth finish on the surfaces for the friction disk. This type of iron contains enough graphite to provide

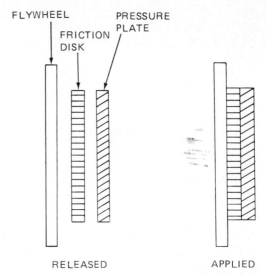

FLYWHEEL   PRESSURE PLATE
FRICTION DISK

RELEASED   APPLIED

Fig. 2-8 Basic clutch elements, showing clutch action. (*Left*) Clutch released. The pressure plate and friction disk have moved away from the flywheel. (*Right*) Clutch engaged. The pressure plate clamps the friction disk to the flywheel so that all parts must rotate together.

some lubrication as the friction disk slips during engagement. When force is applied against the disk by the pressure plate, the friction between the disk facing and the metal surfaces causes the flywheel, the disk, and the pressure plate to lock up and rotate as a solid unit (Fig. 2-9).

To keep the friction disk in proper alignment with the flywheel, the clutch shaft is supported on each end by bearings. A large bearing is installed on the transmission end of the shaft and fits inside the transmission case. The front support for the clutch shaft is the *pilot bearing*. It fits in the center of the crankshaft or flywheel (Figs. 2-6 and 2-8).

The clutch-shaft bearing, which is located ahead of the gear on the clutch shaft, is held in position in the transmission case by a bearing retainer (Fig. 2-2). A release bearing, or throwout bearing, is installed on the front extension, or support, of the bearing retainer, and

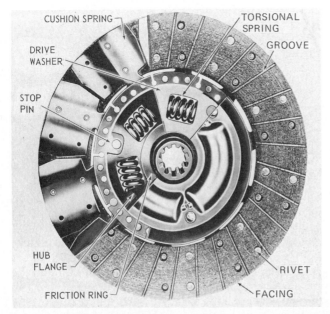

CUSHION SPRING   TORSIONAL SPRING
DRIVE WASHER   GROOVE
STOP PIN
HUB FLANGE   RIVET
FRICTION RING   FACING

Fig. 2-7 Typical friction disk, or driven plate. Facings and drive washer are partly cut away to show springs. (*Buick Motor Division of General Motors Corporation*)

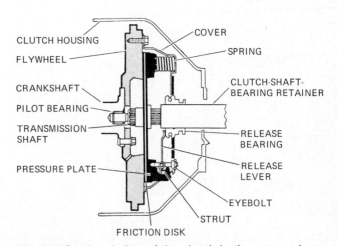

CLUTCH HOUSING   COVER
FLYWHEEL   SPRING
CRANKSHAFT
PILOT BEARING   CLUTCH-SHAFT-BEARING RETAINER
TRANSMISSION SHAFT   RELEASE BEARING
PRESSURE PLATE   RELEASE LEVER
EYEBOLT
STRUT
FRICTION DISK

Fig. 2-9 Sectional view of the clutch in the engaged position.

can slide on it. The travel of the release bearing is controlled by a clutch lever (or *yoke* or *fork*), which is attached to the clutch linkage on one end and to the release bearing on the other.

Most friction disks are made of spring steel in the shape of a single flat disk consisting of several flat segments or waved cushion springs. Friction linings, or facings, are attached to each side of the segments or cushion springs by brass rivets. Facing material must be heat resistant, since the friction of engagement and disengagement produces heat.

Cotton and asbestos fibers woven or molded together and impregnated with resins or other binding agents have been widely used. In many disks of this type, copper wires are woven or pressed into the facing material to give it added strength. However, asbestos is now being replaced with other materials. Some clutch friction disks have metallic-ceramic facings (Fig. 2-10) instead of asbestos or other facings.

To make clutch engagement as smooth as possible and eliminate chatter, several methods have been used to give a little flexibility to the disk. One type of disk is dished so that the inner and outer edges of the facings make contact with the driving members first. Then the rest of the facing makes contact gradually as the spring force increases and the disk flattens. In another type of disk, the flat steel segments attached to the splined hub are replaced with waved cushion springs. These also cause the facings to make contact gradually as the disk flattens.

The friction disk usually has a flexible center or damping device. This is needed to absorb the *torsional vibration* of the crankshaft, which would be transmitted to the power train if it was not eliminated. Torsional vibration is the twist-untwist motion given to the crankshaft by the firing impulses of the engine cylinders.

The damping device has a series of heavy coil springs placed between the drive washers, which are riveted to the cushion springs, and the hub flange, which is attached to the disk hub (Fig. 2-7). The disk hub is driven through the coil springs. They absorb the torsional vibration. Stop pins limit the relative motion between the hub flange and the drive washers. This provides frictional damping that prevents oscillation between the hub flange and the drive washers.

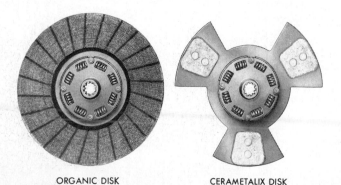

ORGANIC DISK   CERAMETALIX DISK

Fig. 2-10 Organic or asbestos friction disk compared with a ceramic-metallic (Cerametalix) friction disk. (*Chevrolet Motor Division of General Motors Corporation*)

✿ 2-6 Operation of coil-spring clutch

When the clutch pedal is pushed down, the clutch fork moves against the release bearing, forcing the bearing inward (to the left in Fig. 2-2). This operates release levers that overcome the spring force and pull the pressure plate slightly away from the friction disk. With the force against the friction disk removed, the disk then slides away from the flywheel face along the clutch-shaft splines. When this happens, the friction disk and the clutch shaft begin to slow and then stop. As the clutch pedal is released, the springs again force the pressure plate to clamp the friction disk against the flywheel. Now the disk rotates with the flywheel once more.

The basic type of coil-spring clutch is also known as the *Borg and Beck clutch*. It is shown in Figs. 2-2, 2-9, and 2-11. Its operation is based on the frictional contact between two smooth metallic surfaces and the facings riveted to the friction disk. One of the metal surfaces is on the flywheel and the other is on the clutch pressure plate (Figs. 2-4 and 2-11).

When the clutch is in the engaged position (Fig. 2-11), spring force between the clutch cover and pressure plate clamps the disk tightly between the pressure plate and flywheel face. When force is applied, the friction between the flywheel, friction disk, and pressure plate causes the disk to rotate with the flywheel and pressure plate. This continues as long as the engine is running and the clutch is engaged. Because the hub of the disk is splined to the clutch shaft, the shaft rotates with the disk.

To uncouple the engine and the transmission, the clutch pedal is depressed. This action, in turn, through a series of levers, operates the clutch yoke, or fork, assembly, causing the yoke or fork to move the throwout, or release, bearing in toward the flywheel. Movement of the throwout bearing in this direction releases the spring force that holds the flywheel, friction disk, and pressure plate together. As a result the pressure plate moves away from the friction disk (Fig. 2-12). This action permits the flywheel and pressure plate to rotate independent of the friction disk and clutch shaft. Figure 2-3 shows a typical linkage between the clutch pedal and fork.

Figure 2-13 shows one type of clutch fork assembly. The fork goes through a hole in the side of the clutch housing. A dust seal covers the space between the hole and fork, while allowing it freedom of movement. Inside the housing, a *ball stud* (a stud with a round end) or other device provides a pivot point. The fork must pivot about this point when the outer end of the fork is moved back and forth by operation of the clutch pedal. This movement causes the release bearing, attached to the large inner end of the fork, to slide back and forth along the front extension of the front-bearing retainer.

As the throwout bearing slides in toward the clutch, the bearing makes contact with a series of three or more release levers that are equally spaced around the clutch. The release levers pivot about pins, taking up the spring force and causing the pressure plate to move away from the friction disk. Figure 2-14 illustrates the two positions of the release lever: the engaged position and the disengaged position.

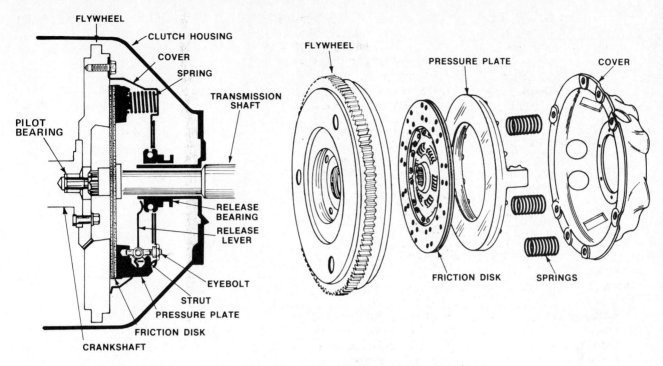

Fig. 2-11 Coil-spring clutch in the engaged position. Note how the pressure plate has clamped the friction disk against the flywheel. The major parts are shown to the right.

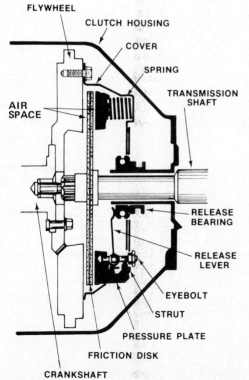

Fig. 2-12 Coil-spring clutch in the disengaged position.

The force of the throwout bearing against the inner end of the release lever causes it to move to the left (in Fig. 2-14). This action causes the lever to pivot about the pin in the eyebolt. The outer end of the release lever moves to the right (in Fig. 2-14). This exerts force through the strut against the pressure plate and causes the plate to move to the right. The movement compresses the coil springs and relieves the spring force on the pressure plate, which has been clamping the friction disk to the flywheel.

As the pressure plate moves to the right (in Fig. 2-14), clearance appears between the pressure plate and friction disk, and between the friction disk and flywheel. These clearances allow the flywheel and the pressure plate to rotate independent of the friction disk.

In the basic clutch of this type (Fig. 2-12), the force against the pressure plate is provided by a series of coil

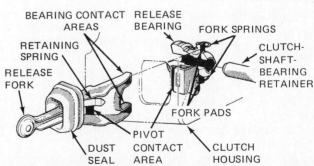

Fig. 2-13 The fork fits onto the release bearing to move it toward and away from the flywheel. (Chrysler Corporation)

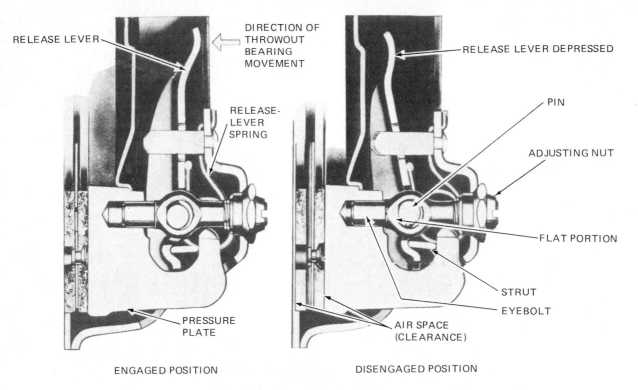

RELEASE LEVER

DIRECTION OF THROWOUT BEARING MOVEMENT

RELEASE-LEVER SPRING

RELEASE LEVER DEPRESSED

PIN

ADJUSTING NUT

FLAT PORTION

STRUT

EYEBOLT

PRESSURE PLATE

AIR SPACE (CLEARANCE)

ENGAGED POSITION

DISENGAGED POSITION

Fig. 2-14 The two limiting positions of the pressure plate and release lever. (*Oldsmobile Division of General Motors Corporation*)

springs located between the clutch cover and the pressure plate (Fig. 2-5). The springs can exert up to about 3000 pounds per square inch (psi) force [20,684 kPa] against the friction disk. However, this much force can make the clutch pedal too stiff for some drivers to operate comfortably.

✿ 2-7 Semicentrifugal clutch One method of allowing a lighter foot-pedal force and still having sufficient force applied against the friction disk is to add a series of rollers between the clutch cover and the pressure plate (Fig. 2-15). At low speeds the pedal must only overcome the springs. But as speed increases, centrifugal force moves the rollers out. The rollers wedge themselves between the pressure plate and the clutch cover with a force that increases with engine speed. This wedging action of the rollers aids the springs and locks the clutch tighter at higher speeds. The addition of centrifugally operated rollers makes the clutch a "semicentrifugal" clutch.

At low and medium speeds, a semicentrifugal clutch works well. It requires low pedal force at low speeds and applies greater force against the disk as engine speed increases. However, at high engine speeds, the rollers tend to hold force against the pressure plate even though the clutch pedal is depressed. This causes the disk to drag and makes gear clash likely during high-speed shifting. Figure 2-5 shows this type of clutch in disassembled view.

The *Long* type of semicentrifugal clutch is shown in Figs. 2-3 and 2-16. In this clutch, each release lever has a weight on its outer end which extends through the clutch cover. These weights are acted on by centrifugal

force to assist the springs in applying force against the pressure plate. However, at high speeds, the pedal force required to operate the clutch is great and also may be considered excessive by some drivers.

✿ 2-8 Diaphragm-spring clutch A diaphragm-spring clutch is widely used on cars with small- to medium-size engines (Fig. 2-17). It has a diaphragm spring (Fig. 2-18) that supplies the force to hold the friction disk against the flywheel. The diaphragm spring also acts as the release lever to take up the spring force when the clutch is disengaged. Figure 2-19 shows the diaphragm-spring clutch in sectional view.

The diaphragm spring is actually a Belleville spring that has a solid ring on the outer diameter. It has a series of tapering fingers pointing inward toward the center of the clutch (Fig. 2-20). The action of the clutch diaphragm is like the action that takes place when the bottom of an oil can is depressed. It "dishes" inward. When the throwout bearing moves in against the ends of the fingers, the entire diaphragm is forced against a pivot ring, causing the diaphragm to dish inward. This raises the pressure plate from the friction disk.

Figures 2-21 and 2-22 illustrate the two positions of the diaphragm spring and clutch parts. In the engaged position (Fig. 2-21), the diaphragm spring is slightly dished, with the tapering fingers pointing slightly away from the flywheel. This position places spring force against the pressure plate around the entire circumference of the diaphragm spring. The diaphragm spring is shaped to exert this initial force.

When the throwout bearing is moved inward against the spring fingers (as the clutch pedal is depressed), the

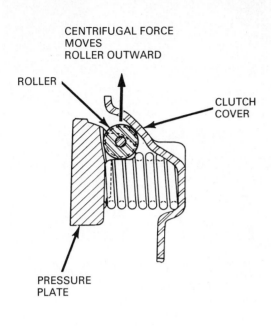

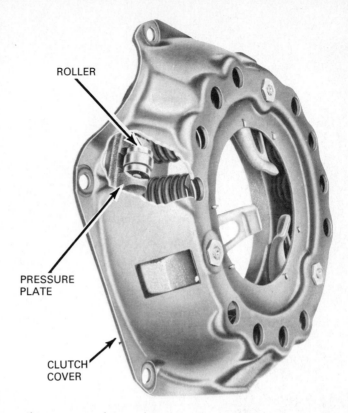

Fig. 2-15 A semicentrifugal clutch, with rollers added between the pressure plate and the clutch cover to assist the springs at higher engine speeds. (*Chrysler Corporation*)

spring is forced to pivot about the inner pivot ring, dishing in the opposite direction. The outer circumference of the spring now lifts the pressure plate away through a series of retracting springs placed about the outer circumference of the pressure plate (Fig. 2-22).

Figure 2-21 is a sectional view of a diaphragm-spring clutch that has a differently shaped spring than the clutch shown in Fig. 2-19. On the diaphragm spring shown in Fig. 2-19, the inner ends of the tapered fingers are bent outward toward the throwout bearing. In Fig. 2-21, the fingers are flat. However, both clutches func-

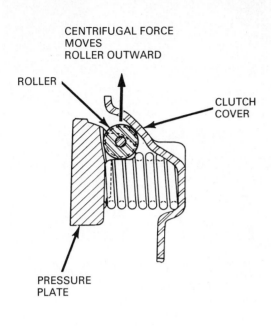

Fig. 2-16 Cutaway view of a centrifugal clutch.

tion in a similar manner. Spring force varies according to the size and thickness of the diaphragm spring.

❀ 2-9 Double-disk clutches Sometimes a clutch with greater holding power is required. This may be because more power must go through the clutch but size is limited so that a clutch with a larger diameter cannot be installed. Then a clutch with two friction disks may be used. Figure 2-23 shows a clutch of this type in exploded view. It has two friction disks separated by an intermediate, or center, drive plate.

Use of the second friction disk adds clutch-plate area, thereby providing greater torque-carrying capacity. Figure 2-24 shows a double-disk clutch in sectional view. When the clutch is engaged, each friction disk transmits half of the flywheel torque to the clutch shaft. These clutches are operated and work in the same way as single-disk coil-spring clutches (❀ 2-6).

Figure 2-25 is a sectional view of a diaphragm-spring clutch using two friction disks and an intermediate pressure plate. The added pressure plate and friction disk give the double-disk clutch greater holding power. This makes it suitable for use with higher-output engines. Double-disk clutches are widely used in medium and heavy trucks.

❀ 2-10 Clutch for transaxle The transaxle is a combined transmission and differential, with a clutch on manually shifted transmissions. In front-wheel-drive cars, the transaxle is attached to the engine (Fig. 1-2). Figure 1-21 shows in simplified form the power flow

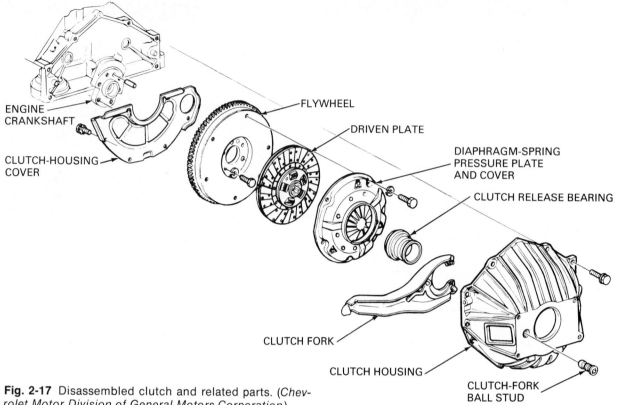

Fig. 2-17 Disassembled clutch and related parts. (*Chevrolet Motor Division of General Motors Corporation*)

Labels in figure 2-17:
- ENGINE CRANKSHAFT
- CLUTCH-HOUSING COVER
- FLYWHEEL
- DRIVEN PLATE
- DIAPHRAGM-SPRING PRESSURE PLATE AND COVER
- CLUTCH RELEASE BEARING
- CLUTCH FORK
- CLUTCH HOUSING
- CLUTCH-FORK BALL STUD

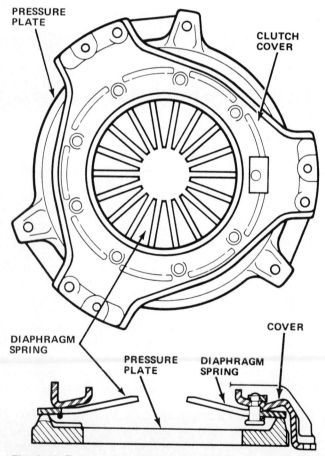

Fig. 2-18 Top and sectional view of a diaphragm-spring clutch. (*American Motors Corporation*)

Labels in figure 2-18:
- PRESSURE PLATE
- CLUTCH COVER
- DIAPHRAGM SPRING
- PRESSURE PLATE
- DIAPHRAGM SPRING
- COVER

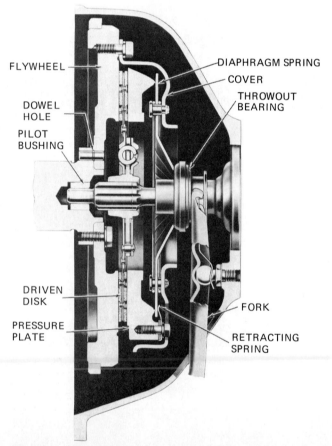

Fig. 2-19 Diaphragm-spring clutch with bent tapering fingers. (*Chevrolet Motor Division of General Motors Corporation*)

Labels in figure 2-19:
- FLYWHEEL
- DOWEL HOLE
- PILOT BUSHING
- DRIVEN DISK
- PRESSURE PLATE
- DIAPHRAGM SPRING
- COVER
- THROWOUT BEARING
- FORK
- RETRACTING SPRING

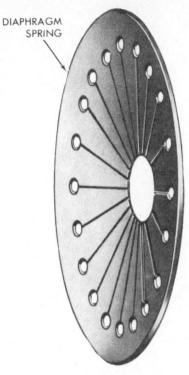

Fig. 2-20 A diaphragm spring, as used in the diaphragm-spring clutch.

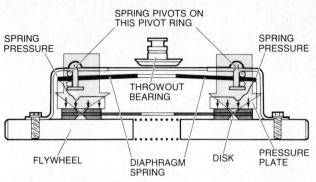

Fig. 2-21 Diaphragm-spring clutch in the engaged position. (*Chevrolet Motor Division of General Motors Corporation*)

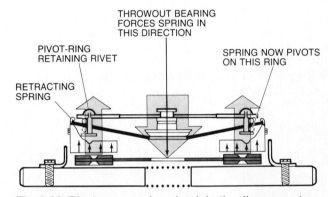

Fig. 2-22 Diaphragm-spring clutch in the disengaged position. (*Chevrolet Motor Division of General Motors Corporation*)

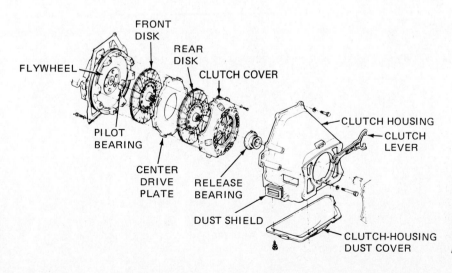

Fig. 2-23 A coil-spring clutch having two friction disks and a center drive plate. (*Ford Motor Company*)

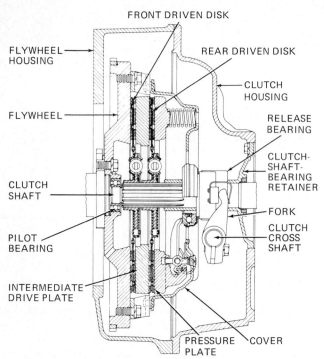

FRONT DRIVEN DISK

FLYWHEEL HOUSING

FLYWHEEL

CLUTCH SHAFT

PILOT BEARING

INTERMEDIATE DRIVE PLATE

REAR DRIVEN DISK

CLUTCH HOUSING

RELEASE BEARING

CLUTCH-SHAFT-BEARING RETAINER

FORK

CLUTCH CROSS SHAFT

PRESSURE PLATE

COVER

Fig. 2-24 Sectional view of a coil-spring clutch with two friction disks and an intermediate drive plate. (*Chevrolet Motor Division of General Motors Corporation*)

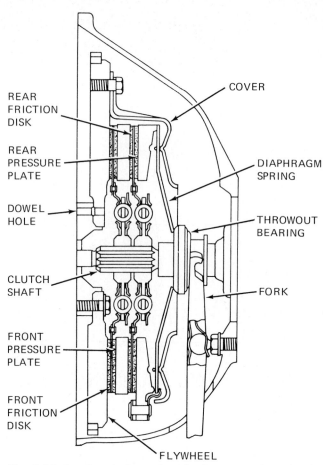

REAR FRICTION DISK

REAR PRESSURE PLATE

DOWEL HOLE

CLUTCH SHAFT

FRONT PRESSURE PLATE

FRONT FRICTION DISK

COVER

DIAPHRAGM SPRING

THROWOUT BEARING

FORK

FLYWHEEL

Fig. 2-25 Sectional view of a diaphragm-spring clutch using two pressure plates. (*Chevrolet Motor Division of General Motors Corporation*)

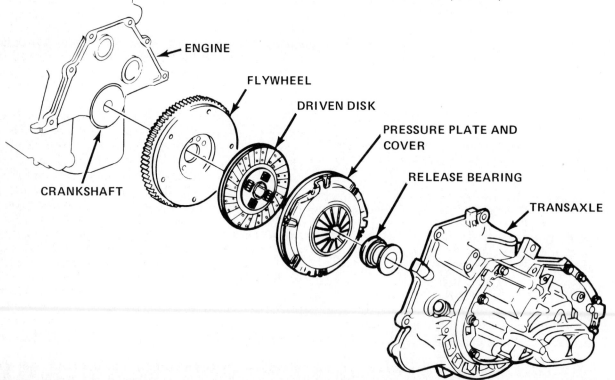

ENGINE

FLYWHEEL

DRIVEN DISK

PRESSURE PLATE AND COVER

RELEASE BEARING

TRANSAXLE

CRANKSHAFT

Fig. 2-26 Disassembled clutch, showing its position between the engine and the transaxle. (*Pontiac Motor Division of General Motors Corporation*)

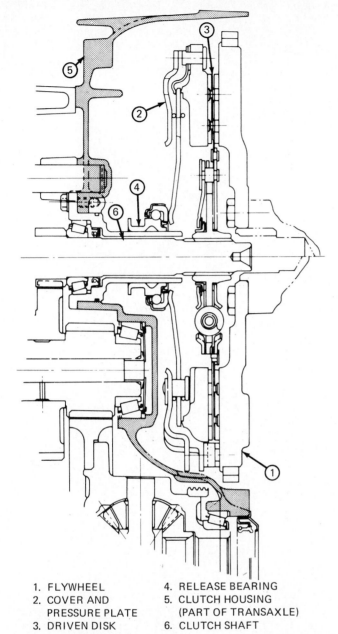

1. FLYWHEEL
2. COVER AND
   PRESSURE PLATE
3. DRIVEN DISK
4. RELEASE BEARING
5. CLUTCH HOUSING
   (PART OF TRANSAXLE)
6. CLUTCH SHAFT

Fig. 2-27 Sectional view of the clutch used with the transaxle. (*Pontiac Motor Division of General Motors Corporation*)

from the engine crankshaft, through the clutch, to the transaxle. Transaxles are covered in detail in ✿ 4-14 and 4-15 and in Chap. 11.

The clutch in most small cars with a transaxle is a typical single-disk diaphragm-spring clutch (✿ 2-8). Figure 2-26 shows its location between the transaxle and the engine. Figure 2-27 is a sectional view of the clutch. Instead of being contained within a clutch housing, the clutch has been adapted to fit into the transaxle assembly.

✿ 2-11 Clutch linkage   Control of the clutch is maintained by the driver through the foot pedal and suitable linkage. Three types of clutch linkages are used in cars. These are the rod type, the cable type, and the hy-

draulic type. Although the operation of each type of clutch linkage varies, the purpose is the same: to convert a light force applied to the clutch pedal (which travels a relatively long distance) into a greatly increased force that moves the pressure plate a very short distance.

Figure 2-28 shows a rod-type clutch linkage. Notice that the system includes an overcenter spring. The purpose of this spring is to reduce the force the driver must apply to the foot pedal to operate it.

A typical cable-operated clutch linkage is shown in Fig. 2-29. On many cars, it is easier for the manufacturer to install a cable system than to try to develop a workable rod arrangement.

✿ 2-12 Hydraulic clutch linkage   A hydraulically operated clutch linkage (Fig. 2-30) is used on vehicles in which the clutch is located so that it would be difficult to run rods or cable from the foot pedal to the clutch. This type of linkage is also used on high-output engines, which require heavy pressure-plate springs. When a clutch is designed to transmit high torque, strong springs are used to provide sufficient force on the friction disk. With insufficient force, the pressure plate and flywheel would slip on the friction disk, quickly ruining it.

However, heavy spring force increases the force that must be applied to the clutch fork. This, in turn, increases the force that the driver must apply to the clutch pedal. To reduce the force required to operate the clutch pedal, a hydraulic system is used.

Figure 2-31 shows a hydraulically operated clutch in disassembled view. Figure 2-30 shows the system schematically. The clutch pedal does not work the release lever directly through rods or cable. Instead, when the driver pushes down on the clutch pedal, a push rod is forced into a master cylinder. This forces hydraulic fluid out of the master cylinder, through a tube, and into a servo, or "slave," cylinder. This is similar to the action in a hydraulic brake system when the brakes are applied.

As the fluid is forced into the servo cylinder, the fluid forces a piston and push rod out. This movement causes the clutch fork to move, pushing the throwout bearing against the release levers on the pressure plate.

The hydraulic system can be designed to multiply the driver's efforts so that a light force applied to the foot pedal produces a much greater force on the clutch fork. A small piston in the master cylinder travels a relatively long distance with only a low input force. This moves a larger piston in the servo cylinder a short distance, transmitting a greater force. There is the additional advantage that no mechanical linkage between the two is required. Only hydraulic lines are required. These lines may be preformed to any angle, or they may be sections of flexible tubing. Many trucks have hydraulically operated clutches.

✿ 2-13 Clutch safety switch   Late-model cars using clutches have a clutch safety switch that prevents starting if the clutch is engaged. The clutch pedal must be depressed at the same time that the ignition switch is

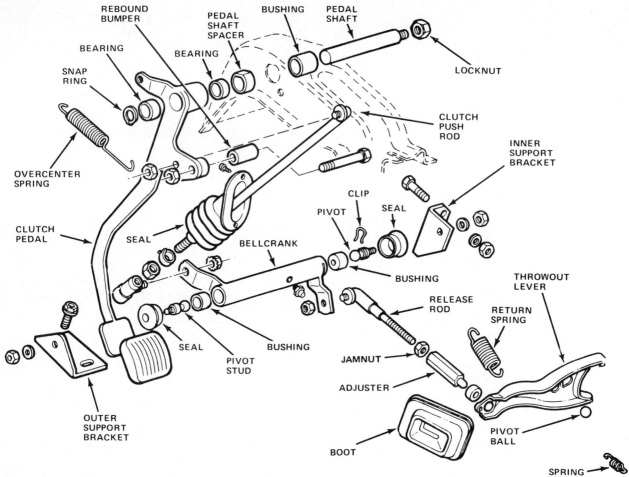

Fig. 2-28 A rod-type clutch operating linkage that includes an overcenter spring to reduce the force required to operate the foot pedal. (*American Motors Corporation*)

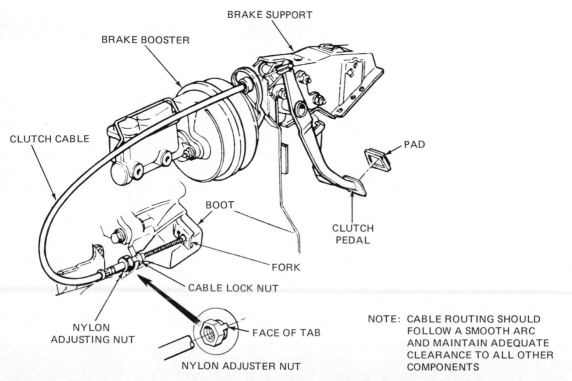

Fig. 2-29 A clutch fork operated by a cable from the foot pedal. (*Ford Motor Company*)

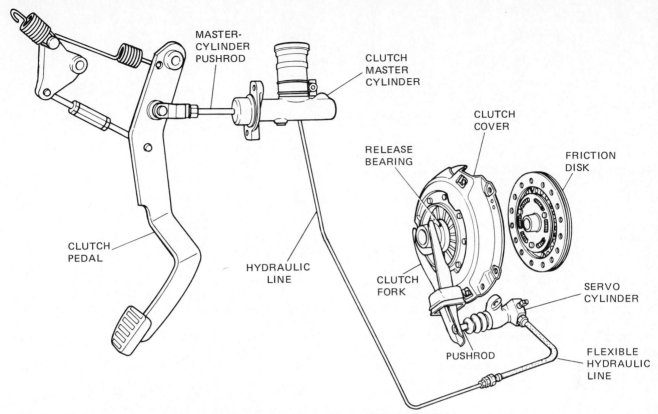

Fig. 2-30 A hydraulically operated clutch linkage. (*Nissan Motor Company, Ltd.*)

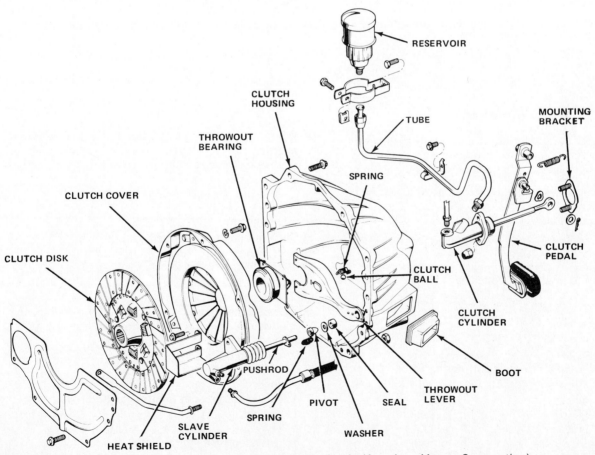

Fig. 2-31 Disassembled view of a hydraulically operated clutch. (*American Motors Corporation*)

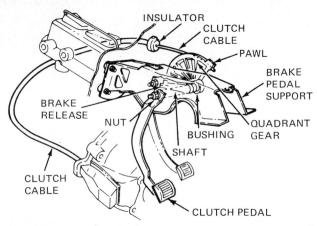

Fig. 2-32 Clutch linkage which includes a self-adjusting device. (*Ford Motor Company*)

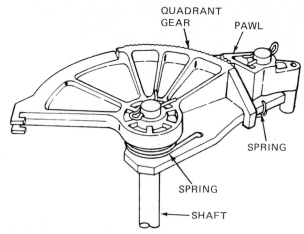

Fig. 2-33 Self-adjusting clutch mechanism. (*Ford Motor Company*)

turned to START. The movement of the clutch pedal closes the safety switch so that the circuit to the starting motor can be completed.

The purpose of the clutch safety switch is to prevent starting with the transmission in gear and the clutch engaged. If this happened, the car might move before the driver is ready. This could lead to an accident.

✿ 2-14 Self-adjusting clutch A self-adjusting clutch (Fig. 2-32) is used on many 1981 and later model cars. This device eliminates the need for routine clutch adjustment.

The adjusting device is a spring-loaded quadrant gear

(Fig. 2-32) that is attached to the clutch pedal through a shaft. A pawl located at the top of the quadrant gear engages the quadrant-gear teeth. As the clutch pedal is depressed, the quadrant gear rotates, pulling the pawl and cable through its travel. This disengages the clutch.

In the released or disengaged position, the spring-loaded quadrant gear takes up excess free play in the clutch mechanism. This compensates for movement of the release lever, clutch cable, and quadrant gear as the clutch disk wears. When sufficient wear occurs, the pawl (Fig. 2-33) engages a new tooth on the quadrant gear. This automatically keeps the clutch pedal in the proper position for proper clutch operation.

Chapter 2 review questions

Select the *one* correct, best, or most probable answer to each question. Then check your answers against the correct answers given at the end of the book.

1. The friction disk is splined to a shaft which extends into the:
   a. transmission
   b. driveshaft
   c. differential
   d. engine

2. The friction disk is positioned between the flywheel and the:
   a. engine
   b. crankshaft
   c. pressure plate
   d. differential

3. When the clutch is engaged, spring force clamps the friction disk between the pressure plate and the:
   a. flywheel
   b. differential
   c. reaction plate
   d. clutch pedal

4. When the clutch pedal is depressed, the throwout bearing moves in and causes the pressure plate to release its force on the:
   a. throwin bearing
   b. pressure springs
   c. friction disk
   d. flywheel

5. The clutch cover is bolted to the:
   a. friction disk
   b. flywheel
   c. car frame
   d. engine block

6. To make engagement as smooth as possible, the friction disk has a series of waved:
   a. cushion pads
   b. cushion bolts
   c. cushion springs
   d. disks

7. The release levers in the typical clutch pivot on:
   a. springs
   b. levers
   c. threaded bolts
   d. pins

8. In the typical friction disk, torsional vibration is absorbed by the use of a series of heavy:
   a. cushion bolts
   b. coil springs
   c. waved pads
   d. friction pads

9. In the diaphragm-spring clutch, inward movement of the throwout bearing causes the diaphragm spring to:
   a. dish inward
   b. expand
   c. contract
   d. flatten

10. In the semicentrifugal clutch, the force of the pressure plate against the friction disk increases with vehicle speed because of weights located on the:
    a. pressure plate
    b. flywheel
    c. release levers
    d. clutch shaft

CLUTCH SERVICE

After studying this chapter, and with proper instruction and equipment, you should be able to:

1. Explain the clutch trouble-diagnosis procedure.
2. List nine clutch troubles, and explain the possible causes of each.
3. Adjust the clutch linkage on a car.
4. Check and adjust the "negative" free travel on a car with a constant-contact release bearing.
5. Service the clutch linkage.
6. Remove and replace a clutch.
7. Replace the crankshaft bushing or bearing.
8. Check and correct the clutch-housing alignment.

3-1 Clutch trouble-diagnosis procedure Many different troubles can develop in the clutch, producing various symptoms. To correct the trouble, you must be able to locate the cause with a minimum of wasted time and effort. This requires a logical approach.

Trouble diagnosis is more than following a series of steps in an attempt to find the solution to a problem. It is a way of looking at systems that are not working right. Here are the basic rules.

1. Know the system. This means that you should know how the parts go together, how they work together as a system, and what happens if some part goes bad or the parts fail to work together as they should.
2. Know the history of the system on the car. How old is the system? What sort of treatment has it had? What is its service history? Has it been serviced for the same problem? The answers to these questions might save you a lot of time.
3. Know the history of the condition causing the driver's complaint. Did the trouble start all at once? Or did it come on gradually? Was it related to some other condition, such as an accident, or to a previous service problem?
4. Know the odds. Some troubles happen more often than others. Be aware of what can happen frequently and what happens rarely. Trouble such as the clutch slipping is more likely to be caused by an improperly adjusted linkage than by weak springs in the pressure-plate assembly.
5. Do not cure the symptom and leave the cause. Replacing a worn friction disk may be a temporary fix, but if the trouble is caused by cracks in the flywheel, you have not eliminated the cause of the problem.

6. Be sure you have fixed the basic condition that caused the trouble.
7. Get all the information (items 1, 2, and 3) that you can from the driver. Such information may greatly simplify your search for the cause of a trouble.

NOTE: The trouble-diagnosis procedure outlined above applies equally well to determining the cause of failure of any part or system on the car. Be sure you understand the steps and the need for the service technician to follow a logical procedure based on all available information. This produces the best possible repair and minimizes the chances of a "come-back."

3-2 Clutch trouble-diagnosis chart Several types of clutch troubles may occur. Usually, the trouble falls into one of the following categories: slipping, chattering or grabbing when engaging, spinning or dragging when disengaged, clutch noises, clutch-pedal pulsations, and rapid wear of the friction-disk facing. The chart on page 33 lists possible causes of each of these troubles and gives the number of the section in this book that explains more fully how to locate and eliminate the trouble.

NOTE: The complaints and possible causes are not listed in the order of frequency of occurrence. Item 1 (or item a) does not necessarily occur more often than item 2 (or item b).

3-3 Clutch slips while engaged Clutch slippage is extremely hard on the clutch facings and mating surfaces of the flywheel and pressure plate. The slipping clutch generates excessive heat. As a result, the clutch facings wear rapidly and may char and burn.

Clutch Trouble-Diagnosis Chart

(See ❋3-3 to 3-10 for detailed explanations of the trouble causes and corrections listed below.)

COMPLAINT	POSSIBLE CAUSE	CHECK OR CORRECTION
1. Clutch slips while engaged (❋3-3)	a. Incorrect pedal-linkage adjustment	Readjust
	b. Broken or weak pressure springs	Replace
	c. Binding in clutch-release linkage	Free, adjust, and lubricate
	d. Broken engine mount	Replace
	e. Worn friction-disk facings	Replace facings or disk
	f. Grease or oil on disk facings	Replace facings or disk
	g. Incorrectly adjusted release levers	Readjust
	h. Warped clutch disk	Replace
2. Clutch chatters or grabs when engaged (❋3-4)	a. Binding in clutch-release linkage	Free, adjust, and lubricate
	b. Broken engine mount	Replace
	c. Oil or grease on disk facings or glazed or loose facings	Replace facings or disk
	d. Binding of friction-disk hub on clutch shaft	Clean and lubricate splines; replace defective parts
	e. Broken disk facings, springs, or pressure plate	Replace broken parts
	f. Warped clutch disk	Replace
3. Clutch spins or drags when disengaged (❋3-5)	a. Incorrect pedal-linkage adjustment	Readjust
	b. Warped friction disk or pressure plate	Replace defective part
	c. Loose friction-disk facing	Replace defective part
	d. Improper release-lever adjustment	Readjust
	e. Friction-disk hub binding on clutch shaft	Clean and lubricate splines; replace defective parts
	f. Broken engine mount	Replace
4. Clutch noises with clutch engaged (❋3-6)	a. Friction-disk hub loose on clutch shaft	Replace worn parts
	b. Friction-disk dampener springs broken or weak	Replace disk
	c. Misalignment of engine and transmission	Realign
5. Clutch noises with clutch disengaged (❋3-6)	a. Clutch throwout bearing worn, binding, or out of lubricant	Lubricate or replace
	b. Release levers not properly adjusted	Readjust or replace assembly
	c. Pilot bearing in crankshaft worn or out of lubricant	Lubricate or replace
	d. Retracting spring (diaphragm-spring clutch) worn	Replace pressure-plate assembly
6. Clutch-pedal pulsations (❋3-7)	a. Engine and transmission not aligned	Realign
	b. Flywheel not seated on crankshaft flange or flange or flywheel bent (also causes engine vibration)	Seat properly, straighten, replace flywheel
	c. Clutch housing distorted	Realign or replace
	d. Release levers not evenly adjusted	Readjust or replace assembly
	e. Warped pressure plate or friction disk	Realign or replace
	f. Pressure-plate assembly misaligned	Realign
	g. Broken diaphragm	Replace
7. Friction-disk-facing wear (❋3-8)	a. Driver "rides" clutch	Keep foot off clutch except when necessary
	b. Excessive and incorrect use of clutch	Reduce use
	c. Cracks in flywheel or pressure-plate face	Replace
	d. Weak or broken pressure springs	Replace
	e. Warped pressure plate or friction disk	Replace defective part
	f. Improper pedal-linkage adjustment	Readjust
	g. Clutch-release linkage binding	Free, readjust, and lubricate
8. Clutch pedal stiff (❋3-9)	a. Clutch linkage lacks lubricant	Lubricate
	b. Clutch-pedal shaft binds in floor mat	Free
	c. Misaligned linkage parts	Realign
	d. Overcenter spring out of adjustment	Readjust
	e. Bent clutch pedal	Replace
9. Hydraulic-clutch troubles (❋3-10)	a. Hydraulic clutches can have any of the troubles listed elsewhere in this chart	Inspect the hydraulic system; check for leakage
	b. Gear clashing and difficulty in shifting into or out of gear	Inspect the hydraulic system; check for leakage

When the flywheel face and pressure plate wear, they may groove, crack, and score. The heat in the pressure plate can cause the springs to lose their tension, which makes the situation worse.

Clutch slippage is very noticeable during acceleration, especially from a standing start or in low gear. A rough test for clutch slippage can be made by starting the engine, setting the parking brake, and shifting into high gear. Then slowly release the clutch while accelerating the engine slowly. If the clutch is in good condition, it should hold so that the engine stalls immediately after clutch engagement is completed.

The dynamometer can also be used to detect a slipping clutch. Connect a tachometer to read engine rpm. Run the vehicle at intermediate speed at part throttle. Note the engine rpm and speedometer reading. Then push the accelerator pedal to the floor, using the dynamometer to hold the same car speed while opening the throttle. Any increase in engine rpm at the same vehicle speed is clutch slippage.

Several conditions can cause clutch slippage. The pedal linkage may be incorrectly adjusted. If the incorrect adjustment reduces pedal lash too much, the throwout bearing may be against the release fingers even with a fully released pedal. This condition can take up part of the spring force so that the pressure plate is not locking the friction disk to the flywheel. The remedy for this problem is to readjust the linkage.

Binding linkage or a broken return spring may prevent full return of the linkage to the engaged position. Replace the spring if it is broken. Lubricate the linkage. Much of the clutch linkage is pivoted in nylon or neoprene bushings. These should be lubricated with silicone spray, SAE 10 oil, or multipurpose grease, depending on the manufacturer's recommendation.

NOTE: If the linkage is not at fault, the slippage could be caused by a broken engine mount. This could allow the engine to shift enough to prevent good clutch engagement. The correction is to replace the mount.

If none of the above is causing the slipping, then the clutch should be removed for service. Conditions in the clutch that could cause slipping include worn friction-disk facings, broken or weak pressure-plate or diaphragm springs, grease or oil on the disk facings, incorrectly adjusted release levers, or a warped clutch disk.

The recommendation of most manufacturers is to replace the disk and pressure-plate assembly if there is internal wear or damage or weak springs. Pressure-plate assemblies can be rebuilt, but this usually is a job for a clutch rebuilder.

NOTE: One clue to a slipping clutch is metal and facing material in the bottom of the clutch housing. This condition can be detected by removing the inspection cover from under the clutch and flywheel.

Careful: If the clutch disk and pressure-plate assembly are replaced, the flywheel should be inspected carefully for damage such as wear, cracks, grooves, and checks. Any of these conditions, if well advanced, will require replacement of the flywheel. Putting a new disk facing against a damaged flywheel will lead to rapid facing wear.

✿ 3-4 Clutch chatters or grabs when engaged The cause of clutch chattering is most likely inside the clutch. The clutch should be removed for service or replacement. However, before this is done, check the clutch linkage to make sure it is not binding. If it binds, it could release suddenly to throw the clutch into quick engagement, with a resulting heavy jerk.

A broken engine mount can also cause chattering. The engine is free to move excessively, and this can cause the clutch to grab or chatter when engaged. The remedy is to replace the mount.

Inside the clutch, the trouble could be due to oil or grease on the disk facings or to glazed or loose facings. If this is the case, the disk should be replaced. The trouble could also be due to binding of the friction-disk hub on the splines of the clutch shaft. This condition requires cleaning and lubrication of the splines in the hub and on the shaft.

NOTE: Clutch chatter after removal and reinstallation of an engine may be caused by a misaligned clutch housing. Some clutch housings have small shims that can be lost during engine or clutch-housing removal. These shims must be reinstalled in the original positions to ensure housing alignment. It is also possible for dirt to get between the clutch housing and cylinder block, or either could be nicked or burred. Any of these conditions can throw off the housing alignment.

Other clutch problems, such as glazed or loose facings or oil or grease on the facings, require replacement of the friction disk and pressure plate.

✿ 3-5 Clutch spins or drags when disengaged The clutch friction disk spins briefly after disengagement when the transmission is in neutral. This normal spinning should not be confused with a dragging clutch. When the clutch drags, the friction disk is not releasing fully from the flywheel or pressure plate as the clutch pedal is depressed. Therefore, the friction disk continues to rotate with or rub against the flywheel or pressure plate. The common complaint of drivers is that they have trouble shifting into gear without clashing. This is because the dragging disk keeps the transmission gears rotating.

The first thing to check with this condition is the pedal-linkage adjustment. If there is excessive pedal lash, or free travel, even full movement of the pedal will not release the clutch fully. If linkage adjustment does not correct the problem, the trouble is in the clutch.

Internal clutch troubles could be due to a warped friction disk or pressure plate or to a loose friction-disk facing. One cause of loose friction-disk facings is abuse of the clutch. This abuse includes "popping" the clutch for a quick getaway (letting the clutch out suddenly with the engine turning at high rpm), slipping the clutch for drag-strip starts, and modifying the engine for increased power output.

The pressure-plate release levers may be incorrectly adjusted so they do not fully disengage the clutch. Also, the friction-disk hub may be binding on the clutch shaft. This condition may be corrected by cleaning and lubricating the splines.

NOTE: A broken engine mount can also cause clutch spinning or dragging. The engine is free to move excessively, which can cause the clutch to spin or drag when disengaged. The remedy is to replace the mount.

✿ 3-6 Clutch noise Clutch noises are usually most noticeable when the engine is idling. To determine the cause, note whether the noise is heard when the clutch is engaged, when it is disengaged, or during pedal movement to engage or disengage the clutch.

Noises heard while the pedal is in motion are probably due to dry or dirty linkage pivot points. Clean and lubricate them as noted in ✿ 3-3.

Noises that are heard when the transmission is in neutral but disappear when the pedal is depressed are transmission noise. (These noises could also be due to a dry or worn pilot bearing in the crankshaft.) They are usually rough-bearing sounds. The cause is worn transmission bearings, sometimes caused by clutch-popping and shifting gears too fast. These conditions throw an extra load on the transmission bearings and on the gears.

Noises heard while the clutch is engaged could be due to a friction-disk hub that is loose on the clutch shaft. This condition requires replacement of the disk or clutch shaft, or perhaps both if both are excessively worn. Friction-disk dampener springs that are broken or weak will cause noise. This condition requires replacement of the disk. Misalignment of the engine and transmission will cause a backward-and-forward movement of the friction disk on the clutch shaft. The alignment must be corrected.

The throwout bearing is the most frequently replaced part of the clutch. Noises heard while the clutch is disengaged could be due to a clutch throwout bearing that is worn or binding, or has lost its lubricant. Such a bearing squeals when the clutch pedal is depressed and the bearing begins to spin. The bearing should be lubricated or replaced. If the release levers are not properly adjusted, they will rub against the friction-disk hub when the clutch pedal is depressed. The release levers should be readjusted or the assembly should be replaced.

If the pilot bearing in the crankshaft is worn or lacks lubricant, it will produce a high-pitched whine when the transmission is in gear, the clutch is disengaged, and the car is stationary. Under these conditions, the clutch shaft (which is piloted in the bearing in the crankshaft) is stationary, but the crankshaft and bearing are turning. The bearing should be lubricated or replaced.

In the diaphragm-spring clutch, worn or weak retracting springs will cause a rattling noise when the clutch is disengaged and the engine is idling. Eliminate the noise by replacing the pressure-plate assembly.

✿ 3-7 Clutch pedal pulsates Clutch-pedal pulsations are noticeable when a slight force is applied to the clutch pedal with the engine running. The pulsations can be felt by the foot as a series of slight pedal movements. As pedal force is increased, the pulsations cease. This condition often indicates trouble that must be corrected before serious damage to the clutch results.

One possible cause is misalignment of the engine and transmission. If the two are not in line, the friction disk or other clutch parts will move back and forth with every revolution. The result will be rapid wear of clutch parts. Correction is to detach the transmission, remove the clutch, and then check the housing alignment with the engine and crankshaft. At the same time, the flywheel can be checked for wobble or runout. A flywheel that is not seated on the crankshaft flange will also produce clutch-pedal pulsations. The flywheel should be removed and remounted to make sure that it seats evenly.

If the clutch housing is distorted or shifted so that alignment between the engine and transmission has been lost, it is sometimes possible to restore alignment. This is done by installing shims between the housing and engine block and between the housing and transmission case. Otherwise, a new clutch housing is required.

NOTE: Clutch-pedal pulsations caused by such conditions as a bent flywheel, a flywheel that is not seated on the crankshaft flange, or housing misalignment usually result only from faulty reassembly after a service job. They do not normally develop during operation of the vehicle.

Another cause of clutch-pedal pulsations is uneven release-lever adjustment (so that release levers do not meet the throwout bearing and pressure plate together). Release levers of the adjustable type should be readjusted. Still another cause is a warped friction disk or pressure plate. A warped friction disk must be replaced. If the pressure plate is out of line because of a distorted clutch cover, the cover sometimes can be straightened to restore alignment. However, usually the pressure-plate assembly is replaced.

In the diaphragm-spring clutch, a broken diaphragm will cause clutch-pedal pulsations. The clue here is that the pulsations develop suddenly, just as the diaphragm breaks.

✿ 3-8 Friction-disk facings wear rapidly Rapid wear of the friction-disk facings is caused by slippage between the facings and the flywheel or pressure plate. If the driver has the habit of "riding" the clutch by resting the left foot on the pedal, part of the pressure-plate spring force will be taken up so that slippage may take place. Likewise, frequent use of the clutch, incorrect clutching and declutching, overloading the clutch, and slow clutch engagement and disengagement increase clutch-facing wear. Speed shifting ("snap" shifting), increasing engine output, and drag-strip starts shorten clutch life. Also, the installation of wide, oversize tires increases the load on the clutch. Some manufacturers will not warranty the clutch if oversize tires are installed.

Rapid facing wear after installation of a new friction disk can be caused by heat checks and cracks in the flywheel and pressure-plate faces. The sharp edges act like tiny knives. They shave off a little of the face during each engagement and disengagement. Because of this, the pressure-plate assembly should also be replaced whenever the friction disk is replaced. In addition, the flywheel face should be inspected, and if it is damaged the flywheel should be replaced.

Several conditions in the clutch itself can cause rapid friction-disk-facing wear. For example, weak or broken pressure springs will cause slippage and facing wear. In this case, the springs or the pressure-plate assembly must be replaced. If the pressure plate or friction disk is warped or out of line, it must be replaced or realigned. In addition, an improper pedal-linkage adjustment or binding of the linkage may prevent full spring force from being applied to the friction disk. With less than full spring force, slippage and wear may take place. The linkage must be readjusted and lubricated.

☼ 3-9 Clutch pedal stiff

A clutch pedal that is stiff or hard to depress is likely to result from lack of lubricant in the clutch linkage, from binding of the clutch-pedal shaft in the floor mat, or from misaligned linkage parts that are binding. In addition, the overcenter spring (on cars so equipped) may be out of adjustment or broken. Also, if the clutch pedal has been bent so that it rubs on the floorboard, it may not operate easily. The remedy for each of these troubles is to lubricate or readjust the clutch parts as necessary, or to replace the clutch pedal.

☼ 3-10 Hydraulic-clutch troubles

The hydraulic clutch can have any of the troubles described previously plus several in the hydraulic system. These special troubles include gear clashing and difficulty in shifting into into or out of gear. The cause is usually loss of fluid from the hydraulic system. Fluid loss prevents the system from completely declutching for gear shifting. The hydraulic system should be checked and serviced in the same way as the hydraulic system in hydraulic brakes. Leaks may be in the master cylinder, servo cylinder, or in the line or connections between the two.

☼ 3-11 Clutch service

The major clutch services include clutch-linkage adjustment, clutch removal and replacement, and clutch disassembly, inspection, adjustment, and reassembly. If a clutch defect develops, you must do more than just replace a worn part. You must determine what caused the part to wear and fix the trouble so that the new part will not wear rapidly.

One of the most common causes of rapid disk-lining wear and clutch failure is improper pedal lash, or free travel. If pedal free travel is not sufficient, the clutch will not engage completely. It will slip and wear rapidly. In addition, the throwout bearing will be operating continuously and will soon wear out.

CAUTION: Asbestos is used in the facings of many friction disks because it can stand up under the high pressures and temperatures inside the clutch. How-

ever, authorities claim that breathing asbestos dust can cause lung cancer. For this reason, be careful when working around clutches. Do not blow the dust out of the clutch housing with compressed air. This dust may contain powdered asbestos. The compressed air could send the dust up into the air around you and you could inhale it. Instead, use damp cloths to wipe out the clutch housing. Then, after working on a clutch, always wash your hands thoroughly to remove any trace of asbestos dust.

☼ 3-12 Clutch-linkage adjustment

Clutch-linkage adjustment may be required at intervals to compensate for wear of the facing or lining on the friction disk. In addition, certain points in the linkage or pedal support may require lubrication. The adjustment of the linkage changes the amount of clutch-pedal free travel, or free play. The free travel of the pedal is the distance that the pedal moves before the throwout bearing makes contact with the release levers in the pressure plate. After this happens, there is a noticeable increase in the force required for further pedal movement. From this point on, pedal movement causes release-lever movement and compression of the springs in the pressure plate. In normal operation, free travel is lost as the friction disk wears. Free travel seldom increases.

A test of pedal free travel should be made with your finger rather than your foot. Your finger can detect the increase in force more accurately than your foot.

There are two checks to make on the clutch and linkage:

1. Make sure the clutch fully disengages.
2. Make sure there is adequate free travel.

☼ 3-13 Checking for clutch disengagement

With the engine idling and the brakes firmly applied, hold the clutch pedal about ½ inch [13 mm] from the floor mat. Move the shift lever between first and reverse several times. If this can be done smoothly, the clutch is disengaging fully. If the shifts are not smooth, the clutch is not disengaging and the linkage must be adjusted (☼ 3-14). If the adjustment does not cure the problem, check for the following:

1. Clutch-pedal bushing may be sticking or worn.
2. Fork or yoke may not be properly installed on its pivot or ball stud. Lack of lubrication can cause the fork or yoke to be pulled off.
3. Linkage may be bent or damaged.
4. Loose or broken engine mounts may allow the engine to shift enough to cause the clutch linkage to bind. Make sure there is clearance between the linkage and any mounting brackets.
5. Clearance between the end of the release bearing and release levers in the pressure plate may be insufficient due to worn parts.

☼ 3-14 Free-travel adjustment

Figure 3-1 shows the clutch-pedal free-travel adjustment on several models of cars built by Chevrolet using the diaphragm-spring clutch (☼ 2-8). The procedure is as follows:

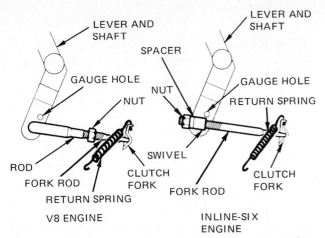

Fig. 3-1 Adjustment of clutch-pedal free travel on various models of cars built by Chevrolet. (*Chevrolet Motor Division of General Motors Corporation*)

1. Disconnect the return spring at the clutch fork.
2. Rotate the clutch lever and the shaft assembly until the clutch pedal is firmly against the rubber bumper.
3. Push the outer end of the clutch fork to the rear until the release bearing lightly touches the fingers on the diaphragm spring.
4. Install the push rod in the gauge hole and increase the length of the linkage rod until all lash is removed from the system.
5. Reinstall the push rod in the lower hole in the lever. Install the retainer and tighten the locknut. Do not change the rod length.
6. Reinstall the return spring and recheck pedal free travel (✿ 3-12).

NOTE: Free-travel adjustment on other makes of cars (including Ford, Chrysler, American Motors, Toyota, Honda, and Datsun) is similar to that for Chevrolet, described above. Refer to the shop manual covering the model of car you are working on for details and specifications. Several examples of adjustment procedures on some of these cars are discussed in ✿ 3-16 to 3-21. In general, the free travel of the clutch pedal should be about 1 inch [25 mm].

✿ 3-15 Post-adjustment check
After the free-travel adjustment is completed, Chevrolet recommends the following procedure to check for clutch slippage:

1. Drive in high gear at 20 to 25 mph [32 to 40 km/h].
2. Depress the clutch pedal to the floor and increase engine speed to 2500 to 3500 rpm.
3. Slip your foot off the clutch pedal to engage the clutch quickly. At the same time push the accelerator pedal to the floor. Engine speed should drop noticeably for a moment as the clutch takes hold, but then the car should accelerate. If the engine speed increases when you simultaneously engage the clutch and open the throttle, the clutch is slipping.

Careful: Do not repeat this procedure more than once or you may overheat the clutch. A clutch that has been slipping prior to adjusting the linkage may continue to slip after correct adjustment because of heat damage. Let the clutch cool for *at least 12 hours* and then repeat the test. If slippage still occurs, there may be enough damage inside the clutch to require overhaul or replacement of the clutch.

✿ 3-16 Chrysler Horizon and Omni clutch free-travel adjustment
Figure 3-2 shows the procedure for adjusting free travel on a front-wheel-drive car with a transversely mounted engine and a transaxle. The adjustment is made with the adjustable sleeve. Pull up on the cable (as shown to the left in Fig. 3-2) and rotate the sleeve downward until it makes snug contact against the grommet. Then rotate the sleeve slightly to allow the end of the sleeve to seat in the rectangular groove in the grommet (as shown to the right in Fig. 3-2). Now, free play (as shown at the bottom in Fig. 3-2) should be about ¼ inch [6.4 mm].

✿ 3-17 Chrysler, Plymouth, and Dodge clutch free-travel adjustment
Figure 3-3 shows the linkage on some models of cars built by Chrysler. Inspect the insulator (lower right in Fig. 3-3) and replace it if it is worn or damaged. To adjust the free travel, allow the insulator to rest lightly in the release fork. Then turn the self-locking adjusting nut until the washer is pushed against the insulator. The linkage should then have $5/32$ inch [4 mm] of free movement at the end of the fork after the linkage has been operated. This provides about 1 inch [25 mm] of free travel at the pedal.

NOTE: If the adjusting nut does not turn easily, the washer is binding on the fork rod. Tap the washer lightly to free it.

✿ 3-18 Ford Pinto and Bobcat clutch free-travel adjustment
Figure 3-4 shows the linkage arrangement for the clutch on the Ford Pinto and Bobcat. To make the adjustment, loosen the clutch-cable locknut on the transmission side of the flywheel housing. Pull the cable toward the front of the car until the tabs on the nylon adjustor nut are clear of the housing boss. Then rotate the nut toward the front of the vehicle about ¼ inch [6.35 mm].

Release the cable to neutralize the system. Then pull the cable forward once more until the free movement in the release lever is eliminated. Rotate the adjusting nut toward the housing until the face of the index tabs touch the housing. Then index the tabs so that they drop into the nearest housing groove. Tighten the locknut at the rear of the housing boss to the specified torque.

✿ 3-19 Ford Fairmont/Zephyr and Mustang/Capri clutch-linkage adjustment
Since 1979, these cars no longer have free travel as in earlier models. Instead, the release bearing is in constant contact with the clutch fingers. Figure 3-5 shows the linkage arrangement for the Fairmont/Zephyr cars with the 3.3-liter [200-cubic-inch] engine and for the Mustang/Capri with this engine both standard and turbocharged. Instead of checking

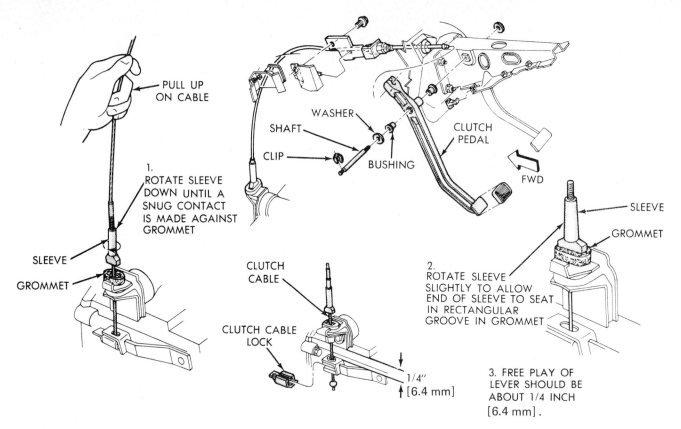

PULL UP ON CABLE

1.
ROTATE SLEEVE DOWN UNTIL A SNUG CONTACT IS MADE AGAINST GROMMET

SLEEVE

GROMMET

WASHER

SHAFT

CLIP

BUSHING

CLUTCH PEDAL

FWD

CLUTCH CABLE

CLUTCH CABLE LOCK

1/4" [6.4 mm]

SLEEVE

GROMMET

2.
ROTATE SLEEVE SLIGHTLY TO ALLOW END OF SLEEVE TO SEAT IN RECTANGULAR GROOVE IN GROMMET

3. FREE PLAY OF LEVER SHOULD BE ABOUT 1/4 INCH [6.4 mm].

Fig. 3-2 Adjusting clutch-pedal free play on a front-wheel-drive car with a transversely mounted engine and a transaxle. (*Chrysler Corporation*)

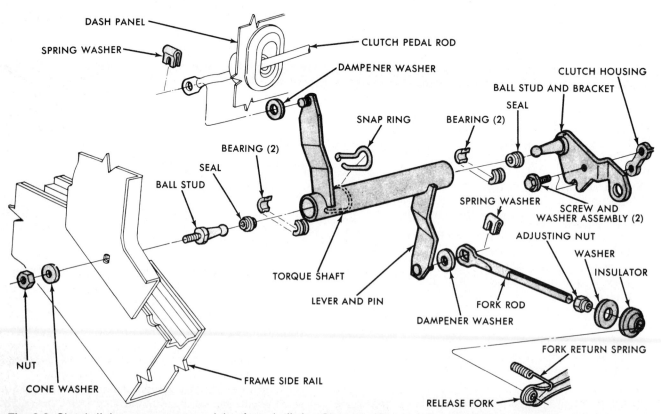

DASH PANEL

SPRING WASHER

CLUTCH PEDAL ROD

DAMPENER WASHER

CLUTCH HOUSING

BALL STUD AND BRACKET

SEAL

BEARING (2)

SNAP RING

BEARING (2)

SEAL

BALL STUD

SPRING WASHER

SCREW AND WASHER ASSEMBLY (2)

ADJUSTING NUT

WASHER

INSULATOR

TORQUE SHAFT

LEVER AND PIN

FORK ROD

DAMPENER WASHER

FORK RETURN SPRING

NUT

CONE WASHER

FRAME SIDE RAIL

RELEASE FORK

Fig. 3-3 Clutch linkage on some models of car built by Chrysler. (*Chrysler Corporation*)

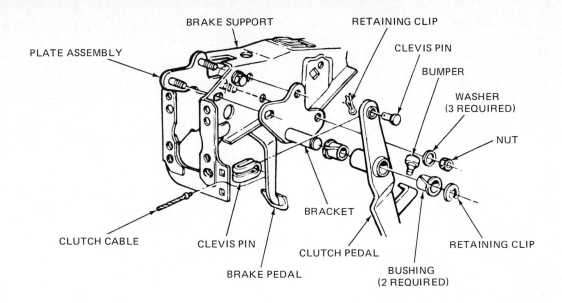

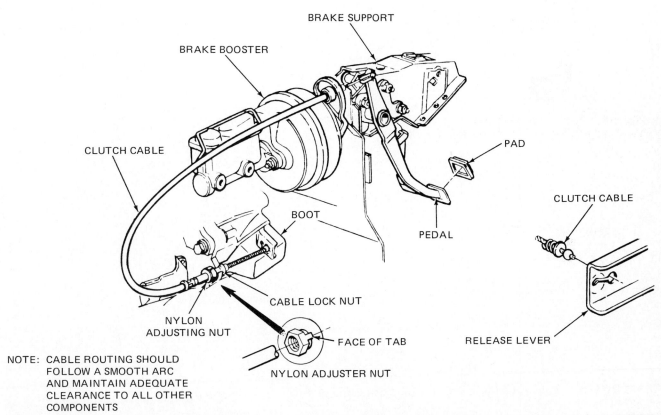

Fig. 3-4 Clutch-pedal and -linkage adjustments on Ford Pinto and Bobcat. (*Ford Motor Company*)

and adjusting the free travel, the clutch-pedal height is adjusted. The procedure is as follows:

1. From under the car, remove the dust shield.
2. Loosen the clutch-cable locknut. If the adjusting nut is turned clockwise, the clutch pedal will be raised. If it is turned counterclockwise, the pedal will be lowered.
3. On all 2.3-liter engines, the total clutch-pedal stroke should be 5.3 inches [135 mm]. For specifications on other models, see the Ford shop manual.

4. After making the adjustment, tighten the locknut, being careful not to change the adjustment. Pump the pedal several times to seat the linkage. Then recheck the pedal height.
5. When the clutch system is properly adjusted on the 2.3-liter models, the clutch pedal can be raised by hand about 2.7 inches [68 mm] from its regular running position before hitting the pedal stop. This "negative" free travel is required to provide for clutch-facing wear.
6. Install the dust shield.

♻ 3-20 Chevrolet clutch-linkage service Figures 3-6 and 3-7 show the linkages and cross-shaft arrangements on recent Chevrolet Malibu, Camaro, and Nova models. The only time these linkages require disassembly is when damaged or worn parts must be replaced. The nylon bushings that support the clutch-pedal pivot shaft may wear after long usage and require replacement. The cross shaft (Fig. 3-7) has nylon ball-stud seats that could wear. After replacement, bushings and

moving parts should be lubricated. Use graphite grease to lubricate the ball-stud and seat.

♻ 3-21 Ford and Chrysler clutch-linkage service Servicing the clutch linkages on other cars besides Chevrolet (covered in ♻ 3-20) is similar to the Chevrolet procedures. Refer to Figs. 3-2 to 3-5 for illustrations of the various linkages. Before you service any linkage or clutch, look up the procedure and specifi-

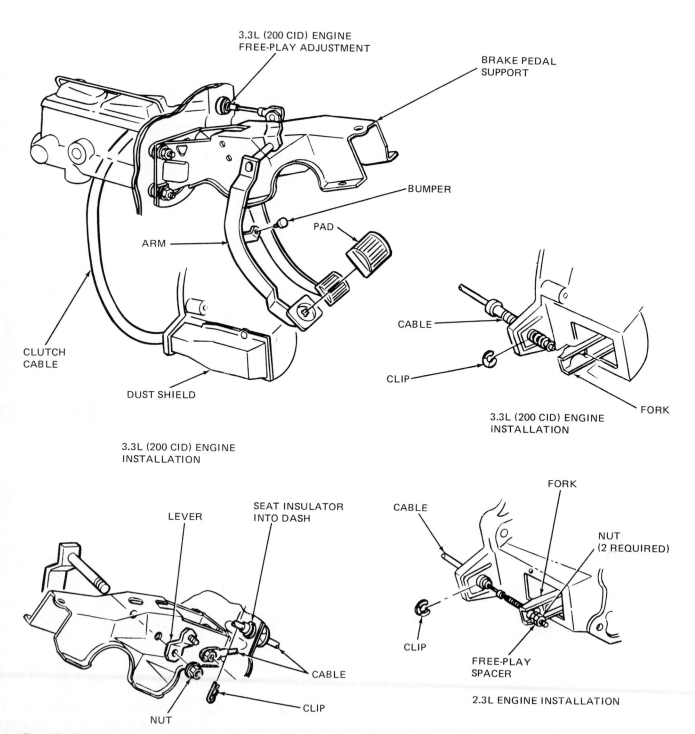

Fig. 3-5 Clutch-pedal and linkage adjustment on some Ford-built cars that have a pedal-height adjustment and a "negative" free-travel adjustment. (*Ford Motor Company*)

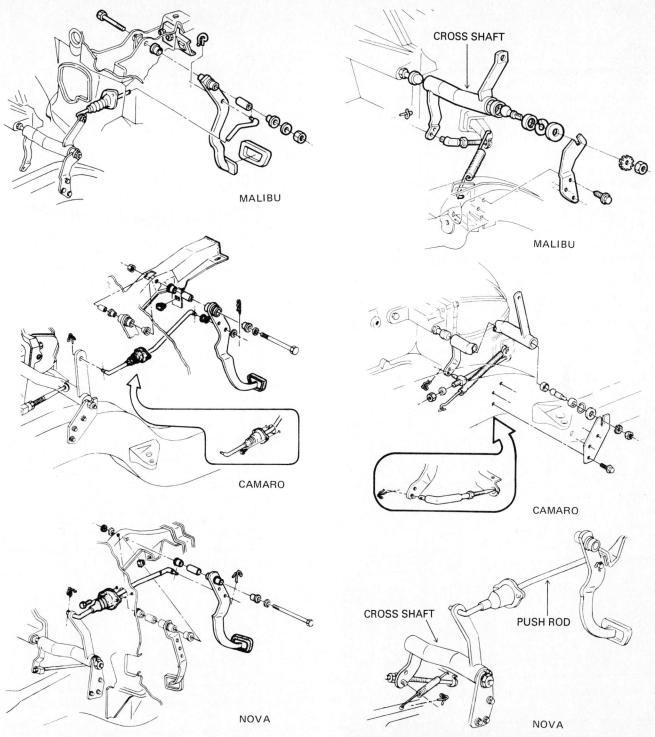

CROSS SHAFT

MALIBU

MALIBU

CAMARO

CAMARO

CROSS SHAFT

PUSH ROD

NOVA

NOVA

Fig. 3-6 Clutch linkages on Chevrolet Malibu, Camaro, and Nova models. (*Chevrolet Motor Division of General Motors Corporation*)

Fig. 3-7 Cross-shaft and push-rod parts of clutch linkages on Chevrolet Malibu, Camaro, and Nova models. (*Chevrolet Motor Division of General Motors Corporation*)

cations in the manufacturer's shop manual for the car you are servicing.

✿ **3-22 Clutch removal and replacement** Variations in construction and design make it necessary to use different procedures and tools when removing and replacing clutches on different cars. As a first step in

clutch removal, the transmission must be removed. Transmission removal and replacement are covered in Chap. 6.

When removing the transmission, pull it straight back from the clutch housing until the clutch shaft is clear of the friction-disk hub. Then the transmission can be lowered from the car. This procedure prevents bending and damage to the friction disk. To protect the friction disk during transmission removal, install two guide pins in place of two of the transmission-attaching bolts (Fig. 3-8). This maintains alignment as the transmission is moved back and prevents the weight of the transmission from springing the hub in the friction disk.

Careful: The pressure-plate cover must be reattached to the flywheel in its original position. If not, dynamic balance may be lost, and vibration and damage will occur. To ensure correct alignment on installation, both the flywheel and the pressure-plate cover are stamped with an X or some similar marking. These markings should align when the clutch pressure-plate assembly is reinstalled on the flywheel. Before removing the pressure plate from the flywheel, if you cannot locate the markings, carefully mark the pressure-plate cover and the flywheel with a hammer and punch (Fig. 3-9). Then you can reinstall the pressure plate in its original position by aligning the marks.

Some clutches use a cross shaft inside the clutch housing as part of the linkage to operate the release bearing. The release fork and cross-shaft must be pulled partly out of the clutch housing to provide room for the pressure plate to clear the cross shaft. This can be done after the clutch release-fork bracket is disconnected at the clutch housing and the release-fork flange cap screws are taken out. On other cross-shaft clutches, it is necessary to detach only the cross shaft so that the fork can be swung up out of the way. On the diaphragm-spring clutch or the clutch using a ball stud on which the fork pivots, snap the fork off its pivot with a screwdriver after removing the clutch linkage from the fork.

Loosen the screws that attach the pressure-plate cover one turn at a time so that the cover will be

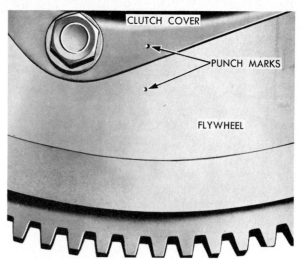

Fig. 3-9 Marking the clutch and flywheel. (*Chrysler Corporation*)

loosened evenly and distortion will not occur. Loosen the screws evenly until the spring force is relieved. Then take the screws out. Use a flywheel turner to rotate the flywheel to get at the upper screws (Fig. 3-10). When the spring force is relieved and the screws are out, the pressure plate and disk can be lowered from the car.

Careful: On some coil-spring clutches, the combined force of the clutch springs may be 3000 pounds [13,344 N] or higher. Always follow the proper procedure when loosening or tightening the screws on the pressure-plate cover.

In general, installation of the clutch is the reverse of removal. Before the clutch is installed, the condition of the pilot bearing in the end of the crankshaft should be checked. It should be lubricated or replaced if

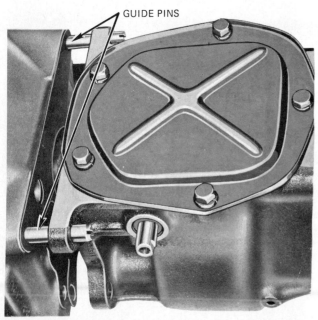

Fig. 3-8 Using guide pins during transmission removal or installation. The pins maintain transmission alignment with the clutch disk as the transmission is moved backward or forward so that the disk will not be damaged. (*Buick Motor Division of General Motors Corporation*)

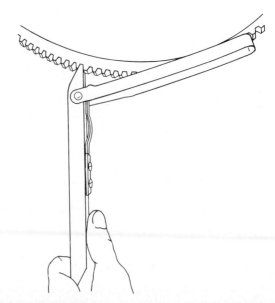

Fig. 3-10 Using a flywheel turner to rotate the flywheel so that the cover upper-attaching bolts can be reached.

necessary (✿ 3-24). In addition, the condition of the throwout bearing and other clutch parts should be checked. Any defective parts should be replaced (✿ 3-24). Figure 3-11 shows a disassembled view of a clutch and housing, together with installation instructions.

NOTE: The clutch-housing alignment should be checked whenever the clutch is removed for service. A misaligned housing can cause improper clutch release, friction-disk failure, front transmission-bearing failure, uneven wear of the pilot bushing in the crankshaft, or clutch noise, vibration, and jumping out of gear. Checking and adjusting clutch-housing alignment is covered in ✿ 3-25.

To proceed with the clutch installation, turn the flywheel until the X or other alignment mark is at the bottom. Then use the clutch aligner to maintain alignment of the friction disk with the pilot bearing in the crankshaft (Fig. 3-12). Or you can use a spare clutch shaft, or transmission drive pinion (Fig. 3-13). Hold the friction disk and clutch cover in place. Turn the cover until the X or other mark on it aligns with the similar mark on the flywheel. Install the attaching bolts, turning them down one turn at a time to take up the spring tension gradually and evenly. Use the flywheel turner (Fig. 3-10) to rotate the flywheel for access to the upper bolts.

As a final step in the procedure, after the transmission has been reinstalled and the clutch linkages reattached, check the clutch-pedal free travel. Make any adjustments that are necessary (✿ 3-12 to 3-21).

✿ 3-23 Clutch inspection and repair Any parts in the pressure plate that are worn or damaged can be replaced by a clutch rebuilder. For example, if the pressure-plate springs lose tension because of overheating, or if the release-lever bearings wear excessively, then the pressure-plate-and-cover assembly can be disassembled and these parts replaced. However, most manufacturers recommend replacing a worn or defective clutch with a complete new one. Shop manuals no longer provide disassembly-assembly instructions for the technician. The maximum disassembly usually covered in current manufacturers' service manuals is shown in Fig. 3-14.

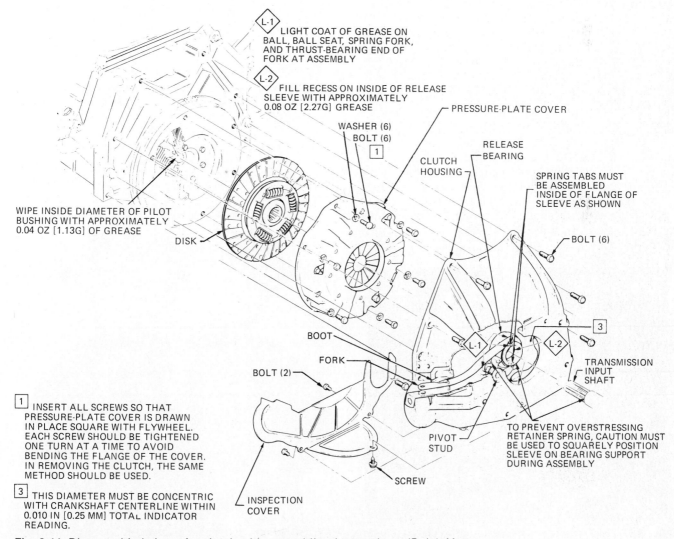

Fig. 3-11 Disassembled view of a clutch with assembling instructions. (*Buick Motor Division of General Motors Corporation*)

43

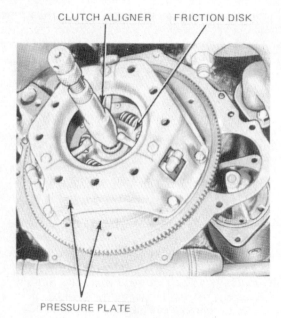

CLUTCH ALIGNER FRICTION DISK

PRESSURE PLATE

Fig. 3-12 Using a clutch aligner to hold the friction disk in position during installation of the pressure-plate assembly. (*Ford Motor Company*)

Some coil-spring clutches have an adjustment that can be made on the release levers. This adjustment requires a clutch-gauge plate and a clutch-lever height gauge (Figs. 3-15 and 3-16). First, place the clutch-gauge plate on the flywheel (Fig. 3-15). Then place the cover assembly on top with the release levers over the machined lands on the gauge. Next, attach the cover assembly to the flywheel. Tighten the screws one turn at a time in rotation to avoid distorting the cover. Depress the release levers several times to seat them. Then measure their height with the height gauge (Fig. 3-16). The height gauge has four settings that can be used for measuring above and below the hub. Figure 3-17 shows details of the adjustment.

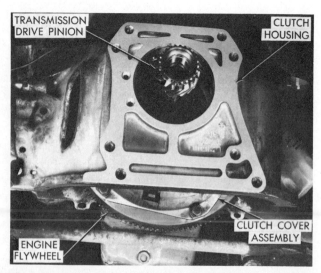

TRANSMISSION DRIVE PINION
CLUTCH HOUSING
CLUTCH COVER ASSEMBLY
ENGINE FLYWHEEL

Fig. 3-13 Using a spare transmission drive pinion (clutch shaft) to align the clutch during installation. (*Chrysler Corporation*)

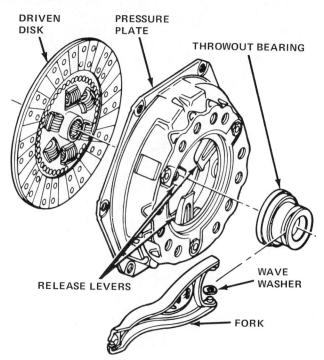

DRIVEN DISK
PRESSURE PLATE
THROWOUT BEARING
RELEASE LEVERS
WAVE WASHER
FORK

Fig. 3-14 Clutch assembly and related parts. (*American Motors Corporation*)

On the indirect-spring-force-type clutch, remove the release clip, loosen the locknut, and turn the adjusting screw until the lever is at the specified height. Tighten the locknut and recheck. If okay, install the release clip.

On the direct-spring-force-type clutch, turn the adjusting nuts until the lever is at the correct height. Work

MACHINED LAND
CLUTCH-GAUGE PLATE

Fig. 3-15 Placing the clutch assembly on a flywheel and clutch-gauge plate to check release-lever adjustments. (*American Motors Corporation*)

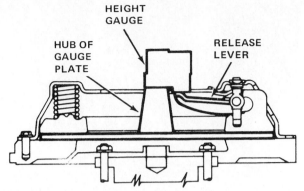

Fig. 3-16 Measuring release-lever adjustment with a height gauge. (*American Motors Corporation*)

the lever several times. Recheck. If okay, stake the nut with a dull punch.

✿ 3-24 Inspecting and servicing clutch parts Clutch parts can be checked as follows after the clutch has been removed from the vehicle.

CAUTION: Never use compressed air to blow the dust out of the clutch housing. This dust may contain asbestos. Asbestos, when breathed into your lungs, can cause lung cancer (✿ 3-11). Instead, wipe out the housing with a damp cloth. Dispose of the cloth so that no one else will use it. Wash your hands after handling the friction disk and other clutch components.

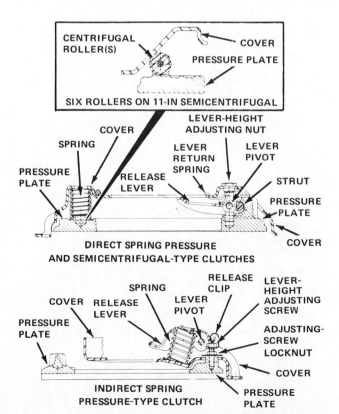

Fig. 3-17 Sectional views of clutches, showing adjustment of release levers. (*American Motors Corporation*)

1. Check for oil leakage through the engine rear-main-bearing oil seal and transmission drive-pinion seal. If leakage is evident, replace the seal.
2. Check the friction face of the flywheel for uniform appearance and for cracks, grooves, and uneven wear. If there is uneven wear, check the flywheel runout with a dial indicator. A warped or damaged flywheel should be replaced. The effect of heat checks and cracks on friction-disk-facing wear is discussed in ✿ 3-8.
3. Check the pilot bushing in the end of the crankshaft. Replace the bushing if it is worn. The bushing may be removed with a bushing puller or by the following method: Fill the crankshaft cavity and bushing bore with heavy grease. Then insert a clutch-aligning tool or spare transmission drive pinion (Figs. 3-12 and 3-13) into the pilot bushing. Tap the end of the tool or drive pinion with a lead hammer. Pressure from the grease will force the bushing out. Install a new bushing with the aligning tool.
4. Check the pilot-bearing surface on the end of the clutch shaft for wear. Replace the clutch shaft if the pilot-bearing surface is rough or worn.
5. Check the friction disk, handling it with care. Do not touch the facings. Any oil or grease will cause clutch slippage and rapid facing wear. Replace the disk if the facings show evidence of oil or grease, are worn to within 0.015 inch [0.38 mm] (Plymouth) of the rivet heads, or are loose. The disk should also be replaced if there is other damage, such as worn splines, loose rivets, or evidence of heat damage.
6. Wipe the pressure-plate face with solvent. Check the face for flatness with a straightedge. Then check the face for burns, cracks, grooves, and ridges.

NOTE: If the friction disk is replaced, then most manufacturers recommend replacing the pressure-plate assembly also.

7. Check the condition of the release levers. The inner ends should have a uniform wear pattern from contact with the throwout bearing.
8. Test the cover for flatness on a surface plate.
9. If any of the pressure-plate parts are not up to specifications, replace the pressure-plate assembly. A new friction disk should be installed at the same time.
10. Examine the throwout bearing. When held in the hand, the bearing should turn freely under a light thrust load. There should be no noise. The bearing should turn smoothly, without roughness. Note the condition of the face where the release levers touch. Replace the bearing if it is not in good condition. Figure 3-18 shows the lubrication points of throwout bearings. Light graphite grease is recommended.

Careful: Never clean the bearing in solvent or carburetor cleaner. The bearing is prelubricated and sealed. Liquid cleaners remove the lubricant and ruin the bearing.

11. Check the fork for wear on throwout-bearing at-

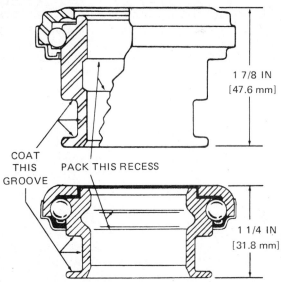

Fig. 3-18 Release-bearing lubrication points. (*Chevrolet Motor Division of General Motors Corporation*)

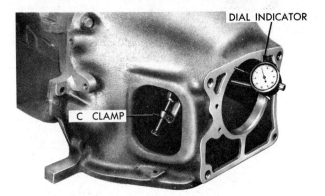

Fig. 3-20 Measuring clutch-housing face for squareness with a dial indicator. (*Chrysler Corporation*)

tachments or other damage. On reassembly, be sure that the dust seal or cover is in good condition.

Careful: Steam cleaning can cause clutch trouble. Steam may enter and condense on the facings of the friction disk, pressure plate, and flywheel. The disk facings will absorb moisture. If the car is allowed to stand with the facings wet, they may stick to either the flywheel or the pressure plate. Then the clutch will not disengage. To prevent this, start the engine immediately after steam cleaning. Slip the clutch slightly to heat and dry the facings. Do not overdo it or you could burn out the clutch.

12. Check the alignment of the clutch housing (✿ 3-25).

✿ 3-25 Checking clutch-housing alignment
Whenever a clutch has been serviced, the clutch hous-

ing should be checked for alignment. This procedure includes checking the housing-bore runout and housing-face squareness (Figs. 3-19 and 3-20).

To check bore runout, substitute a 3-inch [76-mm] bolt for one of the crankshaft bolts. Mount a dial indicator in the bore, as shown in Fig. 3-19. Slowly rotate the engine clockwise to check runout. If the runout is excessive, it can be corrected by installing offset dowels (Fig. 3-21).

Dowels are available with varying amounts of offset. To install the dowels, remove the clutch housing and the old dowels. The dial indicator shows you how much the bore is out of alignment and in which direction. This determines which pair of dowels to select (the pair with the correct amount of offset). The slots in the dowels should align in the direction of maximum bore runout to correct the alignment.

To check housing-face squareness, reposition the dial indicator as shown in Fig. 3-20. Slowly rotate the engine clockwise and note how much the housing face is out of line. To correct alignment, place shim stock of the necessary thickness in the proper positions between the clutch housing and engine block.

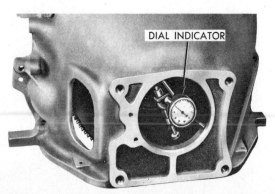

Fig. 3-19 Measuring clutch-housing-bore runout with a dial indicator. (*Chrysler Corporation*)

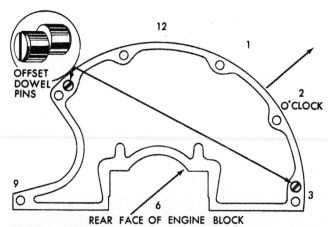

Fig. 3-21 Offset dowel diagram. (*Chrysler Corporation*)

Chapter 3 review questions

Select the *one* correct, best, or most probable answer to each question. Then check your answers against the correct answers given at the end of the book.

1. Clutch slippage while the clutch is engaged is particularly noticeable:
   a. during idle
   b. at low speed
   c. when starting the engine
   d. during acceleration
2. Clutch chattering or grabbing is noticeable:
   a. during idle
   b. at low speed
   c. when engaging the clutch
   d. during acceleration
3. Clutch dragging is noticeable:
   a. when the clutch is disengaged
   b. at road speed
   c. during acceleration
   d. at high speed
4. Clutch noises are usually most noticeable when the engine is:
   a. accelerating
   b. decelerating
   c. idling
   d. being started
5. Clutch-pedal pulsation is noticeable when the engine is running and:
   a. accelerating
   b. a slight force is applied to the pedal
   c. decelerating
   d. the car is moving at steady speed
6. Slippage between the friction-disk facings and the flywheel or pressure plate will cause:
   a. clutch-pedal pulsation
   b. rapid facing wear
   c. excessive acceleration
   d. rapid pressure-plate wear
7. Clutch-pedal free travel, or pedal lash, is the distance the pedal moves before the release bearing comes up against the:
   a. release levers
   b. flywheel
   c. floorboard
   d. stop
8. Heat checks or cracks on the flywheel and pressure-plate faces will cause:
   a. excessive clutch slippage
   b. rapid flywheel and pressure-plate wear
   c. rapid wear of friction-disk facings
   d. excessive pedal pulsation
9. The reason for the caution about the asbestos in the clutch facings is that asbestos dust can:
   a. make you sneeze
   b. cause eye irritation
   c. cause lung cancer
   d. damage the clutch
10. Clutch slippage can be caused by all of the following *except:*
    a. incorrect linkage adjustment
    b. loose friction-disk facings
    c. grease on the facings
    d. broken or weak pressure springs

MANUAL TRANSMISSIONS, TRANSAXLES, AND TRANSFER CASES

After studying this chapter, you should be able to:

1. Discuss the purpose and operation of typical manual transmissions.
2. Explain the difference between three-speed, four-speed, and five-speed transmissions.
3. Describe how shifting is accomplished.
4. Explain the purpose of overdrive and how it is achieved.
5. Explain the purpose and operation of manual transaxles.
6. Discuss the purpose and operation of the transfer case.
7. Describe the operation of a planetary-gear set.

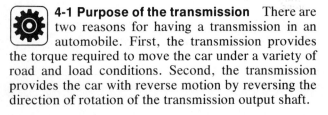

4-1 Purpose of the transmission   There are two reasons for having a transmission in an automobile. First, the transmission provides the torque required to move the car under a variety of road and load conditions. Second, the transmission provides the car with reverse motion by reversing the direction of rotation of the transmission output shaft.

✿ 4-2 Function of the transmission   The primary function of the transmission is to provide a means of varying the speed ratio between the engine and the drive wheels of the car. To do this, the simplest manual passenger-car transmission provides three gear ratios between the engine crankshaft and the drive wheels (✿ 1-6). The crankshaft revolves about 4, 8, or 12 times per drive-wheel revolution. This permits the engine to operate at a fairly high speed and produce greater torque to set the car in motion.

Without the numerically high gear ratio, the engine would turn so slowly that it could deliver little power. The engine crankshaft must turn fairly fast before the engine develops sufficient power to start the car moving.

On first starting, the transmission gears are placed in *low* so that the engine crankshaft turns approximately 12 times for each wheel revolution. The clutch is then engaged so that power is applied to the wheels. Car speed increases with engine speed until the car is moving 5 to 10 miles per hour (mph) [8 to 16 kilometers per hour (km/h)]. At this time the engine crankshaft may be turning as fast as 2000 rpm.

Then the clutch is disengaged and the engine crankshaft speed reduced to permit gear changing. The gears are shifted into *second,* and the clutch is again engaged. Since the ratio is now about 8:1, a higher car speed is obtained as engine speed is again increased.

The transmission is then shifted into *high* (on a three-speed transmission), the clutch being disengaged and engaged for this operation. The ratio between the engine and wheels will then be approximately 4:1. The engine crankshaft will turn four times to cause the wheels to turn once.

After the car begins moving, a numerically lower gear ratio between the engine crankshaft and the drive wheels is needed. This is because in first gear the usable medium-to-maximum engine speed may limit maximum car speed to only 15 to 20 mph [24 to 32 km/h]. Shifting to second gear provides a gear ratio of about 8:1 between the engine crankshaft and the drive wheels. Medium-to-maximum engine speed may produce a car speed above 30 mph [48 km/h]. But higher car speeds may cause overspeeding of the engine, which results in excessive engine noise and wear, poor fuel economy, and possible engine damage.

Shifting to third gear provides a gear ratio of about 4:1 between the engine crankshaft and the drive wheels. This ratio permits higher car speeds at lower engine speeds. It allows the engine to operate at lower, more efficient speeds while the car is traveling at cruising or highway speeds of up to 55 mph [89 km/h].

While the car is in motion and in third gear, normally it is not necessary to shift gears except when the car is slowed or brought to a stop or when additional power at low car speed is required. The procedure of shifting from a higher gear to a lower gear, such as from third gear to second gear, is called *downshifting*. Downshifting provides the additional power that may be needed when climbing a hill or when rapid acceleration from a low car speed is desired. Shifting from a lower gear to a higher gear, such as from second to third after the car is up to highway speed, is called *upshifting*.

NOTE: The fundamental operation of a three-speed sliding-gear transmission is covered in ✿ 1-6. Review this section if you do not thoroughly understand how it works. The modern transmissions discussed in the following sections require that you know how the basic three-speed manual transmission operates.

✿ 4-3 Four-speed and five-speed transmissions
Today more cars are equipped with four-speed transmissions than with the three-speed type. For example, in recent years, about 17 percent of the cars built in the United States were equipped with four-speed transmissions. About ½ percent had five-speed transmissions. A few had three-speed transmissions. The other 82 percent were equipped with automatic transmissions.

Five-speed manual transmissions are like the four-speed units but have a fifth forward speed which is actually an overdrive. In overdrive, there is a speed increase through the transmission instead of speed reduction or direct drive. This allows lower engine rpm to maintain the car at highway speed. Overdrive is covered in ✿ 4-11.

Many cars, especially the smaller ones, now have front-wheel drive. In these cars, typically a four-speed transmission is combined with the differential. This combination is called a *transaxle*. Chapters 11 to 13 cover front-wheel drives and transaxles. Four-wheel-drive vehicles have a transfer case attached to the rear of the transmission (✿ 4-17 to 4-19). Figure 4-1 compares the exterior views of various manual transmissions with a typical transaxle and a transfer case.

✿ 4-4 Types of manual transmissions
In addition to three-, four-, and five-speed manual transmissions, there are several other types. Some transmissions used on heavy-duty trucks and other equipment may have as many as 10 forward speeds and 2 reverse speeds. These are combination units using a five-speed transmission with a two-speed auxiliary transmission mounted in back of the main transmission. By using the two speeds in the auxiliary transmission, the driver obtains twice the number of gear ratios available from the main transmission.

Essentially there is little difference in the basic operation of the various types of manual transmissions. However, those providing more gear ratios have additional gears and shifting positions.

Modern transmissions have synchromesh devices in them to make gear shifting easier. These devices ensure that gears that are about to be meshed are revolving at synchronized speeds. This means that the gear teeth

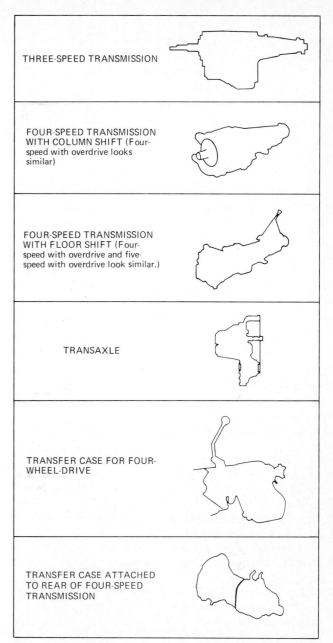

Fig. 4-1 Exterior views of various manual transmissions compared with a transaxle and a transfer case. (*ATW*)

THREE-SPEED TRANSMISSION

FOUR-SPEED TRANSMISSION WITH COLUMN SHIFT (Four-speed with overdrive looks similar)

FOUR-SPEED TRANSMISSION WITH FLOOR SHIFT (Four-speed with overdrive and five-speed with overdrive look similar.)

TRANSAXLE

TRANSFER CASE FOR FOUR-WHEEL-DRIVE

TRANSFER CASE ATTACHED TO REAR OF FOUR-SPEED TRANSMISSION

that are about to mesh are moving at the same speed. As a result, the teeth mesh without any clashing of gears. Following sections describe the action of these devices.

✿ 4-5 Four-wheel drive with transfer case
Many utility vehicles, some trucks, and a few cars have four-wheel drive (Figs. 1-24 and 4-2). Engine power can flow to all four wheels. With all four wheels driving, the vehicle can travel over rugged terrain and up steep grades. It can go through rough or muddy ground where two-wheel-drive cars would stall or get stuck. A *transfer case* is required on vehicles with four-wheel drive.

The transfer case (Fig. 4-2) is an auxiliary transmis-

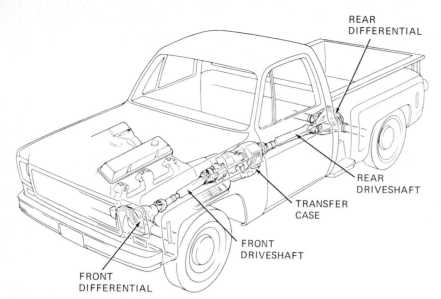

REAR DIFFERENTIAL

REAR DRIVESHAFT

TRANSFER CASE

FRONT DRIVESHAFT

FRONT DIFFERENTIAL

Fig. 4-2 Pickup truck with four-wheel drive. (*Chevrolet Motor Division of General Motors Corporation*)

sion mounted in back of the main transmission. By shifting gears in the transfer case, engine power is divided and transferred to both the front and rear differentials. Transfer cases in automotive vehicles are classed as *full-time* or *part-time,* depending on whether or not the front axle is engaged automatically as soon as the rear wheels begin to spin. With part-time four-wheel drive, the transfer-case shift lever must be moved to engage or disengage the front differential.

Notice in Figs. 4-1 and 4-2 how the design of the transfer case allows the front driveshaft to be placed to one side of the engine crankcase. By not running the front driveshaft under the crankcase, the vehicle ground clearance is increased.

The typical transfer case may be operated in either of two modes. In one, both the front and the rear wheels are driven. In the other, only the rear wheels are driven. In most vehicles, a transfer case also provides the driver with a selection of either of two drive speeds, or *ranges,* high or low. The change from the high-speed range to the low-speed range is made when the driver moves the transfer-case shift lever. As the shift lever is moved, it moves a gear on the main driveshaft in the transfer case from engagement with the high-speed drive gear to engagement with the low-speed drive gear. High speed in the transfer case provides direct drive, or a gear ratio of 1:1. Low speed usually produces a gear ratio of about 2:1.

The front axle is engaged by shifting a sliding gear or a "dog" clutch into engagement with a driven gear on the front-wheel driveshaft inside the transfer case. The sliding gears and clutches are positively driven by splines on the shafts.

Figure 4-3 shows in simplified view the power flow through one type of transfer case. Notice that there are two parts that are moved by the transfer-case shift lever to provide the various gear combinations. A sliding gear on the main shaft locks either the low-speed gear or the high-speed gear to the main shaft. In many transfer cases, this shift cannot be made unless the trans-

mission is in neutral (or in park in an automatic transmission). Otherwise, the transfer-case main shaft and the sliding gear, which is splined to it, will be turning. Gear clash will result.

To engage and disengage the front axle, another sliding gear or a clutch is used to lock the front-axle drive gear to the front-axle driveshaft. In most transfer cases in automotive vehicles today, the front axle can be engaged or disengaged while the vehicle is moving. All the driver must do is release the accelerator pedal to remove the torque load through the gears. Then the shift lever can be moved as desired. However, both the front and rear wheels must be turning at the same speed. If the rear wheels have lost traction and are spinning, or if the brakes are applied and either the front or rear wheels are locked and sliding, gear clashing will occur when engagement of the front axle is attempted.

Various types of transfer cases are installed in automotive vehicles. Some have all gears, as described above. Others are full-time units which have a chain that drives the front driveshaft, instead of gears. This reduces the weight of the transfer case, improving fuel economy. In addition, some full-time transfer cases, such as those used in the American Motors Eagle models, have no low range.

While there is a slight fuel-mileage penalty when full-time four-wheel drive is used, it has certain advantages. For example, when a wheel spins on a vehicle equipped with part-time four-wheel drive, power continues to flow to the spinning wheel. As a result, so little power may reach the other axle that the vehicle may not move, under certain conditions. With full-time four-wheel drive, the transfer case transfers the power from the axle with the spinning wheel to the other axle. This improves traction and keeps the vehicle moving.

The operation of various types of transfer cases is covered in ✿ 4-17 to 4-19. Chapters 15 and 16 explain the trouble diagnosis and service of the various types of transfer cases used in four-wheel-drive vehicles.

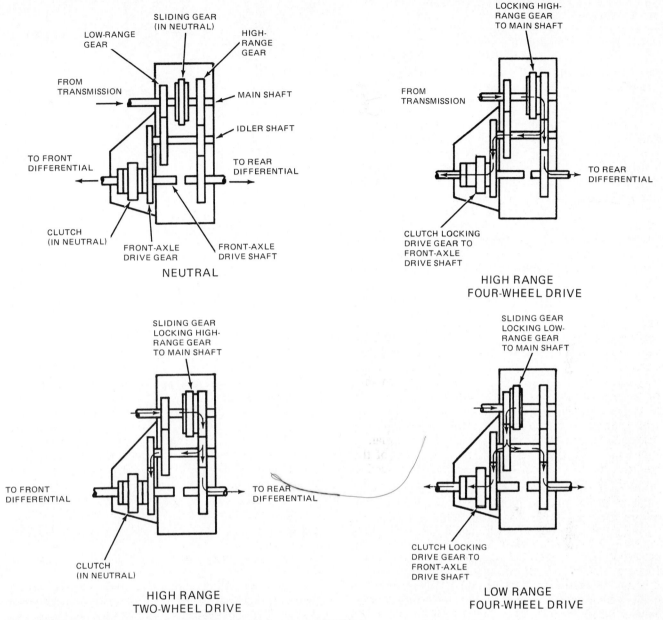

Fig. 4-3 Basic operation of a simplified transfer case. (*ATW*)

✿ 4-6 Steering-column and floor shifts
Years ago, the gearshift lever was always located on the floor of the driver's compartment. The lower end of the lever was attached to the shifting devices in the transmission case. Later, the gearshift lever was moved up to the steering column, where it was more readily accessible to the driver. This position also provided more leg room for the center passenger in the front seat.

When the transmission shift lever is mounted on the steering column, the car is said to have a *column shift*. When a transmission has a floor-shift lever (instead of the column-shift lever), it is called a *floor shift*. A four-speed transmission with a floor shift is often called "four on the floor." Today, many cars with manual transmissions have the gearshift lever back on the floor.

A simple three-speed manual transmission which uses the floorboard-mounted gearshift lever is covered in ✿ 1-6. Figures 1-13 to 1-17 show how the linkage to the shifter yokes or forks works in the transmission. The steering-column-mounted gearshift lever with its linkages is a little more complicated. However, driver operation of each type results in the same action. Movement of the shift lever causes gears to move into or out of mesh (or causes other devices to work) to change the gear ratio by changing the gears through which power flows through the transmission.

Two separate motions of the gearshift lever are required in shifting gears, selection and shifting. The first motion selects the gear to be acted upon. The second moves the gear (or other device) in the proper direction to complete the shift, or change the operating gear ratio. Several different types of selector and shifter devices are in use.

Figures 4-4a and 4-4b show the standard shift patterns for steering-column and floor-shift levers for three-speed transmissions. Below we discuss the linkages for these shift-lever locations. A car with a manual transmission is sometimes referred to as having a *stick shift*.

1. **Steering-column shift** Figures 4-5 and 4-6 show one type of steering-column shift and its linkages to the transmission. To shift into first or reverse, the driver depresses the clutch pedal to momentarily disconnect the engine from the transmission. Then, the driver lifts the shift lever and moves it up for reverse or down for first gear. When the lever is lifted, it pivots on its mounting pin, which forces a tube or rod downward in the steering column. This downward movement pushes downward on a crossover blade at the bottom of the steering column. A slot in the blade engages a pin on the first-and-reverse shift lever (Fig. 4-7).

When the shift lever is moved into first, the first-and-reverse lever is rotated. This movement is carried by the linkage to the transmission (Fig. 4-8). At the transmission, the movement causes the first-and-reverse shift lever to move. This lever is on a shaft that extends through the transmission side cover (Fig. 4-9). There is a lever on the inside end of this shaft, and a shifter fork is mounted on this lever.

When the shaft is rotated by movement of the first-and-reverse shift lever, it causes the shifter fork to move backward or forward inside the transmission. This motion shifts the first-and-reverse sliding gear into mesh.

To shift into second or third, the driver moves the shift lever into the positions shown in Fig. 4-4. This movement causes the slot in the blade to engage the second-and-third shift lever at the bottom of the steering column. The lever is moved to actuate the second-and-third linkage to the transmission.

2. **Floor shift** Many cars have a floor-mounted shift lever. A transmission using a floor-shift lever is shown in Fig. 4-10. Figure 4-11 shows the linkage on an automobile that has a *console*. In the automobile, a

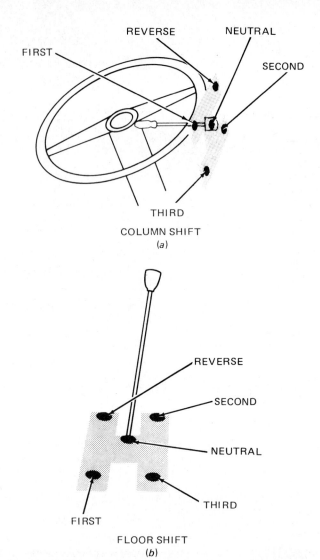

COLUMN SHIFT
(a)

FLOOR SHIFT
(b)

Fig. 4-4 Gearshift patterns for (a)steering-column and (b) floor-shift levers.

console is a small cabinet or raised decorative centerpiece on the floor of the front compartment between the two front bucket seats. It houses the shift lever and sometimes includes a glove compartment, ashtray,

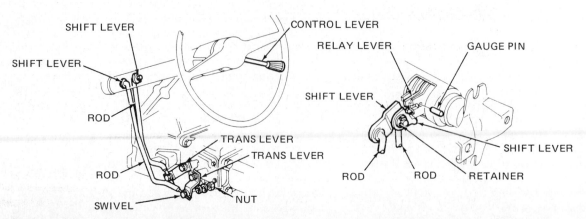

Fig. 4-5 Linkage between steering-column shift (control) lever and transmission. (*Chevrolet Motor Division of General Motors Corporation*)

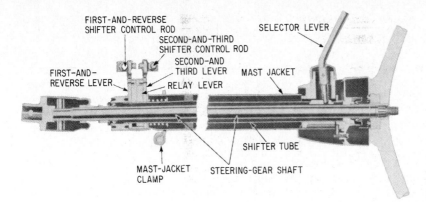

Fig. 4-6 Sectional view of a steering mast, or column, showing gearshift controls. (*Chevrolet Motor Division of General Motors Corporation*)

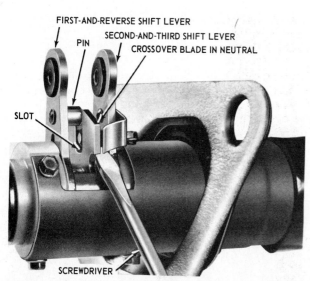

Fig. 4-7 Shift levers and crossover blade at the bottom of the steering column. The screwdriver holds the crossover blade in neutral for an adjustment check. (*Chrysler Corporation*)

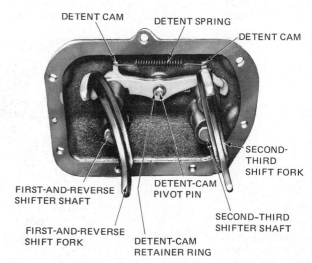

Fig. 4-9 Transmission side cover viewed from inside the transmission. The shift forks are mounted on the ends of levers attached to shafts. The shafts can rotate in the side cover. Detent cams and springs prevent more than one of the shift forks from moving at any one time. (*Chevrolet Motor Division of General Motors Corporation*)

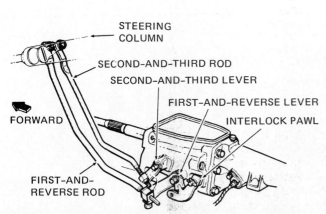

Fig. 4-8 Gearshift linkage between the shift levers at the bottom of the steering column and the transmission levers on the side of the transmission. (*Chrysler Corporation*)

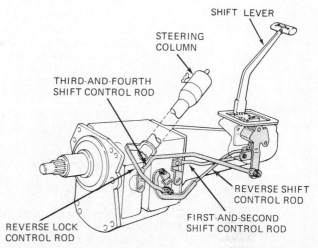

Fig. 4-10 A four-speed floor-shift transmission showing the linkage from the shift lever to the transmission. (*Ford Motor Company*)

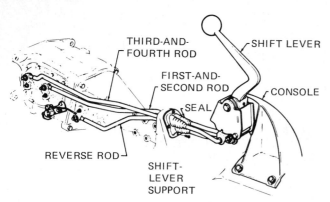

Fig. 4-11 Linkage between a four-speed transmission and a console-mounted floor-shift lever. (*Ford Motor Company*)

gauges, and various controls, such as electric window switches and heater and air-conditioner controls.

In the shift-lever support there are levers attached to each rod. These levers have slots that are selected by a tongue on the lower end of the shift lever as it is moved into the various gear positions. The first movement of the shift lever makes this selection. Then the second movement of the shift lever causes the selected lever and rod to move. This causes the transmission lever to move and thereby shift the selected gear into the selected gear position.

NOTE: Although first and reverse gear ratios may be obtained by moving a sliding gear in some transmissions, the other gears run in constant mesh. It is common practice to describe movement of the shift lever as "shifting gears." However this means only operation of the synchronizers (✿ 4-8) and not actual sliding of the gear into and out of mesh.

Some transmissions do not use linkage rods between the gearshift lever and the transmission. Instead, they use a single-rail or shifter shaft (Figs. 4-12 and 4-13). The lower end of the gearshift lever moves into the bracket on the end of the shifter shaft when a gear is selected. Then further movement of the gearshift lever causes the shifter shaft to move. This then moves the fork that has been selected. The fork and the synchronizer sleeve it surrounds move in the proper direction to produce the selected gear position. Construction and operation of the synchronizer unit is covered in ✿ 4-9.

✿ 4-7 Designations of manual transmissions Manual transmissions are designated according to the number of forward gears (three, four, five) and the distance between the center lines of the main shaft and the countergear shaft. For example, Chevrolet has a

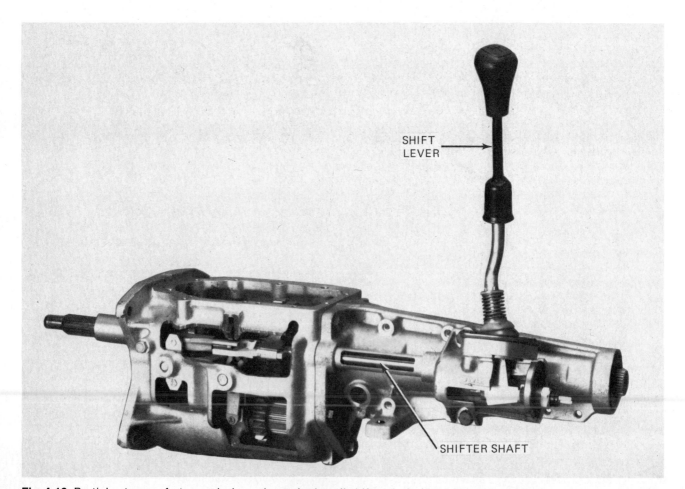

Fig. 4-12 Partial cutaway of a transmission using a single-rail shifter shaft. (*Ford Motor Company*)

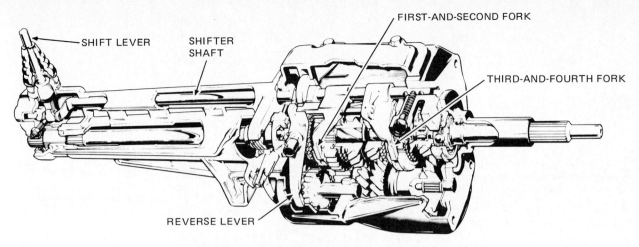

Fig. 4-13 Cutaway view of a single-rail four-speed transmission showing how the shifter shaft is connected to the shifter forks. (*Ford Motor Company*)

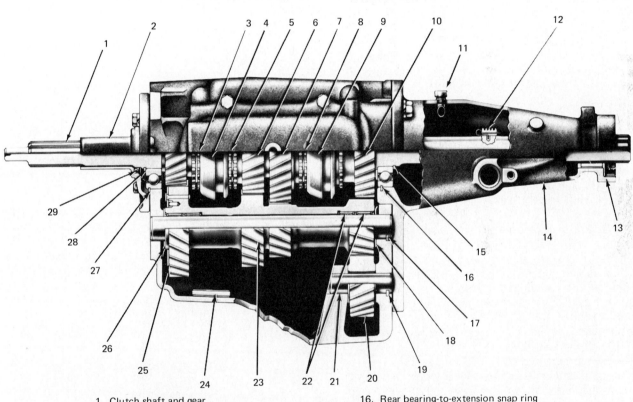

1. Clutch shaft and gear
2. Clutch-gear bearing retainer
3. Third-speed synchronizer ring
4. Second-third-speed-clutch assembly
5. Second-speed synchronizer ring
6. Second-speed gear
7. First-speed gear
8. First-speed synchronizer ring
9. First-reverse-clutch assembly
10. Reverse gear
11. Vent
12. Speedometer gear and clip
13. Rear-extension seal
14. Rear extension housing
15. Rear bearing-to-shaft snap ring

16. Rear bearing-to-extension snap ring
17. Countergear woodruff key
18. Thrust washer
19. Reverse-idler-shaft woodruff key
20. Reverse idler gear
21. Reverse idler shaft
22. Countergear bearings
23. Countergear
24. Case magnet
25. Antilash plate assembly
26. Thrust washer
27. Clutch-gear bearing
28. Snap ring
29. Clutch-gear-retainer lip seal

Fig. 4-14 Sectional view of a three-speed transmission that has synchronizers for all forward gears. (*Chevrolet Motor Division of General Motors Corporation*)

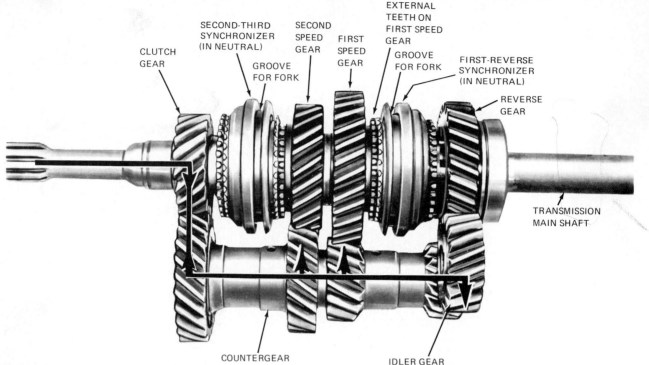

Fig. 4-15 Gear train and shafts in the three-speed transmission. (*Chevrolet Motor Division of General Motors Corporation*)

three-speed 76-mm [2.99 inches] transmission and a four-speed 117-mm [4.64 inches] transmission. The greater the distance between the main shaft and the countergear shaft center lines, the larger the gears. Manual transmissions for larger and more powerful engines usually have larger gears.

✿ 4-8 Gears in a three-speed transmission Now let us look at the operation of the gears in a three-speed transmission (Figs. 4-14 to 4-22). Figure 4-15 shows the gears and shafts in their proper relationship as they are located in the assembled transmission. Figure 4-16 shows the gears only. The gears and other parts in these two illustrations are shown in their position when the transmission is in neutral. No power is being transmitted through the transmission.

The countergear assembly has four gears in it, similar to the countergear assembly in the simplified three-speed transmission shown in Fig. 1-13. The clutch shaft and gear are also similar. However, in the two transmissions, the main shaft and the gears on it are different.

In the transmission shown in Fig. 4-15, the second-speed and first-speed gears do not slide back and forth along the main shaft. They are in constant mesh with their respective gears on the countergear assembly. Also, these two gears are supported on bearings that allow them to rotate independently on the main shaft.

The hubs of the two synchronizer assemblies are splined to the main shaft and rotate with it. However, the synchronizer sleeves of the two assemblies can slide back and forth along the splines on the synchronizer hub. The forks shown in Fig. 4-9 fit in the grooves

of the synchronizers (shown in Fig. 4-15). When the gearshift lever is moved, the linkage selects one of the two synchronizers. The proper fork then moves the selected synchronizer, which then locks one of the gears to the main shaft.

1. **First-speed gear** To shift into first gear, the clutch is disengaged. Then the gearshift lever is moved into first. Movement of the gearshift lever causes the linkage to select the first-reverse synchronizer sleeve and move it to the left (in Fig. 4-17). As the synchro-

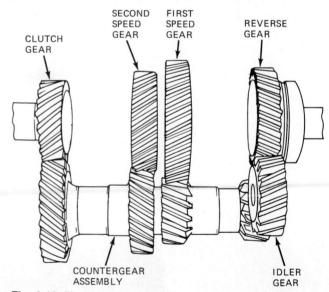

Fig. 4-16 The gears in the three-speed transmission.

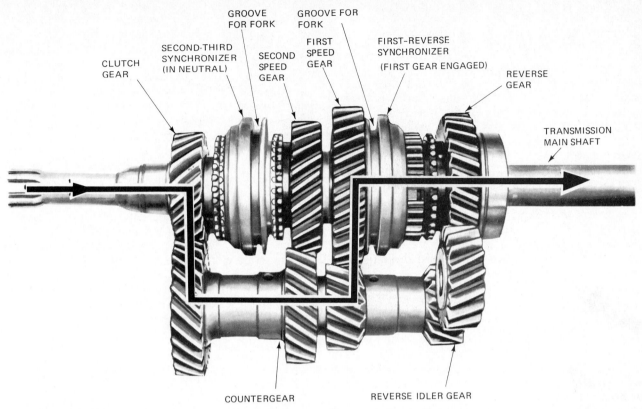

CLUTCH GEAR — SECOND-THIRD SYNCHRONIZER (IN NEUTRAL) — GROOVE FOR FORK — SECOND SPEED GEAR — GROOVE FOR FORK — FIRST SPEED GEAR — FIRST-REVERSE SYNCHRONIZER (FIRST GEAR ENGAGED) — REVERSE GEAR — TRANSMISSION MAIN SHAFT

COUNTERGEAR — REVERSE IDLER GEAR

Fig. 4-17 Power flow through the gear train of a three-speed transmission in first gear. (*Chevrolet Motor Division of General Motors Corporation*)

nizer is moved to the left, the internal splines in the synchronizer engage the external splines on the first-speed gear. This locks the first-speed gear, through the synchronizer, to the main shaft. Figure 4-18 shows the external splines on the first-speed gear. The internal splines on the synchronizer are shown in Fig. 4-19.

When the clutch is engaged and the transmission is in first gear, the power flow through the transmission is as shown in Fig. 4-17. There is gear reduction as the small clutch gear drives the large gear on the countergear. There is also gear reduction as the small first gear on the countergear drives the large first-speed gear on

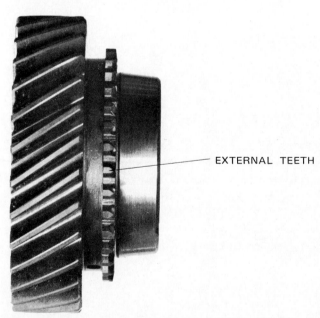

EXTERNAL TEETH

Fig. 4-18 First-speed gear. (*Chevrolet Motor Division of General Motors Corporation*)

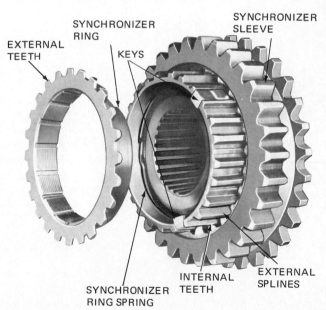

EXTERNAL TEETH — SYNCHRONIZER RING — KEYS — SYNCHRONIZER SLEEVE

SYNCHRONIZER RING SPRING — INTERNAL TEETH — EXTERNAL SPLINES

Fig. 4-19 Synchronizer assembly. (*Chevrolet Motor Division of General Motors Corporation*)

the main shaft. The total gear reduction in first-speed gear is a little less than 3:1 in most three-speed transmissions (2.99:1 in one model).

2. Second-speed gear Figure 4-20 shows the transmission in second-speed gear. The first-reverse synchronizer has been moved to its center or neutral position and out of mesh with the first-speed gear. The second-third synchronizer has been moved to the right so that its internal teeth engage the external teeth on the second-speed gear. The power flow is as shown by the arrow in Fig. 4-20.

The second-speed gear is smaller than the first-speed gear. Also, the countergear that meshes with the second-speed gear is larger than the countergear that meshes with the first-speed gear. Therefore the gear reduction in second gear is less than in first gear. Gear reduction is a little less than 2:1 in most three-speed transmissions (1.83:1 in one model).

3. Third-speed gear In third-speed gear, the first-reverse synchronizer remains in neutral. Power flow is straight through the transmission, as shown in Fig. 4-21. Third gear is achieved by moving the second-third synchronizer to the left, as shown in Fig. 4-21. Its internal teeth engage the external teeth of the clutch gear so that the drive is through these teeth and the synchronizer splines to the main shaft. With the clutch shaft now locked to the main shaft, both shafts turn at the same speed. This is also called *direct drive*. The gear ratio is 1:1.

4. Reverse To obtain reverse, the second-third synchronizer is placed in neutral. Then an extra gear is inserted into the gear train. This gear is called the *reverse idler gear*. It is in constant mesh with the fourth gear on the countergear, as shown in Fig. 4-22.

When the shift lever is moved to reverse, the linkage moves the first-reverse synchronizer to the right, as shown in Fig. 4-22. The internal splines in the synchronizer engage the external splines on the reverse gear. Now the power flow through the transmission is as shown in Fig. 4-22.

Because of the extra gear in the gear train, the main shaft turns in the reverse direction. Therefore the drive wheels also turn in the reverse direction, and the car moves backward. The gear ratio is about 3:1 in most three-speed transmissions (3.00:1 in one model).

5. Interlock To prevent both shift forks from moving in the transmission at the same time, an interlock device is used (Figs. 4-9 and 4-23). This device locks the inoperative shift fork as the other fork is moved into the shifted position. A shift cannot be made into any gear by either fork unless the other fork and its shifter shaft are in their neutral position. This prevents shifting into two gear positions at the same time, which would lock the main shaft and could result in severe transmission damage.

Other types of interlock devices have been used in addition to those shown in Figs. 4-9 and 4-23. One type consists of a spring-loaded ball or plunger that is moved into a notch or hole in the inoperative shifter shaft or shift fork by movement of the shift lever.

⚙ **4-9 Synchronizers** To prevent gear clash during shifting while the car is in motion, synchronizing devices are used in automotive transmissions. These devices ensure that gears that are about to mesh will be

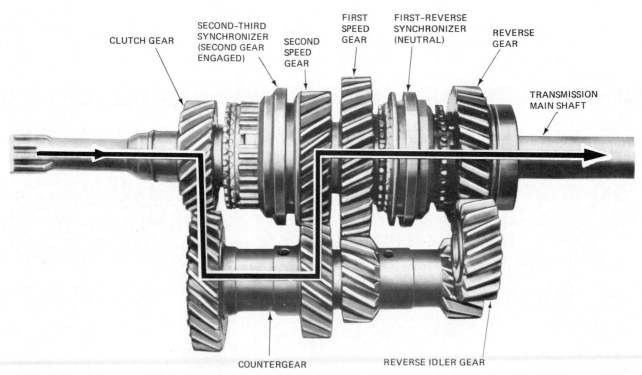

Fig. 4-20 Power flow through a three-speed transmission in second gear. (*Chevrolet Motor Division of General Motors Corporation*)

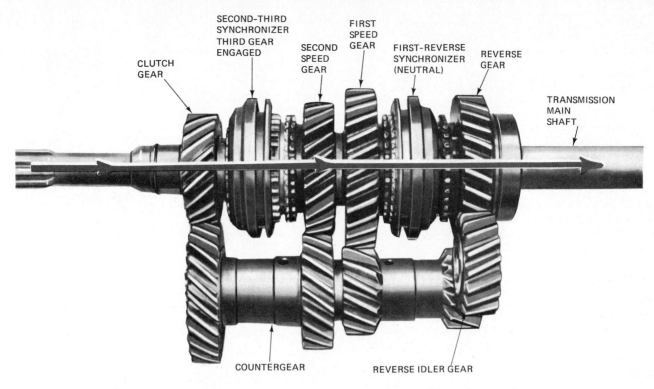

Fig. 4-21 Power flow through a three-speed transmission in third gear. (*Chevrolet Motor Division of General Motors Corporation*)

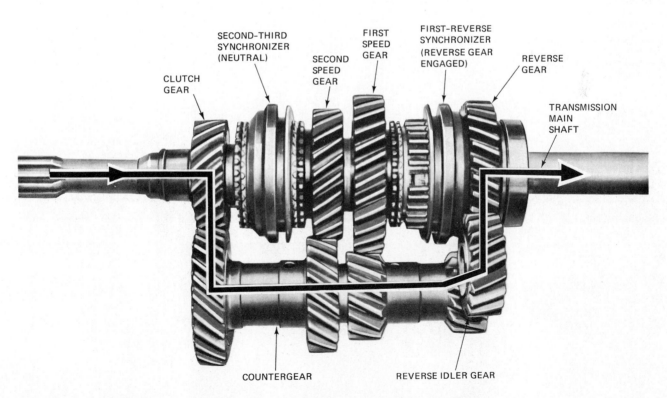

Fig. 4-22 Power flow through a three-speed transmission in reverse. (*Chevrolet Motor Division of General Motors Corporation*)

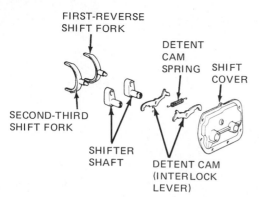

Fig. 4-23 Disassembled view of the shift cover, showing the interlock device. (*Chevrolet Motor Division of General Motors Corporation*)

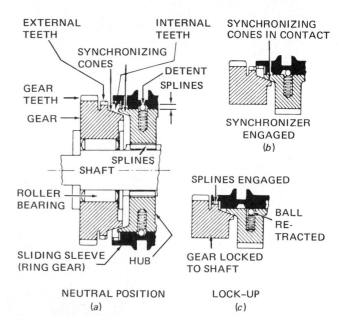

Fig. 4-24 Operation of a transmission synchronizing device that uses cones.

rotating at the same speed, so they will engage smoothly.

One type of synchronizer uses synchronizing cones on the gears and on the synchronizing hub (Fig. 4-24). In the neutral position, the sliding sleeve is held in place by spring-loaded balls resting in detents in the sleeve. When a shift starts, the hub and sleeve, as an assembly, are moved toward the selected gear (to the left in Fig. 4-24). The first contact is between the synchronizing cones on the gear and hub. As the two cones are forced together (upper right in Fig. 4-24), they are brought into synchronization. Both rotate at the same speed.

Further movement of the shift fork forces the sliding sleeve on toward the selected gear. The internal teeth on the sliding sleeve match the external teeth on the gear. With both the gear and the main shaft rotating at the same speed, the sleeve slides over the teeth on the gear without clashing. Now the gear is locked to the

main shaft through the sliding sleeve (lower right in Fig. 4-24) and the shift is completed. Notice that the sliding sleeve moves off center from the hub and over the balls for engagement. This pushes the balls down against their springs.

NOTE: The synchronizer sliding sleeve is called a *synchronizer sleeve* and a *clutch sleeve*.

Another type of synchronizer is shown partly disassembled in Fig. 4-25. Instead of retracting balls, as in Fig. 4-24, this synchronizer has three keys and a pair

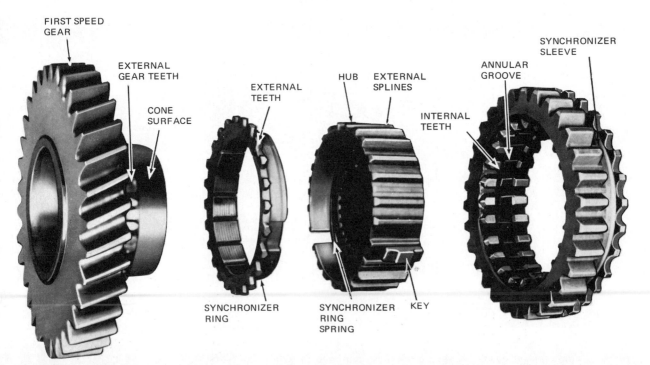

Fig. 4-25 Disassembled view of a synchronizer. (*Chevrolet Motor Division of General Motors Corporation*)

of ring-shaped synchronizing springs. The keys are assembled in slots in the hub, which is splined to the main shaft. Assembled outside the hub is the synchronizing sleeve. The hub has external splines that fit the internal splines of the sleeve. The three keys have raised sections that fit in the annular groove of the sleeve.

NOTE: The sleeve shown in Fig. 4-25 has external teeth, but the synchronizing sleeves in the transmission shown in Fig. 4-15 do not. However, the action is the same.

Synchronizing is a four-stage action: First, the sleeve is moved toward the first-speed gear (when shifting to first). The sleeve slides on the hub splines and carries the three keys with it. Second, the keys move up against the synchronizer ring and push the ring toward the first-speed gear. The ring presses against the cone of the

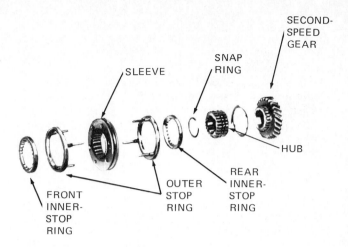

Fig. 4-26 A pin-type synchronizer. (*Chrysler Corporation*)

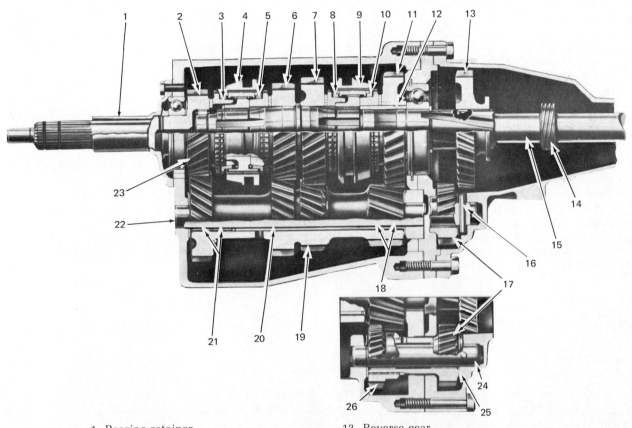

1. Bearing retainer
2. Main drive gear
3. Fourth-speed synchronizing ring
4. Third-and-fourth-speed-clutch assembly
5. Third-speed synchronizing ring
6. Third-speed gear
7. Second-speed gear
8. Second-speed synchronizing ring
9. First-and-second-speed-clutch assembly
10. First-speed synchronizing ring
11. First-speed gear
12. First-speed-gear sleeve
13. Reverse gear
14. Speedometer drive gear
15. Main shaft
16. Reverse-idler-shaft roll pin
17. Reverse idler gear (rear)
18. Countergear bearing roller
19. Countergear
20. Countershaft-bearing-roller spacer
21. Countershaft bearing roller
22. Countergear shaft
23. Oil slinger
24. Reverse idler shaft
25. Thrust washer
26. Reverse idler gear (front)

Fig. 4-27 Sectional view of a four-speed transmission. (*Chevrolet Motor Division of General Motors Corporation*)

first-speed gear. Third, further sleeve movement causes the keys to be pressed out of the annular groove in the sleeve. The sleeve continues to move toward the first-speed gear. The friction between the synchronizing ring and the first-speed gear brings the two into synchronous rotation. Fourth, movement of the sleeve allows the internal teeth of the sleeve to engage the external teeth of the first-speed gear. Meshing is completed. Similar actions take place in the shifts to second and third.

The pin-type synchronizer (Fig. 4-26) has a pair of stop rings. Each has three pins which pin it to the sleeve. The hub is splined to the main shaft. External teeth on the hub mesh with internal teeth on the sleeve. Therefore, the hub, sleeve, and two stop rings are always rotating with the main shaft.

When a shift is made into second, for example, the main shaft and associated parts may be rotating at a different speed from the second-speed gear. However, as the sleeve is moved toward the second-speed gear, the rear inner stop ring moves against the face of the second-speed gear.

This brings the stop ring into synchronous rotation with the sleeve. It permits alignment of the external teeth on the hub and the teeth on the small diameter of the second-speed gear. Now, the sleeve can slip over the teeth of the second-speed gear to lock together the second-speed gear and the hub. Then, when the driver releases the clutch pedal and the engine again delivers power, the second-speed gear drives the main shaft through the synchronizer hub and sleeve.

The action in a shift to third is very similar.

✿ 4-10 Four-speed transmission
The four-speed transmission has four forward speeds, neutral, and reverse. In some four-speed transmissions, fourth speed or "gear" is an overdrive ratio. This means that the main shaft is overdriven. It turns more than one com-

plete revolution for every complete revolution of the transmission input shaft.

One type of four-speed transmission which has synchromesh action in all four forward gears in shown in Fig. 4-27. The gears and shafts are shown in neutral in Fig. 4-28. In this transmission, fourth gear is direct drive, or 1:1.

1. **First gear** Figure 4-29 shows the power flow through the transmission in first gear. The first-second synchronizer has been moved to the right so that its internal splines engage the external splines of the first-speed gear.

2. **Second gear** Figure 4-30 shows the power flow through the transmission in second gear. The first-second synchronizer has been moved to the left so that its internal splines engage the external splines of the second-speed gear.

3. **Third gear** Figure 4-31 shows the power flow through the transmission in third gear. The third-fourth synchronizer has been moved to the right so that its internal splines engage the external splines of the third-speed gear.

4. **Fourth gear** Figure 4-32 shows the power flow through the transmission in fourth gear. The third-fourth synchronizer has been moved to the left so that its internal splines engage the external splines of the clutch gear.

5. **Reverse** In reverse, both synchronizers are in the neutral position. The reverse gear, which is a sliding gear splined to the transmission main shaft, has been moved to the left (as shown in Fig. 4-33) so that it engages the rear reverse idler gear. Now, the rear reverse idler gear causes the main shaft to turn in the reverse direction, and so the car moves backward.

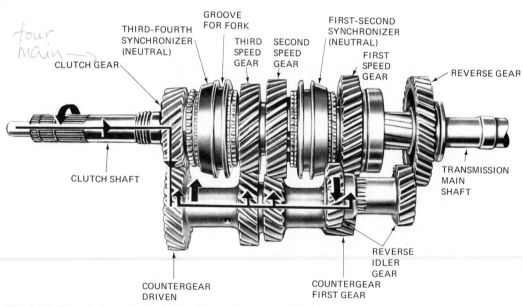

Fig. 4-28 Gear train and shafts of a four-speed transmission, shown in neutral. (*Chevrolet Motor Division of General Motors Corporation*)

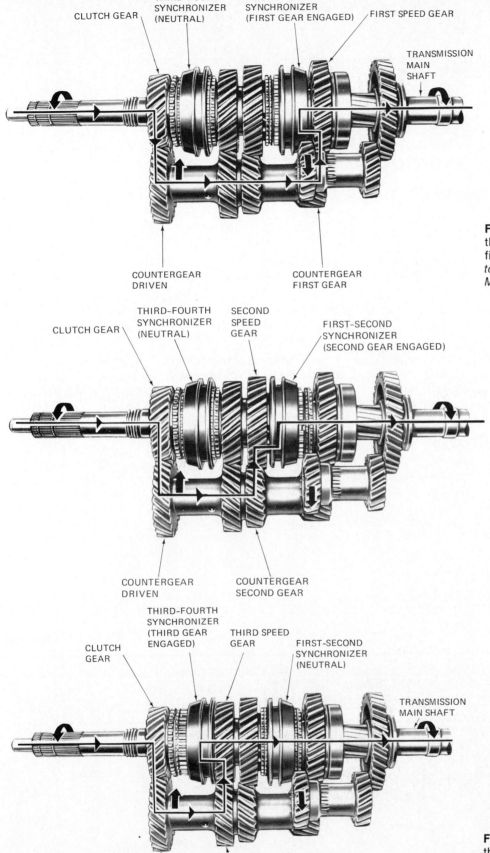

CLUTCH GEAR

THIRD-FOURTH SYNCHRONIZER (NEUTRAL)

FIRST-SECOND SYNCHRONIZER (FIRST GEAR ENGAGED)

FIRST SPEED GEAR

TRANSMISSION MAIN SHAFT

COUNTERGEAR DRIVEN

COUNTERGEAR FIRST GEAR

Fig. 4-29 Power flow through the gear train in first gear. (*Chevrolet Motor Division of General Motors Corporation*)

CLUTCH GEAR

THIRD-FOURTH SYNCHRONIZER (NEUTRAL)

SECOND SPEED GEAR

FIRST-SECOND SYNCHRONIZER (SECOND GEAR ENGAGED)

COUNTERGEAR DRIVEN

COUNTERGEAR SECOND GEAR

Fig. 4-30 Power flow through the gear train in second gear. (*Chevrolet Motor Division of General Motors Corporation*)

CLUTCH GEAR

THIRD-FOURTH SYNCHRONIZER (THIRD GEAR ENGAGED)

THIRD SPEED GEAR

FIRST-SECOND SYNCHRONIZER (NEUTRAL)

TRANSMISSION MAIN SHAFT

COUNTERGEAR DRIVEN

COUNTERGEAR THIRD GEAR

Fig. 4-31 Power flow through the gear train in third gear. (*Chevrolet Motor Division of General Motors Corporation*)

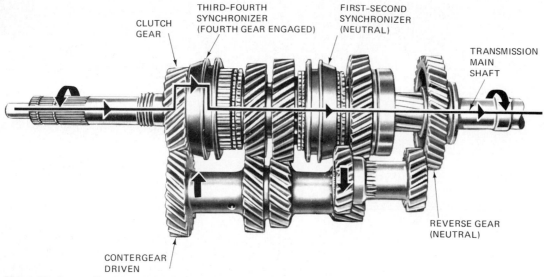

CLUTCH
GEAR

THIRD-FOURTH
SYNCHRONIZER
(FOURTH GEAR ENGAGED)

FIRST-SECOND
SYNCHRONIZER
(NEUTRAL)

TRANSMISSION
MAIN
SHAFT

REVERSE GEAR
(NEUTRAL)

CONTERGEAR
DRIVEN

Fig. 4-32 Power flow through the gear train in fourth gear. (*Chevrolet Motor Division of General Motors Corporation*)

6. Gearshift linkage   Figure 4-11 shows a typical linkage between the floor-mounted gearshift lever and the transmission. This is the linkage arrangement on a four-speed transmission. Note that there is an extra rod—the reverse rod—linking the shift lever and transmission. This rod carries the movement to the reverse fork that moves the reverse gear.

✿ **4-11 Overdrive**   In the transmissions discussed so far, the high-gear position produces a 1:1 ratio between the clutch gear and the transmission output shaft. There is neither gear reduction nor gear increase through the transmission. This is direct drive.

At intermediate and high car speeds, it is sometimes desirable to have the transmission output shaft turn faster than the clutch gear and engine crankshaft.

Therefore, some transmissions are designed with gears that provide an *overdrive* ratio. A transmission is in overdrive when the transmission output shaft is turning faster than the transmission input shaft, or clutch gear.

Years ago, many cars were equipped with a separate overdrive unit which was attached to the rear of the transmission (Fig. 4-34). Today, the overdrive is built into the transmission. Many four-speed and five-speed transmissions have overdrive. The top gear position causes the clutch gear to overdrive the transmission main shaft.

The advantage of overdrive is that it allows engine speed to drop down while still maintaining the car at highway speed. Once the car is moving at a steady speed, it does not require as much power to keep it moving. Therefore, the engine can turn more slowly,

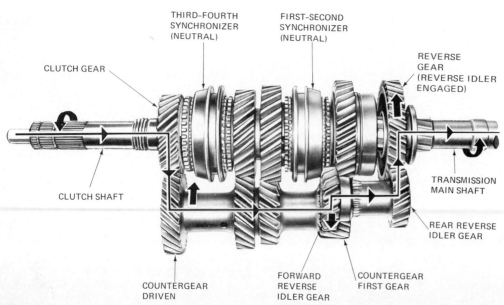

THIRD-FOURTH
SYNCHRONIZER
(NEUTRAL)

FIRST-SECOND
SYNCHRONIZER
(NEUTRAL)

CLUTCH GEAR

REVERSE
GEAR
(REVERSE IDLER
ENGAGED)

CLUTCH SHAFT

TRANSMISSION
MAIN SHAFT

REAR REVERSE
IDLER GEAR

COUNTERGEAR
DRIVEN

FORWARD
REVERSE
IDLER GEAR

COUNTERGEAR
FIRST GEAR

Fig. 4-33 Power flow through the gear train in reverse. (*Chevrolet Motor Division of General Motors Corporation*)

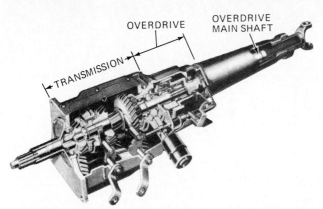

Fig. 4-34 Cutaway view of a three-speed transmission with a separate overdrive unit attached to the rear of the transmission. (*Ford Motor Company*)

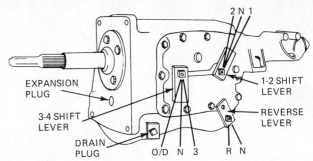

Fig. 4-35 Exterior view of a four-speed transmission with overdrive in fourth gear. (*Chrysler Corporation*)

produce less power, and still maintain car speed. This saves fuel and reduces wear on the engine and accessories. In overdrive, a typical overdrive transmission can maintain a car speed of 55 mph [89 km/h] while allowing the engine to turn at the equivalent of only 44 mph [71 km/h].

✿ 4-12 Four-speed transmission with overdrive
From the outside, a four-speed transmission with over-

drive may look like any other four-speed transmission (Fig. 4-35). Even when you look inside at the gears, you might not be able to see the difference immediately. Figure 4-36 shows a four-speed transmission with overdrive, partly cut away so that the gears can be seen. Notice the different sizes of the clutch gear, the countergear-assembly driven gear, and the overdrive gear. Figure 4-37 is a simplified drawing of the gears. Compare this with Fig. 4-28.

In the overdrive transmission with four forward speeds, the gear ratios in first, second, and third can be compared with the gear ratios of a standard three-speed

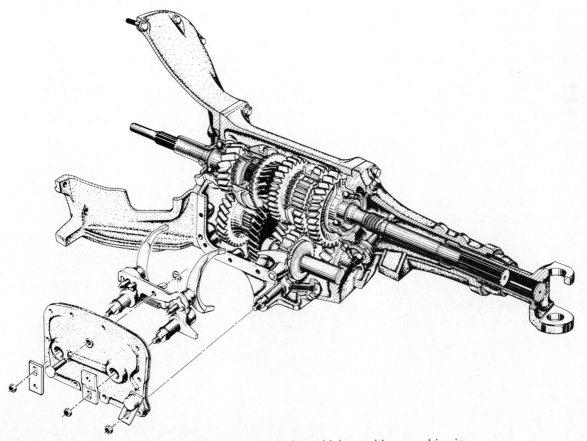

Fig. 4-36 Cutaway view of the four-speed transmission which provides overdrive in fourth gear, showing the gearing. (*Chrysler Corporation*)

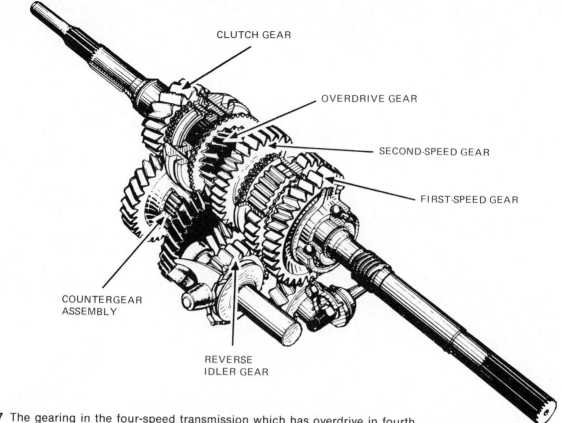

CLUTCH GEAR

OVERDRIVE GEAR

SECOND-SPEED GEAR

FIRST-SPEED GEAR

COUNTERGEAR ASSEMBLY

REVERSE IDLER GEAR

Fig. 4-37 The gearing in the four-speed transmission which has overdrive in fourth gear. *(Chrysler Corporation)*

transmission. First is low, second is intermediate, and third is high, or direct drive with a 1:1 gear ratio. Fourth is overdrive. Figure 4-38 shows the shift pattern for a four-speed transmission with overdrive. Fourth or overdrive (OD) is at the lower right corner.

When the transmission is shifted into overdrive, the clutch synchronizing sleeve locks the overdrive gear to the main shaft. The power flow is then as shown in Fig. 4-39. Note that the clutch gear is smaller than the countergear-assembly driven gear. This means there is gear reduction because the countergear assembly turns more slowly than the clutch gear.

However, the gear on the countergear assembly is larger than the overdrive gear. This means a gear-ratio increase. The overdrive gear turns faster than the gear that is driving it—which is the countergear assembly. The net result of the two gear ratios is that the overdrive gear turns faster than the clutch gear. Therefore the main shaft turns faster than the clutch gear.

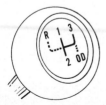

Fig. 4-38 Shift pattern for the four-speed transmission with overdrive.

As an example, let us assume that the four gears in a four-speed transmission with overdrive have the following number of teeth:

Clutch gear	16
Countergear driven gear	24
Overdrive driving gear (on countergear)	28
Overdrive gear (on main shaft)	14

Let's find out how much faster the main shaft will turn than the clutch gear. The gear ratio between the clutch gear and the countergear driven gear is 24:16, or 3:2. This means that the clutch gear turns three times to turn the countergear assembly two times.

Meantime, the gear ratio between the gear on the countergear assembly that is driving the overdrive gear, and the overdrive gear, is 14:28, or 1:2. Therefore, when the countergear turns one time, it turns the overdrive gear on the main shaft two times.

Here is how this works out in terms of revolutions per minute. Suppose the engine speed is 3000 rpm. The clutch gear turns at the same speed of 3000 rpm. The gear reduction between the clutch gear and countergear driven gear is 3:2. Therefore the countergear assembly turns at 2000 rpm.

The gear ratio between the driving gear on the countergear assembly and the overdrive gear is 1:2. This means that when the countergear assembly turns at 2000 rpm, it turns the overdrive gear and the main shaft at 4000 rpm.

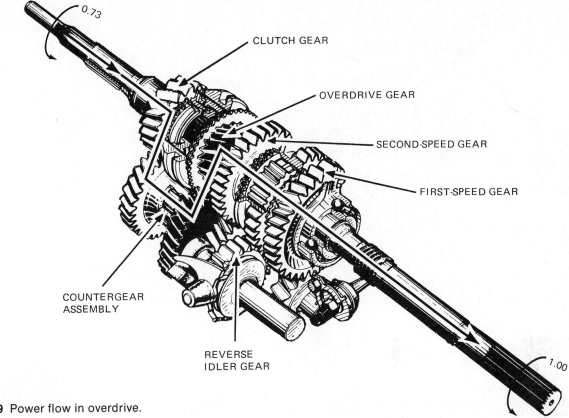

CLUTCH GEAR

OVERDRIVE GEAR

SECOND-SPEED GEAR

FIRST-SPEED GEAR

COUNTERGEAR ASSEMBLY

REVERSE IDLER GEAR

0.73

1.00

Fig. 4-39 Power flow in overdrive.

When the transmission is in overdrive, the overall gear ratio is 3:4 (3000 to 4000). In practical terms, this means the engine speed can drop 25 percent and the car can still maintain highway speed. The only thing that is lost in overdrive is acceleration. With the higher gear ratio through the transmission, there is less torque being delivered to the wheels. However, a downshift can be made into third gear if more torque is desired for passing.

NOTE: The example above is only typical. Gear sizes vary from one transmission to another so that different gear ratios are achieved in overdrive. Figure 4-40 shows the gear ratios of one model of four-speed transmission with overdrive.

GEAR POSITION	GEAR RATIO	
	Input Shaft	Output Shaft
First	3.090	1.000
Second	1.670	1.000
Third	1.000	1.000 (Direct Drive)
Fourth	0.730	1.000 (Overdrive)

Fig. 4-40 Gear ratios of the four-speed transmission in which the fourth speed is an overdrive. (*Chrysler Corporation*)

✪ 4-13 Five-speed transmission with overdrive Several five-speed transmissions with overdrive have been designed. One design widely used by Toyota is built as either a four-speed or a five-speed transmission. Figure 4-41 shows the gear ratios in the five-speed transmission with overdrive. Figure 4-42 shows the gears for both the four-speed and the five-speed transmission. The basic gear arrangement is the same for both. The extra gearing required in the five-speed transmission is shown in boxes at the top and bottom of Fig. 4-42.

✪ 4-14 Transaxle The manual transaxle includes the transmission, clutch, and differential. It is mounted

GEAR POSITION	GEAR RATIO	
	Input Shaft	Output Shaft
First	3.587	1.000
Second	2.022	1.000
Third	1.384	1.000
Fourth	1.000	1.000 (Direct Drive)
Fifth	0.861	1.000 (Overdrive)
Reverse	3.384	1.000

Fig. 4-41 Gear ratios of the five-speed transmission, showing the number of times the output shaft turns for each revolution of the input shaft. (*Toyota Motor Sales Co., Ltd.*)

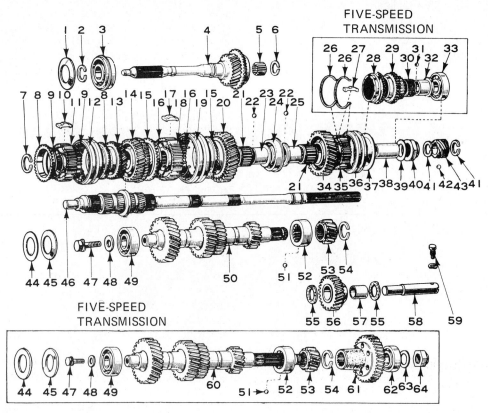

Fig. 4-42 Gears and shafts in the four- and five-speed transmission. The boxes enclose the additional parts in the five-speed transmission. (*Toyota Motor Sales Co., Ltd.*)

1. Gear-thrust-cone spring
2. Shaft snap ring
3. Radial ball bearing
4. Input shaft
5. Roller
6. Hole snap ring
7. Shaft snap ring
8. No. 1 synchronizer ring
9. No. 1 synchromesh-shifting-key spring
10. No. 2 synchromesh-shifting key
11. No. 2 transmission-clutch hub
12. No. 2 transmission-hub sleeve
13. Third-gear subassembly
14. Second-gear subassembly
15. No. 2 synchronizer ring

16. No. 1 synchromesh-shifting-key spring
17. No. 1 synchromesh shifting key
18. No. 1 transmission-clutch hub
19. No. 1 transmission-hub sleeve
20. First-gear subassembly
21. Needle roller bearing
22. Ball
23. First-gear bushing
24. Radial ball bearing
25. Reverse-gear bushing
26. No. 1 synchromesh-shifting-key spring
27. No. 3 synchromesh shifting key
28. No. 1 synchronizer ring
29. Fifth-gear subassembly

30. Needle roller bearing
31. Ball
32. Fifth-gear bushing
33. Radial ball bearing
34. Reverse gear
35. No. 3 transmission-clutch hub
36. No. 3 transmission-clutch hub
37. Spacer
38. Spacer
39. Shim
40. Nut
41. Shaft snap ring
42. Ball
43. Speedometer drive gear
44. Shim
45. Gear-thrust-cone spring
46. Output shaft
47. Bolt with washer
48. Plate washer

49. Radial ball bearing
50. Countergear
51. Ball
52. Cylindrical roller bearing
53. Countershaft reverse gear
54. Shaft snap ring
55. Reverse-idler-gear thrust washer
56. Reverse idler gear
57. Bimetal-formed bushing
58. Reverse-idler-gear shaft
59. Shaft retaining bolt
60. Countergear
61. Fifth-gear countershaft
62. Radial ball bearing
63. Shim
64. Nut

on the engine (Fig. 4-43). The most common arrangement uses the engine and transaxle at the front of the car for front-wheel drive. The engine is mounted crossways, or in a transverse position (Fig. 1-2).

Figure 4-44 is a sectional view of a transaxle. Identify the various gears, the two synchronizers, and the other parts in the transmission. The synchronizers work the same way as those in other manual transmissions (⚙ 4-9).

Figures 4-45 to 4-49 show the power flow through the transmission in first, second, third, fourth, and reverse. Study the power flow shown through each illustration. Note how movement of the synchronizers in one direction or the other locks the different gears to either

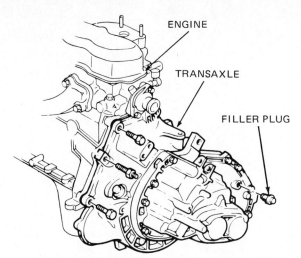

ENGINE

TRANSAXLE

FILLER PLUG

Fig. 4-43 Exterior view of the transaxle. Bolts are removed to detach the transaxle from the engine. (*Chevrolet Motor Division of General Motors Corporation*)

the main shaft or the output shaft (in first, second, third, and fourth). In reverse, the reverse idler gear is introduced into the gear train. This causes the output shaft to rotate in the reverse direction.

Figure 4-50 is a sectional view of a similar transaxle. This view shows the differential and the reverse idler gear. In this transaxle and the transaxle shown in Fig. 4-44, shifting is accomplished in the same way as in other manual transmissions. Movement of the shift lever engages a lever which moves a shift rod. The shift rod moves a lever on the transmission. This movement causes a fork in the transmission to shift a synchronizer one way or the other on a shaft. The synchronizer movement locks a gear to a shaft or to another gear so the desired gear ratio through the transmission is achieved.

⚙ 4-15 Transaxle with dual-speed range A dual-range transaxle provides the usual four forward speeds plus reverse as in other transaxles. In addition, some models of this transmission have an extra set of gears that can supply overdrive in each gear position. This provides two speed ranges which the manufacturer calls the *economy range* and the *power range*. With four forward speeds in each range, this transaxle provides a total of eight forward speeds.

The gear ratios for both the single-range transaxle and the dual-range transaxle are shown in Fig. 4-51. In fourth gear, the power range of the two-speed transaxle provides a gear ratio of 1.105:1. The crankshaft turns 1.105 times to turn the output shaft once. When shifted into the economy range, fourth gear has a gear ratio of 0.855:1, which is overdrive. For each complete revolution of the output shaft, the crankshaft turns less than one complete revolution (only 0.855 of one revolution).

Figures 4-52 and 4-53 are sectional views of the two transaxles. Notice that there is an intermediate gear (item 5 in Fig. 4-52 and item 8 in Fig. 4-53). This is the same as the countergear in standard transmissions. The intermediate gear is located between the input shaft and the output shaft. By changing the design of the

input shaft and the intermediate shaft, it is possible to use the same case for either the single-range or the dual-range gear set.

Basically, the dual-range transaxle has an additional gear and a synchronizer on the input shaft. This can be seen by comparing the dual-range input shaft shown in Fig. 4-53 with the input shaft for the single-range transaxle shown in Fig. 4-52.

Shifting between the two ranges requires a separate lever. Figure 4-54 shows the shift patterns for both the range-selector lever and the gearshift lever. By moving the range-selector lever, the range-selector synchronizer (Fig. 4-54) can be shifted back and forth along the input shaft. This locks either the input low gear or the input high gear to the shaft. The range-selector lever is placed in either low range for power or high range for economy.

⚙ 4-16 Differential and driveshafts for transaxles The driveshafts for front-engine cars with a transaxle are covered in ⚙ 17-7 to 17-10. These driveshafts have universal joints that permit them to carry rotary motion through angles. Slip joints allow the effective length of the driveshaft to change. These joint actions allow the front wheels to move up and down as they meet irregularities in the road and to be swung from side to side for steering.

The differential allows the two front wheels to rotate at different speeds as the car travels around a curve or makes a turn. The outer wheel always has to travel farther and therefore must turn faster (Fig. 4-55). The differential in the transaxle allows this to happen.

When both wheels are turning at the same speed, the differential acts like a solid coupling (⚙ 1-11). The differential case turns both driveshafts and wheels at the same speed. However, when the car makes a turn, the differential pinions rotate on the pinion shaft. This causes the side gear that is splined to the outer-wheel driveshaft to turn faster.

⚙ 4-17 Types of transfer cases Transfer cases have been used with a variety of manual and automatic transmissions. They provide full-time or part-time four-wheel drive, a low range which doubles the number of gear ratios in the transmission, and a power-takeoff point to operate auxiliary equipment, such as a winch.

The transfer case is used in all vehicles with four-wheel drive. Figure 1-24 shows its location in a vehicle. Figure 4-2 shows its location in a pickup truck. The main purpose of the transfer case is to provide a means of sending engine power to both the front and rear wheels (⚙ 4-5). There are two general types, part-time or full-time.

NOTE: Four-wheel-drive vehicles may be designated 4WD vehicles.

In the full-time transfer case, power is available to all four wheels at any time. The transfer case has a gearshift which provides for either direct drive through the transfer case (high range) or gear reduction (low range). Gear reduction means torque increase at the wheels. The transfer case is shifted into low range by

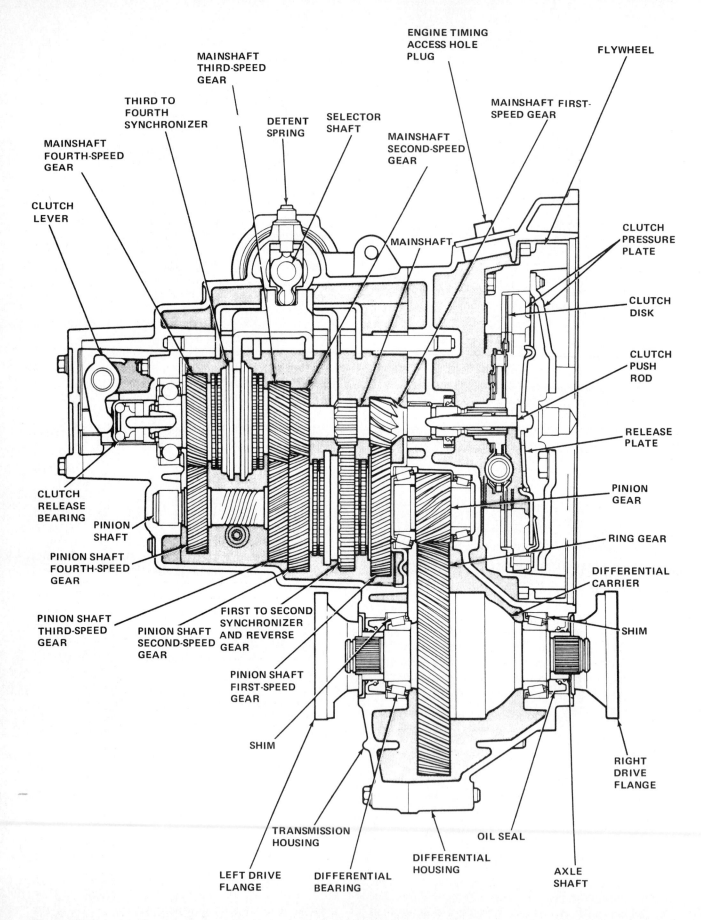

Fig. 4-44 Four-speed transaxle for a front-wheel-drive car. (*Chrysler Corporation*)

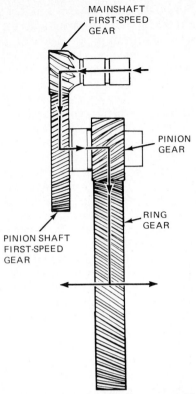

Fig. 4-45 Power flow through the four-speed transaxle in first gear.

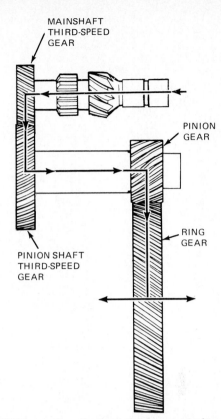

Fig. 4-47 Power flow through the four-speed transaxle in third gear.

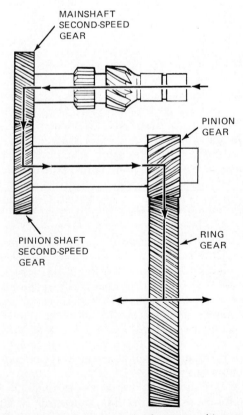

Fig. 4-46 Power flow through the four-speed transaxle in second gear.

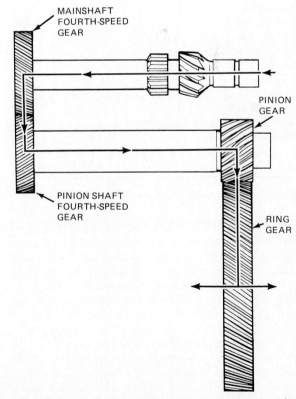

Fig. 4-48 Power flow through the four-speed transaxle in fourth gear.

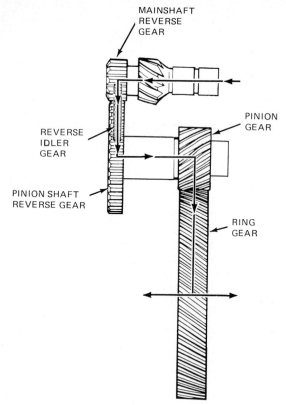

Fig. 4-49 Power flow through the four-speed transaxle in reverse.

the driver when additional torque is needed, such as for climbing steep hills. Some cars with four-wheel drive (for example, the American Motors Eagle) have a transfer case with only a high range.

The part-time transfer case can be shifted into gear reduction, just as in the full-time unit. In addition, the part-time unit also has a gear shift that sends power to only the rear wheels, or to both the front and rear wheels (✸ 4-5).

The gear positions in a typical transfer case for the various transfer-case shift-lever positions are shown in Figs. 4-56 to 4-59. The transfer-case shift lever is located on the floor of the passenger compartment. Figure 4-60 shows two different shift patterns, which vary according to the design of the transfer case.

✸ 4-18 Full-time transfer case with differential If a vehicle has full-time four-wheel drive, a controllable differential is built into the transfer case (Fig. 4-61). The purpose of this interaxle differential is to compensate for any difference in front-wheel and rear-wheel travel while the transfer case is in high range and four-wheel drive. The differential allows the front and rear axles to operate at their own speeds, without forcing wheel slippage in normal dry-road driving.

When differential action is not wanted, such as for maximum engine braking or maximum wheel torque, the driver moves the transfer-case shift lever either to low range or to the LOCK position. This locks together the output shafts in the transfer case for the front and

rear axles. Now differential action cannot occur. Equal torque is delivered to both the front and rear axles.

During turns, the front wheels travel a greater distance than the rear wheels. This is because the front wheels move through a wider arc than the rear wheels (Fig. 4-55). With full-time four-wheel drive, it is the differential in the transfer case (Fig. 4-61) that allows the front wheels to travel farther or turn faster than the rear wheels without slipping. As a result, power-train and tire wear are reduced while the advantages of full-time four-wheel drive are retained.

This differs from the operation of a part-time transfer case. It delivers equal power to each axle while in four-wheel drive. During turns made on a dry surface, one of the axles must slip. Therefore, the driver must shift the part-time transfer case out of four-wheel drive on returning to a dry surface.

There is a disadvantage to a transfer case with a simple differential in it. If the wheels on either axle lose traction and begin to spin, the transfer-case differential continues to deliver maximum torque to the axle with the minimum traction. As a result, insufficient torque to move the vehicle may be provided to the wheels that still have traction.

To overcome this problem, most full-time transfer cases have some type of limited-slip differential in them. The slip-limiting device may be a viscous coupling, brake cones, or clutch plates. Regardless of type, its job is to divert torque from the spinning axle. This causes more torque to be supplied to the wheels with the most traction. A widely used transfer case of this type is the Quadra-Trac. It is built by the Warner Gear Division of the Borg-Warner Corporation. The operation of limited-slip differentials is covered in Chap. 19.

✸ 4-19 Part-time transfer case with planetary gears In 1980, a new design of part-time transfer case for automotive vehicles was introduced by the New Process Gear Division of Chrysler Corporation (Fig. 4-62). This transfer case has an aluminum case, chain drive, and a planetary-gear set for reduced weight and increased mechanical efficiency. The internal parts of the transfer case do not rotate in two-wheel drive. This leaves most of the lubricant undisturbed and reduces the power loss caused by dragging parts through it.

An internal oil pump turns with the rear output shaft to maintain lubrication to critical bearings and bushings whenever the rear driveshaft is turning. This provides improved lubrication during normal operation. In addition, the oil pump allows towing of the vehicle with the transfer case in neutral at any safe speed up to 55 mph [89 km/h] and for any distance. It is not necessary to disconnect the driveshafts. However, the front-wheel locking hubs should be in UNLOCK or AUTO (depending upon the hub type) to prevent unnecessary rotation of the drive-train components. Locking hubs are discussed in Chap. 19.

Figure 4-63 shows the power flow through the transfer case for each drive contition. In neutral, rotation of the input shaft spins only the planetary gears (✸ 4-20) and the ring gear around them. With both the planetary

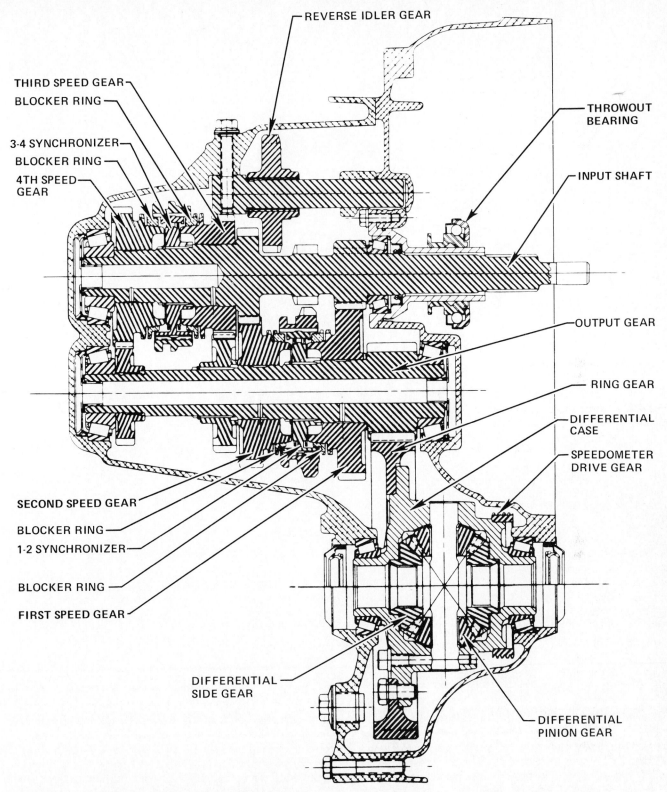

REVERSE IDLER GEAR

THIRD SPEED GEAR
BLOCKER RING

3-4 SYNCHRONIZER
BLOCKER RING

4TH SPEED
GEAR

THROWOUT
BEARING

INPUT SHAFT

OUTPUT GEAR

RING GEAR

DIFFERENTIAL
CASE

SPEEDOMETER
DRIVE GEAR

SECOND SPEED GEAR

BLOCKER RING

1-2 SYNCHRONIZER

BLOCKER RING

FIRST SPEED GEAR

DIFFERENTIAL
SIDE GEAR

DIFFERENTIAL
PINION GEAR

Fig. 4-50 Sectional view of a four-speed transaxle. (*Chevrolet Motor Division of General Motors Corporation*)

		STANDARD TRANSAXLE (SINGLE RANGE)	DUAL RANGE TRANSAXLE	
			Power range	Economy range
Gear ratios	1st	4.226	4.226	3.272
	2nd	2.365	2.365	1.831
	3rd	1.467	1.467	1.136
	4th	1.105	1.105	0.855
	Reverse	4.109	4.109	3.181

Fig. 4-51 Gear ratios of a standard single-range transaxle compared with gear ratios in each range of a dual-range transaxle. (*Chrysler Corporation*)

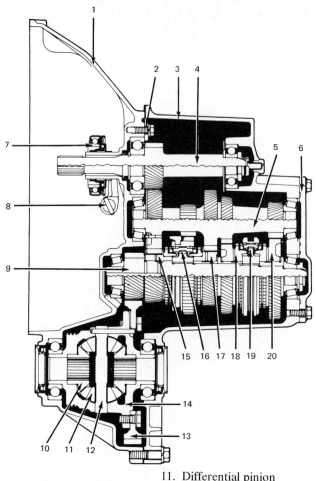

1. Clutch housing
2. Bearing retainer
3. Transaxle case
4. Input shaft
5. Intermediate gear
6. Rear cover
7. Clutch release bearing
8. Clutch release fork
9. Output shaft
10. Differential side gear
11. Differential pinion
12. Pinion shaft
13. Differential drive gear
14. Differential case
15. Fourth-speed gear
16. Third- and fourth-speed synchronizer assembly
17. Third-speed gear
18. Second-speed gear
19. First- and second-speed synchronizer assembly
20. First-speed gear

Fig. 4-52 Sectional view of a single-range four-speed transaxle. (*Chrysler Corporation*)

gears and the ring gear spinning freely, no power is transmitted through the planetary-gear set.

⚙ 4-20 Planetary-gear set operation In its simplest form, a planetary-gear set is made up of three gears (Fig. 4-64). In the center is the sun gear. All gears in the set revolve around the sun gear. Meshing with the sun gear are two or more planet-pinion gears (only one is shown in Fig. 4-64). In any set of two or more gears, the smallest gear is often called the *pinion gear*. The planet pinions are fastened together by the planet-pinion carrier, or planet carrier (Fig. 4-65). This holds the gears in place while allowing them to rotate around their pins or shafts. Usually there are three or four pinion gears fastened to the carrier.

On the outside of the planet pinions is the internal gear, or ring gear (Figs. 4-65 and 4-66). It has teeth

around its inside circumference that mesh with the teeth on the planet pinions.

Figure 4-66 shows how the planet pinions are held in place in the internal gear. Each planet pinion is mounted on a pin, and the pins are set in the planet-pinion carrier. Notice in Figs. 4-65 and 4-66 that the carrier is mounted on a separate shaft from the sun-gear shaft. The planetary gear used in the transfer case discussed in ⚙ 4-19 has four planetary pinions instead of the three shown in Figs. 4-65 and 4-66. Additional planetary pinions provide the strength needed to handle the higher torque flow through the transfer case.

A planetary-gear set can provide any of five conditions. These are:

1. A speed increase with a torque decrease (overdrive)
2. A speed decrease with a torque increase (reduction)

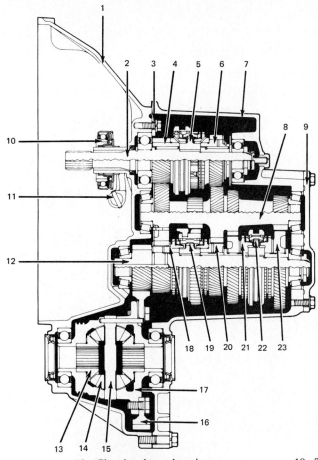

1. Clutch housing
2. Input shaft
3. Bearing retainer
4. Input low gear
5. Synchronizer assembly
6. Input high gear
7. Transaxle case
8. Intermediate gear
9. Rear cover

10. Clutch release bearing
11. Clutch release fork
12. Output shaft
13. Differential side gear
14. Differential pinion
15. Pinion shaft
16. Differential drive gear
17. Differential case
18. Fourth speed gear

19. Third and fourth speed
    synchronizer assembly
20. Third speed gear
21. Second speed gear
22. First and second speed
    synchronizer assembly
23. First speed gear

Fig. 4-53 Sectional view of a dual-range transaxle. (*Chrysler Corporation*)

3. Direct drive (lockup)
4. Neutral
5. Reverse

In the transfer case, the two drive conditions used are speed reduction and direct drive. Speed reduction provides the low range. Direct drive provides the high range. To get speed reduction, the internal gear, or ring gear, is locked in a stationary position. For direct drive, the internal gear and the planet-pinion carrier are locked together.

Gear reduction is provided when the sun gear on the input shaft is the driving gear (Fig. 4-67). In the transfer case, this shaft is coupled directly to the transmission. When the internal gear is held stationary and the sun gear is driving, the planet pinions rotate on their pins. As the planet pinions rotate, they must "walk around"

the internal gear since they are meshed with it. This action causes the pinion carrier to rotate in the same direction as the sun gear.

However, the planet-pinion carrier turns more slowly than the sun gear. This is because of the gear reduction between the planet pinions and the sun gear. As the pinions move around the inside of the internal gear, the shaft attached to the planet carrier is driven in the same direction as the sun gear, but at a lower speed. This action provides the transfer case with low range. The output shaft turns more slowly than the input shaft, but with increased torque.

NOTE: Figure 4-67 shows the internal gear held stationary by the clamping action of a brake band wrapped around it. In the transfer case using a planetary-gear set (☀ 4-19), when the selector lever is moved to low

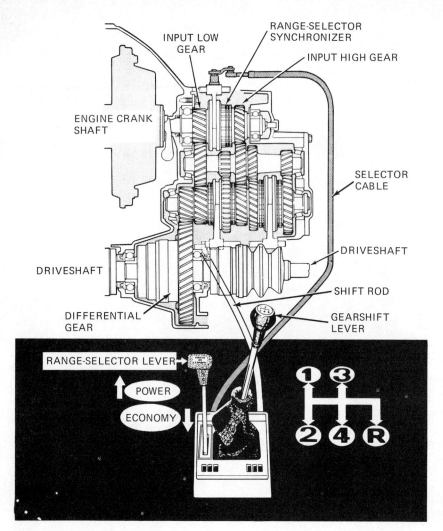

RANGE-SELECTOR SYNCHRONIZER

INPUT LOW GEAR

INPUT HIGH GEAR

ENGINE CRANK SHAFT

SELECTOR CABLE

DRIVESHAFT

DRIVESHAFT

SHIFT ROD

DIFFERENTIAL GEAR

GEARSHIFT LEVER

RANGE-SELECTOR LEVER

POWER

ECONOMY

Fig. 4-54 Control levers and linkage for the "Twin Stick" dual-range trans-axle. (*Chrysler Corporation*)

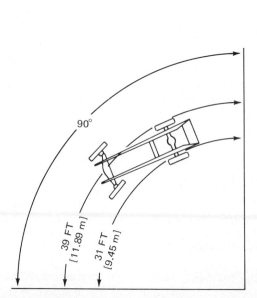

90°

39 FT [11.89 m]

31 FT [9.45 m]

Fig. 4-55 During a turn, the front wheels travel farther than the rear wheels.

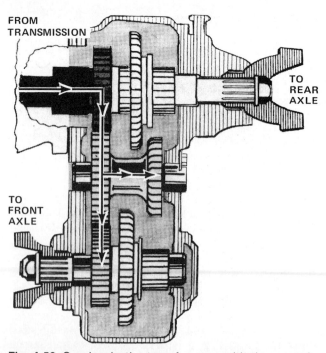

FROM TRANSMISSION

TO REAR AXLE

TO FRONT AXLE

Fig. 4-56 Gearing in the transfer case with the gears in neutral. (*American Motors Corporation*)

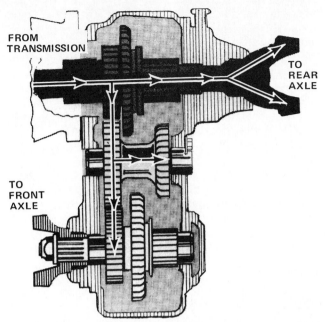

Fig. 4-57 Gearing with the transfer case in two-wheel drive, high range. Only the rear wheels are being driven. (*American Motors Corporation*)

range, the planetary-gear assembly slides forward on its shaft. In this position, a locking plate bolted to the case engages the teeth of the internal gear to hold it stationary.

To get direct drive, the internal gear and the planet-pinion carrier are locked together. Now, the whole planetary-gear assembly turns as a solid unit because the planet pinions cannot rotate on their pins. This provides direct drive. The output shaft turns at the same speed as the input shaft.

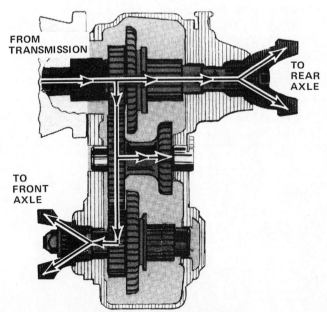

Fig. 4-58 Gearing with the transfer case in four-wheel drive, high range. All four wheels are being driven. (*American Motors Corporation*)

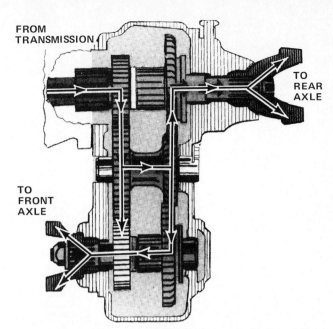

Fig. 4-59 Gearing with the transfer case in four-wheel drive, low range. (*American Motors Corporation*)

☼ 4-21 Gear lubricants Manual transmissions, transaxles, and transfer cases are all various types of gearboxes. They are very similar in three ways. All have:

1. Gears that transmit power
2. Splined shafts that rotate while other parts are sliding on them
3. Bearings that support the shafts and transfer the load to the case or housing

In the gearbox, the moving metal parts must not touch each other. They must be continuously separated by a thin film of lubricant to prevent excessive wear and premature failure.

As gear teeth mesh, there is a sliding or wiping action between the contact faces. This action produces friction and heat. Without lubrication, the gears would wear quickly and fail. However, lubrication provides a fluid film between the contact faces. This prevents metal-to-metal contact. Therefore, all gearboxes in the automotive vehicle have some type of lubricant or *gear oil* in them.

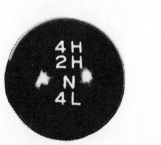

Fig. 4-60 Shift patterns for transfer cases. (*American Motors Corporation*)

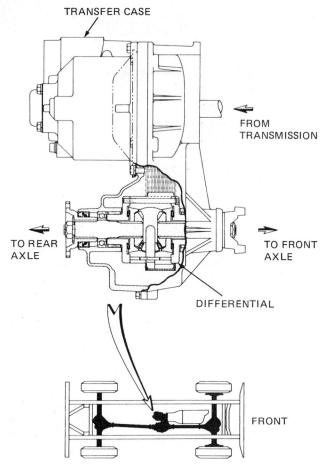

Fig. 4-61 Transfer case for full-time four-wheel drive, which includes a controllable differential. (*American Motors Corporation*)

A gear oil has five jobs to do. These are:

1. To lubricate all moving parts and prevent wear
2. To reduce friction and power loss
3. To protect against rust and corrosion
4. To keep the interior clean
5. To cool the gearbox

In addition, the oil must have adequate load-carrying capacity to prevent puncturing of the oil film. Chemical additives are mixed with gear oil to improve its load-carrying capacity. An oil that has an additive in it to increase the load-carrying capacity is called an *extreme-pressure* (EP) lubricant. Other additives are also added to the oil to improve the viscosity (thickness), to prevent channeling (solidifying), to improve stability and oxidation resistance, to prevent foaming, to prevent rust and corrosion, and to prevent damage to the seals.

The typical gear oil is a straight mineral oil (refined from crude oil) with the required additives in it. Today, some gear oils are made from synthetic oil. Regardless of type, gear oil for use in most cars and light trucks has a classification of SAE 75W, 75W-80, 80W-90, 85W-90, 90, or 140. The higher the number, the thicker the oil.

Gear oil is *not* recommended for use in all gearboxes by all manufacturers. Gears which are lightly loaded, such as the planet-pinion gears in a planetary-gear set, do not require gear oil. Therefore, some transfer cases are filled with SAE 10W-30 engine oil. Other transfer cases use automatic-transmission fluid (ATF).

Automatic-transmission fluid is also used as the factory fill in manual transmissions built by Chrysler. If excessive gear rattle is heard at idle or during acceleration in direct drive or in overdrive, the automatic-transmission fluid may be drained out and the transmission filled with a multipurpose gear oil, such as SAE 85W-90. Some manual transaxles are also filled with automatic-transmission fluid.

To prevent the lubricant from leaking out, the gearbox has an oil-tight case. Seals are used around each cover and shaft. In addition, seals are provided around the input shafts and the output shafts. The clutch shaft on many transmissions does not have a separate seal. Instead, an oil slinger is used to throw back any oil that reaches it. Other designs have a passage in the clutch-shaft-bearing retainer that returns to the case any oil passing through the bearing.

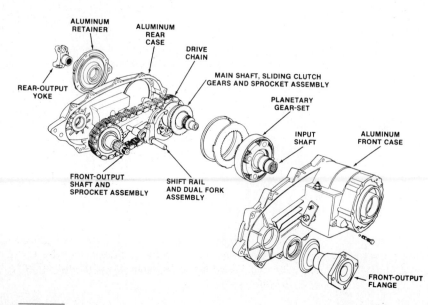

Fig. 4-62 A partially disassembled part-time four-wheel-drive transfer case, which uses a planetary-gear set to achieve gear reduction. (*Chrysler Corporation*)

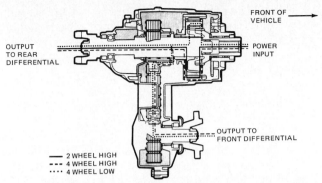

Fig. 4-63 Power flow through the part-time transfer case with planetary gears. (*Chrysler Corporation*)

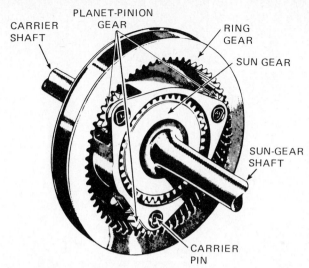

Fig. 4-66 An assembled planetary-gear set.

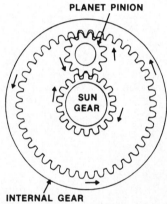

Fig. 4-64 The three gears that make up the basic planetary-gear set.

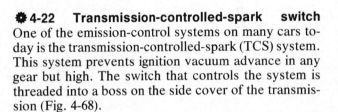

✿ 4-22 Transmission-controlled-spark switch
One of the emission-control systems on many cars today is the transmission-controlled-spark (TCS) system. This system prevents ignition vacuum advance in any gear but high. The switch that controls the system is threaded into a boss on the side cover of the transmission (Fig. 4-68).

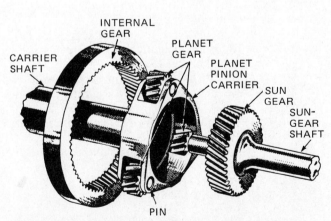

Fig. 4-65 Exploded view of a planetary-gear set, showing how the planet-pinion gears rotate on pins that are part of the carrier.

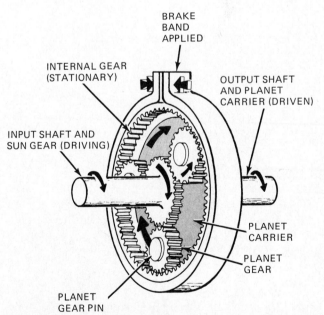

Fig. 4-67 With the internal gear held stationary, the planet carrier and output shaft turn more slowly than the sun gear. This provides the transfer case with low range. (*Ford Motor Company*)

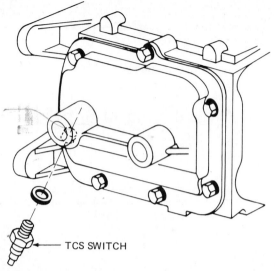

Fig. 4-68 Location of the transmission-controlled-spark switch on a transmission. (*Pontiac Motor Divison of General Motors Corporation*)

The TCS switch is open in all gears but high, thereby preventing vacuum advance. In high, the switch is closed, allowing vacuum advance. Vacuum advance in other gear positions increases the engine exhaust emissions coming from the tail pipe.

✿ **4-23 Backup lights** When the gearshift lever is moved to reverse, the linkage closes a switch that connects the backup lights to the battery. With this arrangement the lights come on automatically. This is a warning that the car is about to back up. The backup lights also allow the driver to see behind the car at night.

✿ **4-24 Speedometer drive** On most cars, the speedometer is driven by a pair of gears in the transmission-extension housing. One of these gears is mounted on the transmission main shaft. The other gear is mounted on the end of the flexible shaft connecting the speedometer to the transmission gear.

Chapter 4 review questions

Select the *one* correct, best, or most probable answer to each question. Then check your answers against the correct answers given at the end of the book.

1. Synchronizing devices in the transmission synchronize gears about to be:
   a. meshed
   b. de-meshed
   c. stopped
   d. started
2. Synchronizing devices are normally used when shifting into:
   a. first
   b. second
   c. third
   d. all of the above
3. The gearshift lever requires two separate motions to shift gears, and the first movement:
   a. moves the gear assembly
   b. selects the gear
   c. meshes the gears
   d. operates the clutch
4. On the main shaft of the four-speed transmission described in this chapter, there are:
   a. three gears
   b. four gears
   c. five gears
   d. six gears
5. The three-speed transmission has:
   a. one shifter fork
   b. two shifter forks
   c. three shifter forks
   d. four shifter forks

6. How many shift-lever positions, including neutral and reverse, are there in the four-speed transmissions described in this chapter?
   a. four
   b. five
   c. six
   d. seven and reverse
7. Including neutral and reverse, how many shift-lever positions are there in the three-speed passenger-car transmission?
   a. three
   b. four
   c. five
   d. six
8. In the standard transmission, the countershaft gears turn:
   a. faster than the clutch gear
   b. slower than the clutch gear
   c. at the same speed as the clutch gear
   d. twice as fast as the clutch gear
9. In overdrive, the:
   a. main shaft turns faster than the crankshaft
   b. clutch gear turns more slowly than the main shaft
   c. the main shaft overdrives the engine crankshaft
   d. all of the above
10. To get overdrive in the transmission, the overdrive gear on the main shaft is made:
   a. larger than the meshing gear on the countershaft
   b. smaller than the meshing gear on the countershaft
   c. the same size as the meshing gear on the countershaft
   d. none of the above

11. In the four-speed transmission with overdrive, the first three speeds are:
    a. low, intermediate, and high
    b. first, second, and third
    c. first, intermediate, and direct drive
    d. all of the above
12. The manual transaxle includes the transmission,
    a. clutch, and axles
    b. clutch, and differential
    c. axles, and wheels
    d. engine suspension, and axles
13. The dual-range transaxle with four forward speeds really offers:
    a. four forward speeds
    b. six forward speeds
    c. eight forward speeds
    d. only two forward speeds
14. The purpose of the transfer case is to transfer engine power:
    a. to the rear wheels
    b. to the rear and front wheels
    c. to the power takeoff
    d. all of the above
15. The gears in the part-time transfer case can be shifted to provide two-wheel-drive high, four-wheel-drive high, and:
    a. front-wheel drive only
    b. front-wheel drive only low
    c. rear-wheel drive low
    d. four-wheel drive low
16. The shifter mechanism (gearshift lever) on the steering column is normally connected to the transmission by:
    a. a single link
    b. two or three linkages
    c. four linkages
    d. a shifter rod
17. Lifting the shift lever toward the steering wheel selects:
    a. first-reverse
    b. first-second
    c. second-third
    d. third-reverse
18. Moving the shift lever down away from the steering wheel selects:
    a. first-reverse
    b. first-second
    c. high-reverse
    d. second-high
19. The first movement of the gearshift lever selects the gear and the second lever movement:
    a. releases the gear
    b. shifts the gear
    c. engages the clutch
    d. releases the clutch
20. The purpose of the planetary-gear set in the transfer case is to allow the vehicle to operate with:
    a. increased mechanical efficiency
    b. part-time four-wheel drive
    c. only the front axle engaged
    d. only the rear axle engaged

MANUAL-TRANSMISSION TROUBLE DIAGNOSIS

After studying this chapter, you should be able to:

1. Explain why the car should be road tested with the customer.
2. Discuss three types of transmission noise and the causes of each.
3. List nine transmission troubles.
4. Explain the possible causes of each trouble.
5. Explain how to correct the possible causes of each trouble.
6. Describe the conditions that can result from adding the wrong lubricant to the transmission.
7. List the places where oil can leak from a transmission.

 5-1 Manual-transmission diagnosis procedure The type of trouble a transmission has is often a clue to the cause of that trouble. Before attempting any repairs, try to determine the cause. Driver complaints, their possible causes, and the checks or corrections to be made are listed and discussed in later sections. Internal transmission troubles are fixed by disassembling the transmission. Or the old transmission can be replaced with a new, rebuilt, or used transmission.

To accurately diagnose a complaint about a manual transmission, a procedure must be followed. Road test the car with the owner to verify that the complaint exists (Fig. 5-1). Road testing with the owner gives you the opportunity to identify the condition that the owner wants corrected. There are two general types of manual-transmission troubles. They are (1) noise and (2) improper operation.

During the road test, determine any related symptoms that may be occurring. Get all the facts and service history possible from the owner. Then, as you drive, determine when, where, and how the symptoms occur.

Immediately begin to analyze the symptoms. Now you are performing the diagnosis step by answering the question "What is wrong?" As soon as you know, tell the owner what to expect. Then, with the owner's permission, perform the required adjustments or repairs. When the job is completed, road test the car again. This time make sure that the trouble you found on the first road test no longer exists.

When proper operation has been restored, you know that the trouble has been corrected. This second road test serves as a quality-control check. As far as possible, it assures both you and the car owner that the job has been done right the first time. This helps prevent shop "come-backs" because of the installation of defective parts or faulty workmanship.

NOTE: See ✿ 3-1 for a discussion of diagnosis theory.

✿ 5-2 Manual-transmission trouble diagnosis Most internal transmission problems can be accurately diagnosed before disassembling the transmission. For example, there are three general types of noise from a manual transmission (Fig. 5-2). The noise provides you

Fig. 5-1 When possible, road test the car with the customer.

NOISE | CAUSE

PERIODIC CLUNK = BROKEN TEETH

GROWL OR WHINE = DEFECTIVE BEARING OR WORN TEETH

GEAR CLASH = DEFECTIVE SYNCHRONIZER

Fig. 5-2 Three types of transmission noise and their causes.

with information about what is taking place inside the transmission. A periodic clunking noise indicates broken teeth. A growl or whine indicates a defective bearing or worn contact faces on the gear teeth. Gear clash during shifting or when shifting is attempted indicates a defective synchronizer.

NOTE: Certain clutch problems produce symptoms similar to the symptoms of transmission problems. Follow the trouble-diagnosis procedures for the transmission you are servicing to determine the actual cause of the problem before attempting any repair. It may be that what you thought was transmission trouble is actually a trouble located in some other part of the car.

☼ 5-3 Manual-transmission trouble-diagnosis chart The chart that follows lists the various manual-transmission troubles together with their possible causes, and checks and corrections to be made. Most transmission troubles can be listed under a few headings, such as "hard shifting," "slips out of gear," and "noises." In the chart, section numbers are given where fuller explanations are found on how to locate and correct the troubles.

NOTE: The troubles and possible causes are not listed in the chart in the order of frequency of occurrence. Item 1 (or item a) does not necessarily occur more often than item 2 (or item b).

☼ 5-4 Hard shifting into gear Difficulty in shifting into gear might be caused by improper linkage adjustment between the gearshift lever and the transmission. Improper adjustment might greatly increase the force necessary for gear shifting. The same trouble could result when the linkage is rusted and in need of lubrication or jammed at any of the pivot points. Adjustment

Manual Transmission Trouble-Diagnosis Chart

(See ☼ 5-4 to 5-11 for detailed explanations of the trouble causes and corrections listed below.)

COMPLAINT	POSSIBLE CAUSE	CHECK OR CORRECTION
1. Hard shifting into gear (☼ 5-4)	a. Gearshift linkage out of adjustment	Adjust
	b. Gearshift linkage needs lubrication	Lubricate
	c. Clutch not disengaging	Adjust (☼ 3-11 to 3-21)
	d. Excessive clutch-pedal free play	Adjust (☼ 3-11 to 3-21)
	e. Shifter fork bent	Replace or straighten
	f. Sliding gears or synchronizer tight on shaft splines	Replace defective parts
	g. Gear teeth battered	Replace defective gears
	h. Synchronizing unit damaged or springs improperly installed (after a service job)	Replace unit or defective parts; install spring properly
	i. Shifter tube binding in steering column	Correct tube alignment
	j. End of transmission input shaft binding in crankshaft pilot bushing	Lubricate; replace bushing
2. Transmission sticks in gear (☼ 5-5)	a. Gearshift linkage out of adjustment or disconnected	Adjust; reconnect
	b. Gearshift linkage needs lubrication	Lubricate
	c. Clutch not disengaging	Adjust (☼ 3-11 to 3-21)
	d. Detent balls (lockouts) stuck	Free; lubricate
	e. Synchronizing unit stuck	Free; replace damaged parts
	f. Incorrect or insufficient lubricant in transmission	Replace with correct lubricant and correct amount
	g. Internal shifter components damaged	Remove transmission to inspect and service shifter parts
3. Transmission slips out of gear (☼ 5-6)	a. Gearshift linkage out of adjustment	Adjust
	b. On floor shift, shift boot stiff or shift-lever binding	Replace boot; adjust console to relieve binding
	c. Insufficient lockout-spring force	Replace
	d. Bearings or gears worn	Replace
	e. End play of shaft or gears excessive	Replace worn or loose parts
	f. Synchronizer worn or defective	Repair; replace
	g. Transmission loose on clutch housing or misaligned	Tighten mounting bolts; correct alignment
	h. Clutch housing misaligned	Correct alignment
	i. Pilot bushing in crankshaft loose or broken	Replace
	j. Input-shaft retainer loose or broken	Replace

COMPLAINT	POSSIBLE CAUSE	CHECK OR CORRECTION
4. No power through transmission (✿ 5-7)	a. Clutch slipping	Adjust (✿ 3-11 to 3-21)
	b. Gear teeth stripped	Replace gears
	c. Shifter fork or other linkage part broken	Replace
	d. Gear or shaft broken	Replace
	e. Drive key or spline sheared off	Replace
5. Transmission noisy in neutral (✿ 5-8)	a. Gears worn or tooth broken or chipped	Replace gears
	b. Bearings worn or dry	Replace; lubricate
	c. Input-shaft bearing defective	Replace
	d. Pilot bushing worn or loose in crankshaft	Replace
	e. Transmission misaligned with engine	Realign
	f. Countershaft worn or bent, or thrust plate or washers damaged	Replace worn or damaged parts
6. Transmission noisy in gear (✿ 5-8)	a. Clutch friction disk defective	Replace
	b. Incorrect or insufficient lubricant	Replace with correct lubricant and correct amount
	c. Rear main bearing worn or dry	Replace or lubricate
	d. Gears loose on main shaft	Replace worn parts
	e. Synchronizers worn or damaged	Replace worn or damaged parts
	f. Speedometer gears worn	Replace
	g. Any condition noted in item 5	See item 5
7. Gears clash in shifting (✿ 5-9)	a. Synchronizer defective	Repair or replace
	b. Clutch not disengaging; pedal lash incorrect	Adjust
	c. Hydraulic system (hydraulic clutch) defective	Check cylinder; add fluid, etc.
	d. Idle speed excessive	Readjust
	e. Pilot bushing binding	Replace
	f. Gearshift linkage out of adjustment	Adjust
	g. Lubricant incorrect	Replace with correct lubricant
8. Transmission noisy in reverse (✿ 5-10)	a. Reverse idler gear or bushing worn or damaged	Replace
	b. Reverse gear on main shaft worn or damaged	Replace
	c. Countergear worn or damaged	Replace
	d. Shift mechanism damaged	Repair, replace defective parts, readjust
9. Oil leaks (✿ 5-11)	a. Foaming due to incorrect lubricant	Replace with correct lubricant
	b. Oil level too high	Use proper amount, no more
	c. Gaskets broken or missing	Replace
	d. Oil seals damaged or missing	Replace
	e. Oil slingers damaged, improperly installed, or missing	Replace correctly
	f. Drain plug loose	Tighten
	g. Transmission retainer bolts loose	Tighten
	h. Transmission or extension case cracked	Replace
	i. Speedometer-gear retainer loose	Tighten
	j. Side cover loose	Tighten
	k. Extension-housing seal worn or drive-line yoke worn	Replace

and lubrication of linkages are discussed in ✿ 6-4 and 6-5.

Another cause of this trouble could be failure of the clutch to disengage completely. If the clutch linkage is out of adjustment or if other conditions, as outlined in ✿ 3-11 to 3-21, prevent full clutch disengagement, then it will be difficult to shift gears into or out of mesh. Gear clashing will probably result since the engine will still be delivering at least some power through the

clutch to the transmission. See ✿ 3-11 to 3-21 for correction of this sort of clutch trouble.

Inside the transmission, hard gear shifting could be caused by a bent shifter fork, a sliding gear or synchronizer that is tight on the shaft splines, battered gear teeth, or a damaged synchronizing unit. A bent shifter fork, which might make it necessary to exert greater force in order to shift gears, should be replaced.

The splines in the gears or on the shaft may become

gummed up or battered from excessive wear so that the gear will not move easily along the shaft splines. If this happens, the shaft and gears should be cleaned or, if worn, replaced. If the gear teeth are battered, they will not slip into mesh easily. Nothing can be done to repair gears with battered teeth. New gears will be required. The synchronizing unit could be tight on the shaft, or it could have loose parts or worn or scored cones. Any of these conditions would increase the difficulty of meshing. To clear up troubles in the transmission, the transmission must be removed and disassembled, as explained in later chapters.

Another condition that can cause hard shifting is binding of the shifter tube in the steering column. The steering column must be partly disassembled so that the binding can be relieved.

❁ 5-5 Transmission sticks in gear

A number of the conditions that cause difficulty shifting into gear can also cause the transmission to stick in gear. For example, improper linkage adjustment between the gearshift lever and the transmission, or lack of lubrication in the linkage, could make it hard to shift out of mesh. Adjustment and lubrication of linkages are discussed in ❁ 6-4 and 6-5.

Another cause could be failure of the clutch to disengage completely. Improper clutch-linkage adjustment, and other conditions outlined in ❁ 3-5 that prevent full disengagement of the clutch, could make it hard to shift out of mesh. See ❁ 3-11 to 3-21 for correction of this type of clutch trouble.

If the detent balls (or the lockout mechanism in the transmission) stick and do not unlock readily when shifting is attempted, it will be hard to shift out of gear. They should be freed and lubricated.

If the synchronizers do not slide freely on the shaft splines, then it will be hard to come out of mesh. The shaft and synchronizers should be cleaned or, if worn, replaced. Lack of lubricant or use of the wrong lubricant in the transmission can cause gears to stick in mesh (Fig. 5-3). Transmission removal, disassembly, reassembly, and installation procedures are covered in following chapters.

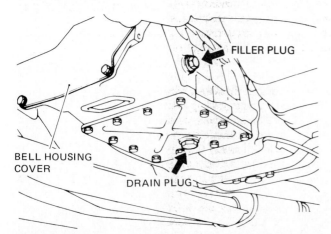

Fig. 5-3 When the transmission sticks in gear, check the lubricant level. It should reach the bottom of the filler-plug hole. *(Chrysler Corporation)*

❁ 5-6 Transmission slips out of gear

Improperly adjusted linkage between the gearshift lever and the transmission might produce force on the linkage in such a way that gears would work out of mesh. Linkage adjustment is outlined in ❁ 6-4 and 6-5. An excessively stiff boot on the shift lever or a binding shift lever may pull the shift lever back to neutral from any gear position (Fig. 5-4). To check the boot, squeeze it. If it is too stiff, replace it. Adjust the console to relieve any binding.

Worn gears or gear teeth may also increase the chances of gears coming out of mesh. If the detent balls lack sufficient spring force, there will be little to hold the gears in mesh and they may slip out. Worn bearings or synchronizers loose on the shaft tend to cause excessive end play, or free motion, that allows the gears to de-mesh.

In addition, if the transmission slips out of high gear, it could be due to misalignment between the transmission and the engine. This condition is serious. It can soon damage the clutch and transmission parts. Misalignment can often be detected by the action of the clutch pedal. It causes clutch-pedal pulsations, as explained in ❁ 3-7. The procedure for checking clutch-housing alignment is described in ❁ 3-25. If the clutch housing is out of line, then the transmission will also be out of line.

❁ 5-7 No power through transmission

If the transmission is in mesh and the clutch is engaged, but no power passes through the transmission, then the clutch could be slipping. Various causes of clutch slippage are described in ❁ 3-3. If the clutch is not slipping, then the trouble is in the transmission. The indication is that something serious has taken place which will require complete transmission overhaul.

Conditions inside the transmission that would prevent power from passing through include gear teeth stripped from gears, a shifter fork or some other linkage part broken, a gear or shaft broken (Fig. 5-5), and a drive key or spline sheared off. The transmission must be taken out and disassembled as explained in following chapters, so that the damaged or broken parts can be replaced.

❁ 5-8 Transmission noisy

Several types of noise may be encountered in transmissions. Whining or

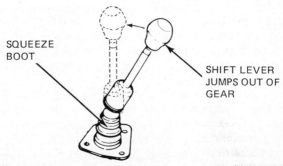

Fig. 5-4 On some transmissions, an excessively stiff shift-lever boot can force the shift lever to jump out of gear. *(Ford Motor Company)*

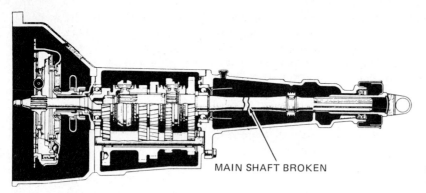

Fig. 5-5 A broken main shaft prevents power flow through the transmission. (*Toyota Motor Sales Co., Ltd.*)

MAIN SHAFT BROKEN

growling, either steady or intermittent, may be due to worn, chipped, rough, or cracked gears. As the gears continue to wear, the noise may take on a grinding characteristic, particularly in the gear position that throws the greatest load on the worn gears. Bearing trouble often produces a hissing noise that will develop into a bumping or thudding sound as bearings wear badly.

Metallic rattles could be due to worn or loose shifting parts in the linkage or to gears loose on the shaft splines. Sometimes, if the clutch friction-disk cushion springs or the engine torsional-vibration dampener are defective, the torsional vibration of the engine will carry back into the transmission. This noise would be apparent only at certain engine speeds.

In analyzing noise in the transmission, first note whether the noise is obtained in neutral with the car stationary or in certain gear positions. If the noise is evident with the transmission in neutral with the car stationary, disengage the clutch. If this does not stop the noise, then the trouble probably is not in the transmission (provided the clutch actually disengages and does not have troubles such as outlined in ✿ 3-6). In this case, the noise is probably in the engine or clutch. But if the noise stops when the clutch is disengaged, then the trouble is probably in the transmission.

A squeal when the clutch is disengaged usually means that the clutch throwout bearing needs lubrication or is defective. Also, a worn or dry pilot bushing in the crankshaft can become noisy. Noise can occur because the crankshaft continues to turn with the clutch disengaged, but the clutch shaft itself (which pilots in the crankshaft bushing) stops turning.

Noise heard when the transmission is in neutral and the clutch engaged, could come from transmission misalignment with the engine, worn or dry bearings, worn gears, a worn or bent countershaft, or excessive end play in the countershaft. Notice that these are the parts which are in motion when the clutch is engaged and the transmission is in neutral.

Noise heard with the transmission in gear could result from any of the conditions noted in the previous paragraph. Also, it could be due to a defective friction disk in the clutch or a defective engine torsional-vibration dampener. In addition, the rear main bearing of the transmission could be worn or dry, gears could be loose on the main shaft, gear teeth could be worn, or synchronizers could be worn or damaged. Another

cause of noise could be worn speedometer gears. Careful listening to notice the particular gear position in which the most noise is obtained is often helpful in pinpointing the worn parts that are producing the noise.

Worn transmission parts should be replaced after transmission removal and disassembly, as outlined in later chapters.

✿ 5-9 Gears clash in shifting   Gear clashing that accompanies shifting may be due to failure of the synchronizing mechanism to operate properly (Fig. 5-6). This condition might be caused by a broken synchronizer spring, incorrect synchronizer end play, or defective synchronizer cone surfaces. It could also be due to gears sticking on the main shaft or failure of the clutch to release fully.

Gear clash can be obtained in low or reverse on many cars if a sudden shift is made to either of these gears while gears are still in motion. In some transmissions these two gear positions do not have synchromesh devices. To prevent gear clash when shifting into either of these positions, it is necessary to pause long enough to allow the gears to stop turning. If the clutch is not disengaging fully, then the gears will still be driven and may clash when the shift is made. Conditions that can prevent the clutch from disengaging fully are discussed in ✿ 3-5.

A worn or dry pilot bushing can keep the clutch shaft spinning even when the clutch is disengaged. This condition can cause gear clash when shifting, as can incorrect lubricant in the transmission.

Following chapters describe transmission removal

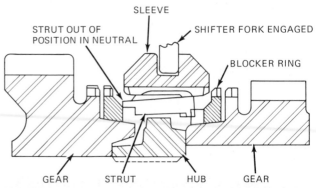

Fig. 5-6 A defective synchronizer causes gear clash during shifts.

and disassembly to replace defective synchromesh parts.

✿5-10 Transmission noisy in reverse
Noise heard when the transmission is in the reverse position is probably due to a damaged or worn reverse idler gear or bushing, reverse gear on the main shaft, countergear, or to a damaged shift mechanism. The transmission should be removed for disassembly and inspection of the parts so that damaged or worn parts can be replaced.

✿5-11 Oil leaks
If the lubricant in the transmission case is not the correct type or if different brands of lubricant are put into the transmission, the lubricant may foam excessively. As it foams, it will completely fill the case and begin to leak out. The same thing might happen if the oil level is too high. The right amount of the recommended oil should be used in the transmission to prevent excessive oil leakage due to foaming.

In addition, if gaskets are broken or missing or if oil seals or oil slingers are damaged or missing, oil will work past the shafts at the two ends of the transmission. Also, if the drain plug is loose or if the transmission-bearing retainer is not tightly bolted to the case, then oil will be lost.

A cracked transmission or extension case will also leak oil. Following chapters explain how to disassemble the transmission so that defective gaskets, oil seals, and slingers can be replaced.

Chapter 5 review questions

Select the *one* correct, best, or most probable answer to each question. Then check your answers against the correct answers given at the end of the book.

1. Hard shifting into gear can be caused by:
   a. gearshift linkage out of adjustment
   b. clutch not disengaging
   c. clutch-pedal free play excessive
   d. all of the above
2. The transmission may stick in gear because:
   a. the gearshift linkage is out of adjustment
   b. the clutch is not disengaging
   c. a synchronizer is stuck
   d. all of the above
3. Noise from the transmission when it is in neutral could be caused by:
   a. failure of the clutch to engage
   b. worn or dry bearings
   c. main-shaft gears having chipped or broken teeth
   d. all of the above
4. The transmission may slip out of gear because:
   a. the gearshift linkage is out of adjustment
   b. the clutch housing is misaligned
   c. the lockout springs have insufficient force
   d. all of the above
5. Noise from the transmission when it is in gear could be caused by:
   a. a slipping clutch
   b. excessive lubricant
   c. worn or damaged gears
   d. all of the above

6. Gear clash while shifting could be caused by:
   a. gears loose on the main shaft
   b. clutch not disengaging
   c. gearshift linkage disconnected
   d. all of the above
7. Transmission oil leaks could be caused by:
   a. foaming due to incorrect lubricant
   b. excessive lubricant
   c. damaged or missing oil seals or slingers
   d. all of the above
8. Noise from the transmission in reverse could be caused by:
   a. worn or damaged reverse idler gear
   b. defective synchronizer
   c. clutch not disengaging
   d. all of the above
9. Hard shifting into gear could be caused by any of the following *except:*
   a. clutch not disengaging
   b. synchronizer damaged
   c. bearings worn
   d. shifter fork bent
10. Gear clash during shifts can be caused by any of the following *except:*
    a. synchronizer defective
    b. idle speed too low
    c. clutch not disengaging
    d. incorrect gearshift-linkage adjustment

MANUAL-TRANSMISSION REMOVAL AND INSTALLATION

After studying this chapter, and with proper instruction and equipment, you should be able to:

1. Remove and install a manual transmission in a car.
2. Service and adjust the shift linkage on a car with a column-mounted shift lever.
3. Service and adjust the shift linkage on a car with a floor-mounted shift lever.

6-1 Preparing for manual-transmission service This chapter covers the removal and installation of manual transmissions used on vehicles with rear-wheel drive. Typical gearshift-linkage adjustments also are covered.

Transmission troubles can be repaired in two ways. These are (1) disassembling the transmission and replacing defective parts and (2) replacing the old transmission with a new, used, or rebuilt unit. If the problem is in the linkages between the shift lever and the transmission, it can usually be fixed with a linkage adjustment or repair. You should have the manufacturer's service manual for the specific model of car you are working on. Different linkage arrangements and different adjustment procedures are required for different cars. However, when the transmission trouble cannot be fixed by in-car service, then the transmission must be removed.

6-2 Manual-transmission removal A variety of arrangements are used to mount manual transmissions in vehicles. Basically, they are very similar. Several bolts attach the transmission to the clutch housing (Fig. 6-1). The rear end of the transmission output shaft or transmission main shaft is attached to the driveshaft (Fig. 6-2). A cross member supports the transmission (Fig. 6-3). Shift rods connect the selector lever and the shift levers on the side of the transmission (Fig. 6-4).

Following is a manual-transmission removal procedure that applies, in general, to all vehicles. However, because of variations from car to car, refer to the manufacturer's shop manual that covers the specific vehicle you are working on. Manuals have step-by-step procedures, together with specifications for checking parts and tightening attaching bolts and nuts.

NOTE: You may have to remove the catalytic converter and its support bracket to give yourself enough room to remove the transmission.

1. On floor-shift cars (Fig. 6-4), remove the shift lever.
2. Raise the vehicle on a lift.
3. Mark the rear-axle flange and driveshaft so that the driveshaft can be reinstalled in the same position when reconnected.
4. Some manufacturers recommend draining the lubricant from the transmission before removing it. Refer to the manufacturer's service manual.
5. Disconnect the speedometer cable from the transmission. Some manufacturers specify that you remove the speedometer-cable adapter and gear from the transmission housing at this time. If the trans-

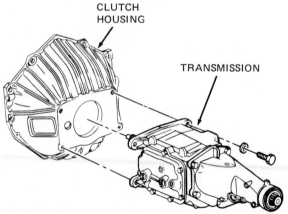

Fig. 6-1 Typical transmission mounting on the clutch housing. *(Chevrolet Motor Division of General Motors Corporation)*

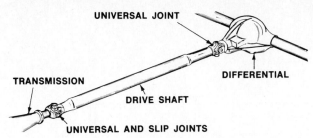

Fig. 6-2 The front end of the driveshaft attaches to the output shaft or main shaft of the transmission.

mission lubricant has not been drained, plug the hole to prevent leakage during transmission removal.

6. Disconnect the wiring to the backup and transmission-controlled-spark switches, if present.

7. Disconnect the driveshaft. Some manufacturers recommend removing the driveshaft completely. If the lubricant has not been drained, install a sealing plug in the transmission extension housing to prevent lubricant leakage.

8. Support the engine with a jack or engine support (Fig. 6-5). Remove the bolts attaching the transmission support to the cross member (Fig. 6-6). Then remove the bolts attaching the cross member to the body or frame and remove the cross member (Fig. 6-3).

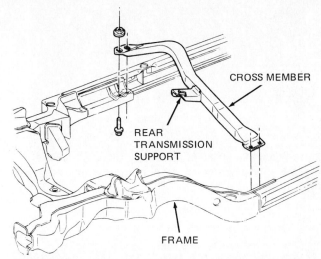

Fig. 6-3 The rear mount on the transmission attaches to a support bracket on the frame or body cross member. *(Chevrolet Motor Division of General Motors Corporation)*

9. Remove the upper bolts attaching the transmission to the clutch housing and install guide pins in the holes (Fig. 3-8). The purpose of these guide pins is to prevent damage to the clutch friction disk. If the transmission were removed hanging down at an angle, it would put stress on the disk hub and probably damage the disk. This is because the clutch

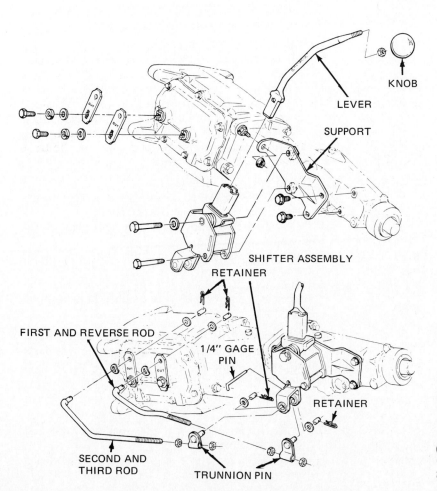

Fig. 6-4 Floor-shift linkage must be disconnected before the transmission can be removed. *(Oldsmobile Division of General Motors Corporation)*

Fig. 6-5 Installing engine support prior to removal of the transmission.

shaft passes through the friction-disk hub and pilots in the end of the crankshaft (Figs. 2-6 and 2-8).

10. Remove the other transmission-attaching bolts. Then slide the transmission rearward until the clutch shaft clears the clutch.

CAUTION: The transmission is heavy. Always use a transmission jack if available (Fig. 6-7). Place the jack under the transmission to support it. If a jack is not available, get another person to help you. Move the transmission to the rear until it is free, then lower it and move it out from under the car.

With the transmission out, inspect the clutch and flywheel condition and tightness. Check the clutch-shaft pilot bushing in the end of the engine crankshaft. If the pilot bushing is worn, it can be replaced as explained in ✿ 3-24, item 3.

✿ 6-3 Manual-transmission installation In general, installation of the transmission is the reverse of removal. Just before installation, shift the transmission into each gear and turn the input shaft to make sure the transmission works properly.

Be sure that the matching faces of the transmission and clutch housing are clean and free of burrs or nicks. Put a small amount of lubricant on the splines of the clutch shaft. Then proceed as follows:

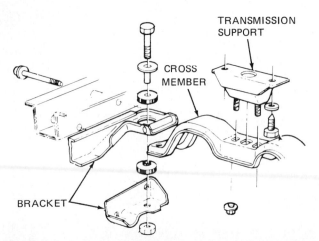

Fig. 6-6 Remove the transmission support from the cross member and from the transmission. *(Ford Motor Company)*

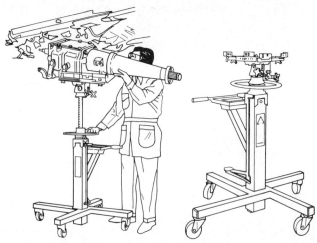

Fig. 6-7 Using a transmission jack to support the transmission and lower it from the engine. *(ATW)*

1. With the guide pins (Fig. 3-8) and engine support (Fig. 6-5) in place, raise the transmission on the transmission jack (Fig. 6-7). Align the splines on the clutch shaft and the friction-disk hub by turning the clutch shaft. This is done by placing the transmission in gear and turning the output shaft.
2. Move the transmission forward, making sure the splines on the clutch shaft and friction-disk hub align, until the transmission seats against the clutch housing. Install the attaching bolts and lockwashers. Torque the bolts to specifications.

Careful: If the transmission does not fit snugly against the clutch housing, or if you cannot move the transmission easily into place, do not force it. The splines may not be properly aligned, or roughness, dirt, or a loose retainer ring in the transmission may be blocking the transmission. If the attaching bolts are tightened under such conditions, the transmission case may break. Move the transmission back or remove it and determine the cause of trouble.

3. Install the cross member and attach the support to the transmission (if removed). Reattach the transmission support to the cross member.
4. Reinstall the speedometer cable and gear (if removed).
5. Reconnect the switch wires disconnected during transmission removal.
6. Install the driveshaft, making sure to align the marks so the driveshaft is reattached in its original position.

NOTE: The driveshaft is attached with special fasteners. If new parts are required, use original-equipment parts or their equivalent. Drive-line service is covered in detail in Chap. 18.

7. Reattach the shift rods. Install the shift lever if it was removed.
8. Fill the transmission with the specified type and quantity of lubricant.
9. Check the transmission shifts. Adjust the linkages as necessary (✿ 6-4 and 6-5).

10. Lower the vehicle to the ground. Start the engine and check the transmission operation. Then road test the car to make sure the transmission operates properly.

✿ 6-4 Column-mounted shift-linkage adjustment

Figure 6-8 shows the adjustments to be made on a column-mounted shift control lever. With rods D and E disconnected, the procedure is as follows:

1. Set levers A and C in reverse. Turn the ignition switch to the lock position. Shift to reverse by moving transmission lever C clockwise to forward detent.
2. Attach rod D to shift lever G with retainer (view A in Fig. 6-8). Slide swivel onto rod D. Insert swivel with clamp into lever C and loosely attach it with nut K and washers.
3. Remove column lash by rotating lever G in downward direction. Complete the attachment of rod D to lever C by tightening nut K to the recommended specification.
4. Turn ignition switch to the unlock position and place levers A, B, and C in neutral. Neutral is obtained by moving levers B and C clockwise to forward detent and then counterclockwise one detent.
5. Align gauge holes in levers F, G, and H. Then insert gauge pin J (view A in Fig. 6-8).

NOTE: The linkage adjustment on many transmissions is checked by the use of various gauge pins.

6. Repeat steps 2 and 3 for rod E and levers B and F.
7. Remove gauge pin J.

NOTE: With the shift lever in reverse the ignition switch must move freely to the lock position. It must not be possible to get the lock position in neutral or in any position other than reverse.

✿ 6-5 Floor-shift linkage adjustment

This adjustment varies with different cars. Figures 4-10 and 4-11 show typical linkage arrangements for three-speed and four-speed floor-shift linkages. Adjustment is as follows:

1. Turn ignition switch to the off position. Raise the vehicle on a lift.
2. Loosen locknuts at the swivels on the shift rods. Rods should pass freely through the swivels.
3. Set the shift levers in neutral at the transmission.
4. Move the shift control lever to the neutral detent and align control assembly levers. Insert the locating gauge into the lever alignment slot.
5. Tighten locknuts at shift-rod swivels. Remove the locating gauge.
6. Shift transmission control into reverse and turn ignition switch to the lock position. Loosen locknut at the back-drive control rod swivel. Then pull down slightly on the rod to remove any slack in the column mechanism. Tighten the clevis jam nut.
7. Check the interlock control. The ignition switch should move freely into and out of the lock position. Readjust the back-drive control rod, if necessary.
8. Check transmission shift operation. Readjust the shift controls, if necessary. Lower the vehicle. Road test the car to make sure the transmission operates properly.

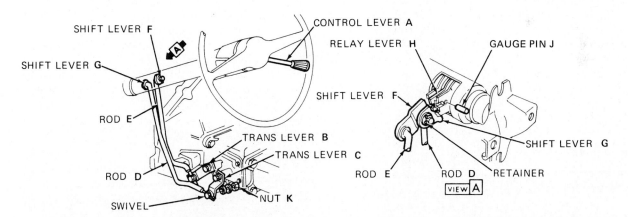

1. Set levers A and C in REVERSE position and turn ignition switch to LOCK position. ⎡NOTE⎤ Obtain REVERSE position by moving trans lever C clockwise to forward detent.
2. Attach rod D to shift lever G with retainer. See View A. Slide swivel onto rod D. Insert swivel with clamp into lever C and loosely assemble with nut K and washers.
3. Remove column lash by rotating lever G in a downward direction and complete attachment of rod D to lever C by tightening nut K using recommended torque.
4. Turn ignition key to UNLOCK position and position levers A, B and C in NEUTRAL. ⎡NOTE⎤ Obtain NEUTRAL position by moving levers B and C clockwise to forward detent then counter-clockwise one detent.
5. Align gauge holes in F, G and H and insert gauge pin J. See View A.
6. Repeat steps 2 and 3 for rod E and levers B and F.
7. Remove gauge pin J.

⎡NOTE⎤ With shift lever in REVERSE the ignition key must move freely to LOCK position. It must not be possible to obtain ignition LOCK position in NEUTRAL or any gear other than REVERSE.

Fig. 6-8 Adjustment of the shift linkage in a car with a steering-column-mounted shift control lever. (Chevrolet Motor Division of General Motors Corporation)

Chapter 6 review questions

Select the *one* correct, best, or most probable answer to each question. Then check your answers against the correct answers given at the end of the book.

1. The purpose of the cross member is to:
   a. carry the shifting action across to the transmission
   b. support the shift mechanism
   c. support the transmission
   d. brace the frame
2. If the catalytic converter is in the way:
   a. push it to one side
   b. remove it
   c. disconnect the front end and bend it down out of the way
   d. remove the transmission from the top of the engine compartment
3. The purpose of marking the driveshaft is so that:
   a. you can adjust its length
   b. you will know which end goes to the front
   c. it can be reinstalled in its original position
   d. you can line it up with the differential
4. The purpose of installing guide pins before removing the transmission is to:
   a. prevent damage to the clutch friction disk
   b. guide the transmission into place on reinstallation
   c. prevent damage to the engine crankshaft
   d. give you room to work when removing the transmission
5. If the transmission does not slide easily into position on reinstallation:
   a. use a soft hammer to tap it into place
   b. force it into place
   c. add more lubricant to the splines
   d. find and eliminate the hang-up

THREE-SPEED MANUAL-TRANSMISSION SERVICE

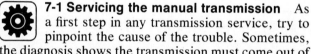

After studying this chapter, and with proper instruction and equipment, you should be able to:

1. Disassemble a three-speed manual transmission.
2. Identify the various parts of a disassembled three-speed transmission.
3. Clean and inspect the bearings and other transmission parts.
4. Reassemble a three-speed transmission and check its operation.

7-1 Servicing the manual transmission As a first step in any transmission service, try to pinpoint the cause of the trouble. Sometimes, the diagnosis shows the transmission must come out of the car for overhaul. Other times, the diagnosis might point to some minor problem that can be fixed without removing the transmission.

Since transmission construction varies considerably from car to car, the removal and disassembly procedures also vary. Before attempting to disassemble a particular transmission, carefully study both the transmission and the transmission section for it in the manufacturer's service manual. If possible, locate the illustrations and the exploded views for the transmission you are working on in the manufacturer's service manual.

In ✿ 4-8 and 4-9, the operation of the gears and synchronizers in a three-speed manual transmission was described. This transmission is shown in partial sectional view in Fig. 4-14. The disassembly and reassembly of this transmission, including inspection and servicing of the transmission parts, are covered in the following sections. Removal of the transmission from the vehicle is discussed in ✿ 6-2.

✿ 7-2 Disassembling the transmission Figure 7-1 shows the transmission completely disassembled. With the transmission on the bench, proceed as follows:

1. Remove the side-cover-attaching bolts and the side-cover assembly (item 62 in Fig. 7-1).
2. Remove the front-bearing retainer (25) and gasket (26).
3. Remove the snap ring holding the bearing in place (28) from the clutch shaft (Fig. 7-2). Then remove the bearing by pulling out on the clutch shaft until a screwdriver can be inserted between the large snap ring around the outside of the bearing and the case. The bearing is a slip fit on the clutch shaft and into the case bore. Removal of the bearing provides clearance to permit removal of the clutch gear and the main-shaft assembly.
4. Remove the speedometer driven gear from the rear extension housing (36).
5. Remove the bolts attaching the extension housing to the transmission case.
6. Remove the reverse-idler-shaft E ring (Fig. 7-3).
7. Remove the drive gear, main shaft, and extension assembly as a unit through the rear of the case. From the main shaft (9), remove the drive gear (31), needle bearings, and synchronizer ring (2).
8. Use snap-ring pliers to expand the snap ring in the extension housing (Fig. 7-4). This snap ring holds the main-shaft rear bearing (20) in place. With the snap ring off, remove the extension housing from the main shaft.
9. Use a short dummy shaft (Fig. 7-5) to drive the countergear shaft, with its woodruff key, out the rear of the case. The dummy shaft will hold the roller bearings in place inside the countergear bore. Remove the gear, bearings, and thrust washer with the dummy shaft.
10. If the reverse idler gear requires removal, use a long punch to drive the reverse idler shaft and woodruff key out through the rear of the case (Fig. 7-6).

✿ 7-3 Disassembling the main shaft The main shaft (9) has all of the items listed in Fig. 7-1 from item

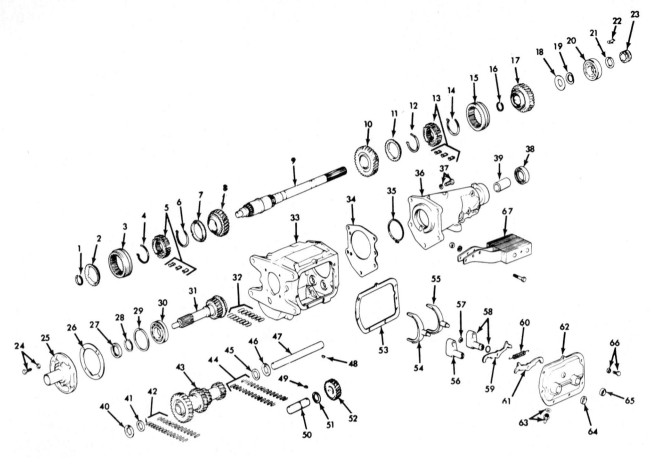

1. Snap ring
2. Synchronizer ring
3. 2-3 synchronizer sleeve
4. Synchronizer key spring
5. Synchronizer hub and keys
6. Synchronizer key spring
7. Synchronizer ring
8. Second gear
9. Main shaft
10. First gear
11. Synchronizer ring
12. Synchronizer key spring
13. Synchronizer hub and keys
14. Synchronizer key spring
15. 1-2 synchronizer sleeve
16. Snap ring
17. Reverse gear
18. Thrust washer
19. Waved washer
20. Rear bearing
21. Snap ring
22. Speedometer gear clip
23. Speedometer drive gear

24. Bearing retainer bolts and washers (4)
25. Front bearing retainer
26. Bearing retainer gasket
27. Bearing retainer oil seal
28. Snap ring
29. Bearing snap ring
30. Front bearing
31. Drive or clutch shaft and gear
32. Pilot bearings
33. Case
34. Extension to case gasket
35. Rear bearing to extension retaining ring
36. Rear extension housing
37. Extension to case retaining bolts and washers
38. Rear extension bushing
39. Rear seal
40. Thrust washer
41. Spacer
42. Countergear shaft bearings
43. Countergear
44. Countergear shaft bearings
45. Spacer

46. Thrust washer
47. Countergear shaft
48. Countergear shaft key
49. Idler shaft key
50. Reverse idler shaft
51. Snap ring
52. Reverse idler gear
53. Side cover gasket
54. 2-3 shift fork
55. 1-rev shift fork
56. 2-3 shifter shaft
57. Retaining E-ring
58. 1-rev shifter shaft with O-ring
59. 2-3 detent cam
60. Detent cam spring
61. 1-rev detent cam
62. Side cover
63. TCS switch and gasket
64. Shifter shaft seal
65. Shifter shaft seal
66. Shift cover bolts and washers
67. Damper assembly

Fig. 7-1 Disassembled three-speed transmission. *(Chevrolet Motor Division of General Motors Corporation)*

1 to item 23, as shown in Fig. 7-4. Snap-ring pliers and the shop press are needed to disassemble the main shaft.

1. Use snap-ring pliers to remove the second-and-third-speed synchronizer-hub snap ring from the main shaft (Fig. 7-7). Then remove the synchronizer assembly and second-speed gear (items 3 to 8) from the main shaft.

2. Remove the speedometer gear from the rear of the main shaft by depressing the retainer clip (22 and 23).

3. Remove the rear-bearing snap ring from the main shaft (Fig. 7-8).

4. Support the reverse gear with a press plate. Press on the rear of the main shaft to push it out of the reverse gear. Remove the gear, thrust washer, spring washer, rear bearing, and snap ring from the rear of the main shaft (items 17 to 21 in Fig. 7-1).

Fig. 7-2 Removing the clutch-gear-bearing retaining ring. (Chevrolet Motor Division of General Motors Corporation)

Fig. 7-5 Removing the countergear shaft. (Chevrolet Motor Division of General Motors Corporation)

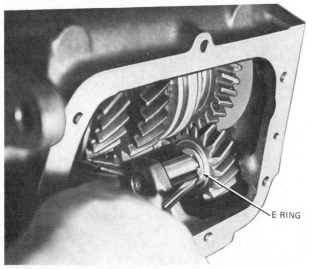

Fig. 7-3 Removing the reverse-idler-shaft E ring. (Chevrolet Motor Division of General Motors Corporation)

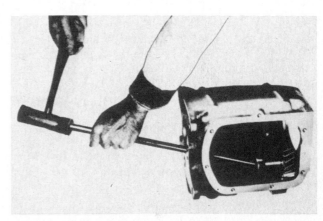

Fig. 7-6 Removing the reverse idler shaft. (Chevrolet Motor Division of General Motors Corporation)

Fig. 7-4 Removing the rear-extension snap ring from the main shaft. (Chevrolet Motor Division of General Motors Corporation)

Fig. 7-7 Removing the second-and-third clutch-hub retaining ring. (Chevrolet Motor Division of General Motors Corporation)

Fig. 7-8 Removing the rear-bearing retaining ring. *(Chevrolet Motor Division of General Motors Corporation)*

5. Remove the first-and-reverse synchronizer-hub snap ring from the main shaft. Slide synchronizer, blocker ring, and first-speed gear from the rear of the main shaft (items 10 to 16).
6. In some transmissions, the hub may be a tight fit and must be pressed from the main shaft.

☼ 7-4 Cleaning and inspecting transmission parts Wash all transmission parts, except for bearings and seals, in solvent. Brush or scrape all dirt from the parts. Do not damage parts with the scraper. Dry parts with compressed air.

CAUTION: When using solvent, wear goggles to protect your eyes. If you get solvent in your eyes, wash them out with water at once (Fig. 7-9).

To clean the bearings, rotate them slowly in clean solvent to remove all lubricant. Then hold the bearing

Fig. 7-9 If you get solvent or other chemicals in your eyes, wash them out at once with water.

assembly stationary so it will not rotate and dry it with compressed air.

Careful: Do not spin the bearing with compressed air. Spinning a dry bearing this way, when it is not lubricated, can damage it.

As soon as the bearing is dry, immediately lubricate it with transmission lubricant. Then wrap each bearing in clean, lint-free cloth or paper for later inspection (☼ 7-6).

Some transmissions have a magnet at the bottom of the case. If present, clean the magnet with solvent and remove any metal particles and dirt.

☼ 7-5 Parts inspection Inspect the transmission case for cracks or worn or damaged bearing bores and threads. Check the front and back of the case for nicks or burrs that could cause misalignment of the case with the flywheel housing or extension housing. Remove all burrs with a fine file.

Check the condition of the shift levers, forks, shift rails, and the lever and shafts. Replace the roller bearings if they are broken, worn, or rough.

Replace the countergear assembly, or any other gears, if they have teeth that are worn, broken, chipped, or otherwise damaged. Replace the countershaft if it is worn, bent, or scored.

The bushing in the reverse gear and the bushing in the reverse idler gear are not serviced separately. If a bushing is worn, the gear-and-bushing assembly must be replaced as a unit.

☼ 7-6 Bearing inspection Bad bearings usually produce a rough growl or grating noise rather than a whine, which is typical of gear noise. Noise is often the clue that points to bad bearings. To inspect a bearing, first clean it as explained in ☼ 7-4. If the bearing has become magnetized, metal particles will be attracted to it. Cleaning the bearing by normal methods will not remove them. The bearing must be demagnetized first.

Bearings fail by lapping, spalling, or locking (Fig. 7-10).

1. Lapping   Lapping is caused by fine particles of abrasive material such as scale, sand, or emery. These particles circulate with the oil and cause wear of the roller or ball and race surfaces. Bearings which are worn enough to be loose but appear smooth without pitting or spalling have been running with dirty oil. This gradually wears away the surfaces.

2. Spalling   Spalling is caused by overloading or by faulty assembly. Bearings that fail by spalling have either flaked or pitted rollers or races. Spalling can be caused by faulty assembly, such as cocking of bearings, misalignment, or excessively tight adjustments.

3. Locking   Locking is caused by large particles of dirt or other material wedging between rollers and races. This usually causes one of the races to spin. If a race spins in the housing, it will wear the housing in which it is assembled so that the housing must be replaced.

HEAVILY SPALLED INNER RACE. UNACCEPTABLE

HEAVY PARTICLE INDENTATION AND LIGHT SPALLING. UNACCEPTABLE

LIGHTLY SPALLED INNER RACE. UNACCEPTABLE

LIGHT PARTICLE INDENTATION. ACCEPTABLE

Fig. 7-10 Inspection of ball bearings. *(Ford Motor Company)*

Here are the four basic checks to be made on ball bearings to determine if they are in good enough condition to be used again.

1. **Inner-ring raceway** Hold the outer ring stationary and rotate the inner ring three times. Examine the raceway of the inner ring for pits or spalling (Fig. 7-10). Note the types of damage and those that require bearing replacement.

2. **Outer-ring raceway** Hold the inner ring stationary and rotate the outer ring three times. Examine the outer ring raceway for damage (Fig. 7-10). Note the types of damage and those that require bearing replacement.

3. **External surfaces** Replace the bearing if there are radial cracks on the front or rear faces of the outer or inner rings. Replace the bearing if there are cracks on the outside diameter or outer ring. Check carefully around the snap-ring groove. Also replace the bearing if the ball cage is cracked or deformed.

4. **Spin test** Lubricate the bearing raceways lightly with clean oil. Turn the bearing back and forth slowly to coat the raceways and balls. Hold the bearing by the inner ring in a vertical position. Some vertical movement between the inner and outer rings is okay.

Spin the outer ring several times by hand. *Do not use an air hose!* If you notice roughness or vibration, or if the outer ring stops abruptly, reclean the bearing, relubricate it, and spin it again. Roughness is usually caused by particles of dirt in the bearing. If the bearing is still rough after cleaning and lubricating it three times, discard it.

Now hold the bearing by the inner ring in a horizontal position with the snap-ring groove up. Spin the outer bearing several times by hand, as described above. If the bearing is still rough after cleaning and relubricating it three times, discard it.

✿ 7-7 Synchronizer assembly service The synchronizer assembly (Fig. 7-11) is a select-fit assembly. The hub and sliding sleeve are selected so that they provide the proper fit. For this reason they should be kept together and not mixed with the parts from another synchronizer. However, the keys and springs can be replaced separately if they are worn or broken.

To disassemble and reassemble the synchronizer, proceed as follows:

1. Mark the hub and sleeve so that they can be matched on reassembly.
2. Push the hub from the sliding sleeve. The keys and springs may now be removed.

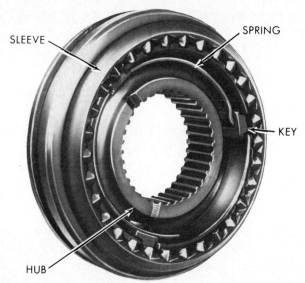

Fig. 7-11 Synchronizer assembly. *(Chevrolet Motor Division of General Motors Corporation)*

3. To reinstall the three keys and two springs, put them in place so that all three keys are engaged by both springs. One spring goes on each side of the sleeve. The tanged end of each spring should be installed in different key cavities on either side. Slide the sleeve on the hub, aligning the marks made before disassembly.

A groove around the outside of the synchronizer hub identifies the end that must be opposite the fork slot in the sleeve when it is assembled. This groove indicates the end of the hub with the greater recess depth.

✿ 7-8 Extension oil seal and bushing If the bushing in the rear of the extension requires replacement, remove the seal and use the specified tool to drive the

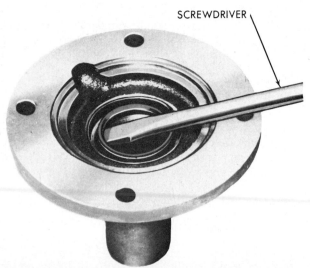

Fig. 7-12 Removing the drive-gear-bearing-retainer seal. *(Chevrolet Motor Division of General Motors Corporation)*

bushing on into the extension. Using the same tool, drive in a new bushing from the rear. Coat the bushing with oil and install a new seal using the special tool.

✿ 7-9 Drive-gear-bearing oil seal If the oil seal in the front-bearing retainer (25 in Fig. 7-1) requires replacement, pry it out (Fig. 7-12). Use the special tool to install a new oil seal.

✿ 7-10 Assembling the main shaft Figure 7-13 shows the main-shaft assembly and drive gear with all gears and other parts. Everything is shown slightly separated so you can see how all the parts go together. See also Fig. 7-1. To reassemble the main shaft:

1. Turn the front of the main shaft up and install the following components.
2. Install the second-speed gear with the clutching splines upward. The rear face of the gear will butt against the flange on the main shaft.
3. Install a blocker ring with the clutching cone downward over the synchronizing surface of the second-speed gear. All three blocker rings used in this transmission are identical.
4. Install the second-and-third synchronizer assembly with the fork slot downward. Press it onto the splines on the main shaft until it bottoms. Both synchronizer assemblies used in this transmission are identical.
5. If the sleeve comes off the second-and-third synchronizer hub, note that the notches on the hub outside diameter face the forward end of the main shaft. Be sure the notches of the blocker ring align with the keys of the synchronizer assembly.
6. Install the snap ring that retains the synchronizer hub in place on the main shaft. Both synchronizer snap rings are identical.
7. Turn the main shaft around so that the rear of the main shaft points up.
8. Install the first-speed gear with the clutching splines up. The front face of the gear will butt against the flange on the main shaft.
9. Install a blocker ring with the clutching cone over the synchronizing surface of the first-speed gear.
10. Install the first-and-reverse synchronizer assembly with the fork slot down. Push the assembly onto the splines on the main shaft.
11. Install the synchronizer-hub-to-main-shaft snap ring. Be sure the notches of the blocker ring align with the keys of the synchronizer assembly.
12. Install the reverse gear with the clutching splines pointing downward.
13. Install the reverse-gear steel thrust washer.
14. Install the reverse-gear spring washer.
15. Install the rear ball bearing with the snap-ring slot downward. Press the bearing onto the main shaft.
16. Install the rear-bearing-to-main-shaft snap ring.
17. Install speedometer drive gear and retaining clip.

✿ 7-11 Assembling the transmission After completing the assembly of the main shaft, you are ready to assemble the transmission.

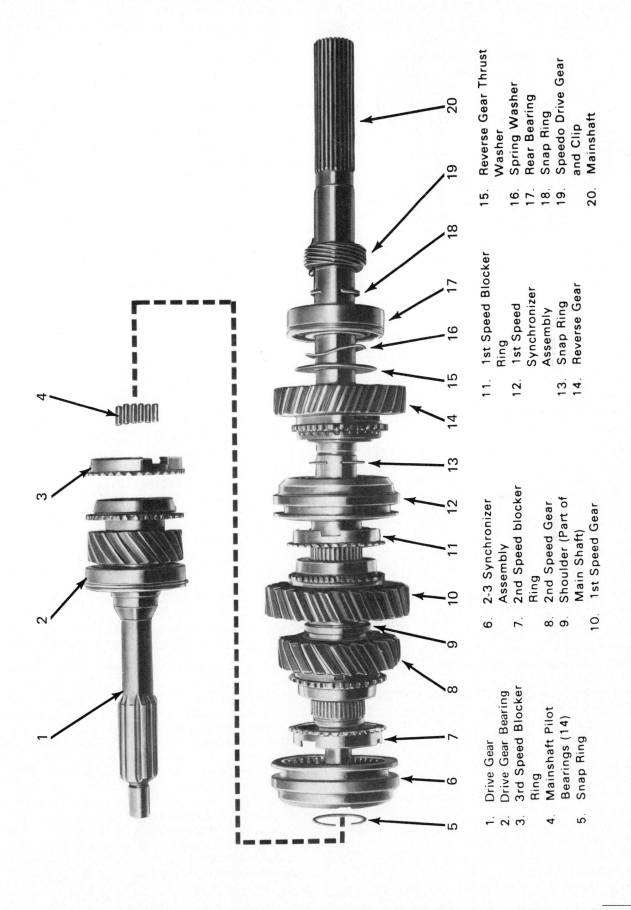

1. Drive Gear
2. Drive Gear Bearing
3. 3rd Speed Blocker Ring
4. Mainshaft Pilot Bearings (14)
5. Snap Ring
6. 2-3 Synchronizer Assembly
7. 2nd Speed blocker Ring
8. 2nd Speed Gear
9. Shoulder (Part of Main Shaft)
10. 1st Speed Gear
11. 1st Speed Blocker Ring
12. 1st Speed Synchronizer Assembly
13. Snap Ring
14. Reverse Gear
15. Reverse Gear Thrust Washer
16. Spring Washer
17. Rear Bearing
18. Snap Ring
19. Speedo Drive Gear and Clip
20. Mainshaft

Fig. 7-13 Clutch-gear and main-shaft assembly. *(Chevrolet Motor Division of General Motors Corporation)*

1. Use a short dummy shaft to install the roller bearings at each end of the countergear (Fig. 7-1). First, insert the dummy shaft. Then apply heavy grease to the rollers and insert them one by one into the space between the shaft and the countergear.

2. Put heavy grease on the two tanged thrust washers and install them in the case so they will be at the two ends of the countergear. The tangs should align with the notches in the case. The grease will hold the washers in place.

3. Put the countergear assembly into the case through the rear opening. Push the countergear shaft, with woodruff key, into the countergear. This pushes out the dummy shaft. Make sure the countergear shaft picks up the two thrust washers as well as the rollers at both ends of the countergear.

4. Install the reverse idler gear and shaft with its woodruff key from the rear of the case. Do not install the E ring.

5. Use snap-ring pliers to expand the snap ring in the extension and put the extension over the rear of the main shaft and onto the rear bearing (20 in Fig. 7-1). Seat the snap ring in the rear-bearing groove (Fig. 7-4).

6. Load the main-shaft pilot bearings into the drive-gear cavity (Fig. 7-13). Assemble the third-speed blocker ring onto the drive-gear clutching surface with its cone toward the gear.

7. Carefully guide the drive gear, pilot bearings, and blocker ring over the front end of the main-shaft assembly (Fig. 7-13). Do not assemble the bearing to the gear yet. Make sure the notches in the blocker ring align with the keys in the second-to-third synchronizer assembly.

8. Stick the extension-to-case gasket to the rear of the case with grease. Then, from the rear of the case, guide the clutch gear, main-shaft assembly, and extension into place. Secure the extension to the case with retaining bolts.

9. Install the front-bearing outer snap ring to the bearing and slide the bearing over the end of the drive-gear shaft and into the front-case bore.

10. Install the snap ring on the drive-gear shaft. Install the bearing retainer and gasket on the case. Make sure the retainer-oil-return hole is at the bottom.

11. Install the reverse-idler-gear-retainer E ring on the shaft (Fig. 7-3).

12. Shift the synchronizer sleeves to neutral positions and install the cover, gasket, and fork assembly in the case. Be sure the forks align with their synchronizer-sleeve grooves.

13. Install the speedometer driven gear in the extension.

14. Tighten all bolts to the specified torques.

15. Rotate the drive-gear shaft and shift the transmission to all gear positions to make sure everything is in proper working order.

16. Fill the transmission with the specified lubricant, up to the level of the filler-plug hole.

17. The transmission is now ready for installation on the vehicle. After installation and adjustment of the linkages, road test the car to make sure the transmission is working properly.

Chapter 7 review questions

Select the *one* correct, best, or most probable answer to each question. Then check your answers against the correct answers given at the end of the book.

1. According to the illustration showing the three-speed transmission completely disassembled, the transmission has:
   a. 37 parts
   b. 57 parts
   c. 67 parts
   d. 73 parts

2. When disassembling the transmission, the first thing to remove is the:
   a. extension housing
   b. clutch gear
   c. side gear
   d. side-cover assembly

3. When disassembling the main shaft, the first thing to remove is the:
   a. second-and-third-speed synchronizer-hub snap ring
   b. idler gear
   c. countergear shaft
   d. second-speed gear

4. After washing the bearings in cleaning solvent, dry them by:
   a. spinning them with compressed air
   b. lubricating them
   c. holding them stationary so they do not spin and drying them with compressed air
   d. warming them in an oven

5. Bad ball bearings usually produce a:
   a. harsh knock
   b. rough growl or whine

c. rough growl or grating noise
d. whine or whistling noise

6. Bearings fail by:
   a. lapping, spalling, or locking
   b. lapping, grinding, or honing
   c. knocking, whining, or grinding
   d. spalling, grinding, or honing

7. The sliding sleeve and hub of a synchronizer:
   a. are interchangeable between synchronizers
   b. are separate service items
   c. are select-fitted during original assembly
   d. cannot be serviced

8. When reassembling the main shaft, the first thing to install is the:
   a. blocker ring
   b. second-speed gear
   c. third-speed gear
   d. speedometer gear

FOUR-SPEED MANUAL-
TRANSMISSION SERVICE

After studying this chapter, and with proper instruction and equipment, you should be able to:

1. Disassemble a standard four-speed manual transmission.
2. Identify the various parts of a disassembled four-speed manual transmission.
3. Reassemble a standard four-speed manual transmission and check its operation.

 8-1 Servicing the four-speed manual transmission This chapter describes in detail the disassembly and reassembly of a standard four-speed manual transmission. The transmission has direct drive in fourth gear. Section 4-10 describes the gearing in this transmission. It is shown in sectional view in Fig. 4-27. Figure 8-1 shows the transmission completely disassembled. This picture is your guide as we describe disassembly and reassembly procedures for the transmission.

Removal of the transmission from the vehicle is discussed in ✿ 6-2. The servicing of the four-speed transmission in which fourth gear is overdrive is covered in the following chapter.

✿ **8-2 Disassembling the four-speed transmission** Figure 8-1 shows the transmission completely disassembled, with all parts named. With the transmission on the bench, proceed as follows to disassemble it:

1. First, remove the transmission side cover. To do this, shift the transmission into second speed by moving the 1-2 shifter lever into forward position. Then remove the cover assembly and allow the oil to drain.
2. Remove the four bolts and two bolt lock strips from the front-bearing retainer (1 in Fig. 8-1) and gasket. Use a special drive-gear wrench to remove the main-drive-gear retaining nut (Fig. 8-2). To do this you must lock the transmission by shifting it into two gears.
3. Put the transmission gears in neutral and drive the lockpin from the reverse-shifter-lever boss (Fig. 8-3). Pull the shifter shaft out about ⅛ inch [3.18 mm] to disengage the shift fork from the reverse gear.

4. Remove the six bolts attaching the case extension to the case. Tap the extension away from the transmission with a soft hammer. When the reverse idler shaft is out as far as it will go, move the extension to the left so that the reverse fork clears the reverse gear. Now, the extension and gasket can be removed.
5. Remove the reverse idler gear, flat thrust washer, shaft, roll spring pin, speedometer gear, and reverse gear.
6. Slide the 3-4 synchronizer clutch sleeve to the fourth-speed position (forward), as shown in Fig. 8-4. Now, carefully remove the rear-bearing retainer and entire main-shaft assembly from the base by tapping the bearing retainer with a soft hammer.
7. Unload the 17 bearing rollers from the main drive gear and then remove the fourth-speed-synchronizer blocker ring. Lift the front half of the reverse idler gear and remove the tanged thrust washer from the case.
8. Use a shop press to press the main drive gear down from the front bearing (Fig. 8-5). From inside the case, tap out the front bearing and snap ring. Also from the front of the case, press out the countershaft with a special tool (Fig. 8-6). This is a short dummy shaft which holds the rollers in place when the countergear assembly is removed. Then remove the countergear assembly and both tanged washers.
9. Remove 112 rollers, six 0.070-inch [1.778-mm] spacers, and the roller spacer from the countergear. Remove the main-shaft front snap rings as shown in Fig. 8-7 and slide the third-and-fourth-speed clutch assembly, third-speed gear, and synchronizing ring from the front of the main shaft.
10. Spread the rear-bearing-retainer snap ring and press the main shaft out of the retainer. Remove

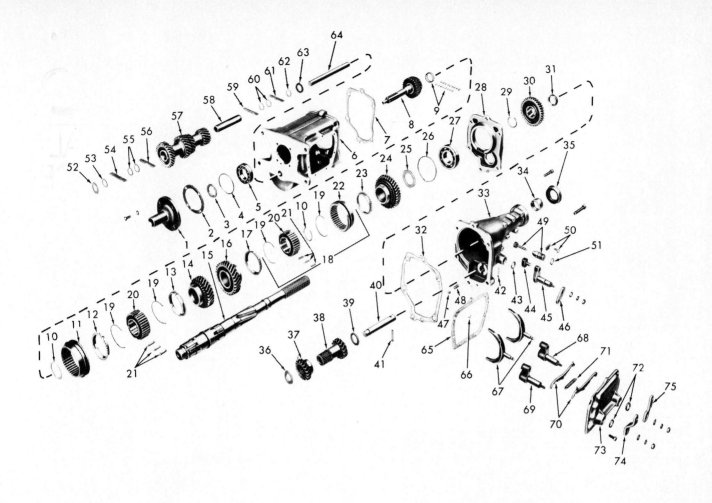

1. Bearing retainer
2. Gasket
3. Bearing retaining nut
4. Bearing snap ring
5. Main-drive-gear bearing
6. Transmission case
7. Rear-bearing-retainer gasket
8. Main drive gear
9. Bearing rollers (17), cage
10. Snap ring
11. Third-and-fourth-speed-clutch sliding sleeve
12. Fourth-speed-gear synchronizing ring
13. Third-speed synchronizing ring
14. Third-speed gear
15. Main shaft
16. Second-speed gear
17. Second-speed-gear synchronizing ring
18. First-and-second-speed-clutch assembly
19. Clutch key spring
20. Clutch hub
21. Clutch keys
22. First-and-second-speed-clutch sliding sleeve
23. First-speed-gear synchronizing ring
24. First-speed gear
25. First-speed-gear sleeve

26. Rear-bearing snap ring
27. Rear bearing
28. Rear-bearing retainer
29. Selective-fit snap ring
30. Reverse gear
31. Speedometer drive, clip
32. Rear-bearing-retainer-to-case extension gasket
33. Case extension
34. Extension bushing
35. Rear oil seal
36. Reverse idler front thrust washer (tanged)
37. Reverse idler gear (front)
38. Reverse idler gear (rear)
39. Flat thrust washer
40. Reverse idler shaft
41. Reverse-idler-shaft roll pin
42. Reverse-shifter-shaft lockpin
43. Reverse-shifter-shaft lip seal
44. Reverse shift fork
45. Reverse shifter shaft and detent plate
46. Reverse shifter lever
47. Reverse-shifter-shaft detent ball
48. Reverse-shifter-shaft ball detent spring
49. Speedometer driven gear and fitting
50. Retainer and bolt
51. O-ring seal

52. Tanged washer
53. Spacer
54. Bearing rollers (28)
55. Spacer
56. Bearing rollers (28)
57. Countergear
58. Countergear roller spacer (seam type)
59. Bearing rollers (28)
60. Spacer
61. Bearing rollers (28)
62. Spacer
63. Tanged washer
64. Countershaft
65. Gasket
66. Detent-cam retainer ring
67. Forward-speed shift forks
68. First-and-second-speed-gear shifter shaft and detent plate
69. Third-and-fourth-speed-gear shifter shaft and detent plate
70. Detent cams
71. Detent-cam spring
72. Lip seals
73. Transmission side cover
74. Third-and-fourth-speed shifter lever
75. First-and-second-speed shifter lever

Fig. 8-1 Disassembled four-speed transmission. (*Chevrolet Motor Division of General Motors Corporation*)

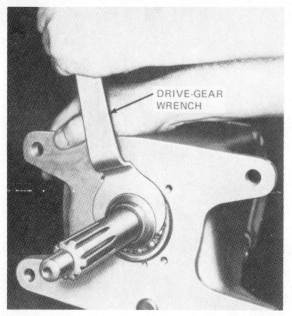

Fig. 8-2 Removing the main-drive-gear retaining nut. *(Chevrolet Motor Division of General Motors Corporation)*

Fig. 8-5 Removing the main drive gear. *(Chevrolet Motor Division of General Motors Corporation)*

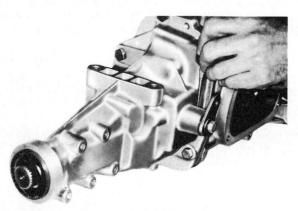

Fig. 8-3 Removing the reverse-shifter-shaft lockpin. *(Chevrolet Motor Division of General Motors Corporation)*

Fig. 8-6 Removing the countershaft with a special tool. *(Chevrolet Motor Division of General Motors Corporation)*

Fig. 8-4 Third-and-fourth-speed synchronizer clutch sleeve in fourth-gear position. *(Chevrolet Motor Division of General Motors Corporation)*

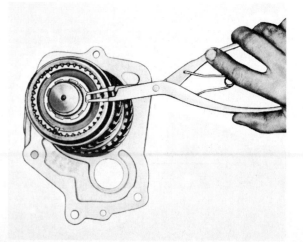

Fig. 8-7 Removing the main-shaft front snap ring. *(Chevrolet Motor Division of General Motors Corporation)*

the main-shaft rear snap ring. Support the second-speed gear and press on the rear of the main shaft to remove the rear bearing, first-speed gear and sleeve, first-speed synchronizing ring, 1-2-speed synchronizing ring, and second-speed gear.

11. Clean and inspect the transmission parts (see ⚙ 7-4 to 7-6).

⚙ **8-3 Reassembling the four-speed transmission** With all parts cleaned and checked to make sure they are good (or replaced with new parts), assemble the main shaft. Next put the countergear assembly together. Then put everything into the transmission case. Proceed as follows:

1. Assemble the main shaft. Install the second-speed gear on the main shaft with the hub toward the rear of the shaft.

   a. Install the 1-2-synchronizer assembly (taper to the rear, hub to the front). Make sure there are synchronizer rings on either side of the assembly with their keyways lined up with the synchronizer keys (Fig. 8-8).

   b. Use a pipe of the correct diameter (Fig. 8-9) to press the first-gear sleeve onto the main shaft.

   c. Install the first-speed gear (hub to front) and use a pipe of the correct diameter to press the rear bearing on. Install the snap ring of the correct thickness to get a maximum distance of 0 to 0.005 inch [0.13 mm] between the ring and rear face of the bearing.

   d. Install the third-speed gear (hub to front) and synchronizer ring (notches to front).

   e. Install the third-and-fourth-speed-gear clutch assembly (hub and sliding sleeve) with both sleeve taper and hub toward the front. Make sure the keys in the hub correspond to the notches in the synchronizer ring.

   f. Install the snap ring in the groove in the main shaft in front of the third-and-fourth-speed clutch

Fig. 8-8 Installing the synchronizing ring. *(Chevrolet Motor Division of General Motors Corporation)*

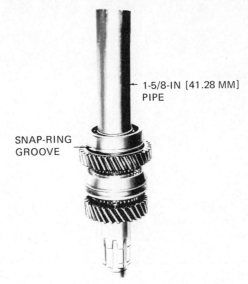

Fig. 8-9 Installing the rear bearing. *(Chevrolet Motor Division of General Motors Corporation)*

assembly, with the ends of the snap ring seated behind the spline teeth.

   g. Install the rear-bearing retainer. Spread the snap ring in the plate to allow the snap ring to drop around the rear bearing. Press on the end of the main shaft until the snap ring engages the groove in the rear bearing.

   h. Install the reverse gear, shift collar to rear, and the two antirattle springs.

   i. Install the retaining clip and speedometer gear.

2. Assemble the countergear. Install the roller spacer in the countergear.

   a. Then install the rollers as follows: Using heavy grease to retain the rollers, install a spacer, 28 rollers, a spacer, 28 more rollers, and then another spacer.

   b. Then at the other end, install a spacer, 28 rollers, a spacer, 28 more rollers, and another spacer.

   c. Insert the special tool into the countergear. This is a dummy shaft to hold the rollers in place when the assembly is installed in the case.

3. Assemble the transmission. Lay the transmission case on its side, with the side cover opening toward you.

   a. Put the countergear tanged washers in place, retaining them with grease. Make sure the tangs are in the case notches.

   b. Put the countergear in the bottom of the case. Turn the case on end, front end down. Lubricate the countershaft and start it into the case. The flat on the end of the shaft should face the bottom of the case. Align the countergear and press the countershaft into the case until the flat on the shaft is flush with the rear of the case. Make sure the thrust washers remain in place.

   c. Check the end play of the countergear by attaching a dial indicator to the case (Fig. 8-10). If the end play is greater than 0.025 inch [0.63 mm], install new thrust washers to reduce the end play.

Fig. 8-10 Checking countergear end play. *(Chevrolet Motor Division of General Motors Corporation)*

d. Into the main drive gear, install the roller bearings, using grease to hold them in place. Install the main drive gear and pilot the bearings through the side-cover opening and into position in front of the case.

e. Put the gasket in position on the front face of the rear-bearing retainer.

f. Install the fourth-speed synchronizing ring on the main drive gear with the notches toward the rear of the case.

g. Slide the 3-4 synchronizing clutch sleeve forward into the fourth-speed detent position (Fig. 8-4). Lower the main-shaft assembly into the case. Make sure the notches on the fourth-speed synchronizing ring correspond to the keys in the clutch assembly (Fig. 8-11). Also, make sure the main drive gear

engages both the countergear and the antilash plate (on standard-ratio models).

h. With the guide pin in the rear-bearing retainer aligned with the hole in the rear of the case, tap the rear-bearing retainer into position with a soft hammer.

i. From the rear of the case, insert the rear reverse idler gear, engaging the splines with the front gear inside the case. Use grease to stick a gasket onto the rear face of the rear-bearing retainer.

j. Install the remaining flat thrust washer on the reverse idler shaft, and install the shaft, roll pin, and thrust washer in the gears and front boss of the case. Make sure the front tanged washer stays in place. The roll pin should be in a vertical position.

k. Pull the reverse-shifter shaft to the left side of the extension and rotate the shaft to bring the reverse-shift fork forward (to reverse detent position). Start the extension onto the transmission shaft (Fig. 8-12) while pushing in on the shifter shaft to engage the shift fork with the reverse-gearshift collar. Then pilot the reverse idler shaft into the extension housing to permit the extension to slide onto the case. Install the six extension-and-retainer-to-case bolts and torque to specifications.

l. Push or pull the reverse-shifter shaft to line up the groove in the shaft with the holes in the boss, and drive in the lockpin. Install the shifter lever.

m. Press the bearing onto the main drive gear (snap-ring groove to front) and into the case until several main-drive-gear threads are exposed. Lock the transmission by shifting into two gears. Install the main-drive-gear retaining nut and torque to specifications. Stake the nut into place at the gear-shaft hole. Do not damage the threads on the shaft.

n. Install the main-drive-gear bearing retainer, gasket, four retaining bolts, and two strip-bolt-lock retainers. Use a sealer on the bolts and torque to specifications.

o. Shift the main-shaft 3-4 sliding clutch sleeve into neutral and the 1-2 sliding clutch sleeve into second. Install the side-cover gasket and put the side cover in place. Be sure the dowel pin aligns. Install attaching bolts and torque to specifications.

p. Put the required amount of the specified lubricant into the transmission. After the transmission has been installed in the vehicle and the linkages properly adjusted, road test the car to make sure the transmission is working properly.

ALIGNMENT PIN

CLUTCH KEY

NOTCH IN SYNCHRONIZING RING

Fig. 8-11 Installing the main-shaft assembly. *(Chevrolet Motor Division of General Motors Corporation)*

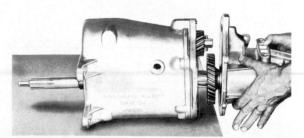

Fig. 8-12 Installing the case extension. *(Chevrolet Motor Division of General Motors Corporation)*

Chapter 8 review questions

Select the *one* correct, best, or most probable answer to each question. Then check your answers against the correct answers given at the end of the book.

1. When disassembling the four-speed transmission, the first thing to remove is the:
   a. front-bearing retainer
   b. side cover
   c. front cover
   d. extension housing
2. The purpose of the short dummy shaft is to hold the:
   a. main-shaft ball bearings in place
   b. main-shaft roller bearings in place
   c. drive-gear bearings in place
   d. countershaft rollers in place
3. During reassembly, the first thing to install on the main shaft is the:
   a. second-speed gear
   b. first-speed gear
   c. fourth-speed gear
   d. synchronizer assembly
4. During reassembly, the countershaft rollers are retained in place by:
   a. clamps
   b. clips
   c. heavy grease
   d. transmission lubricant
5. To check the end play of the countergear, use a:
   a. thickness gauge
   b. dial indicator
   c. depth gauge
   d. Plastigage strip

FOUR-SPEED OVERDRIVE MANUAL-TRANSMISSION SERVICE

After studying this chapter, and with proper instruction and equipment, you should be able to:

1. Disassemble a four-speed manual transmission which has overdrive in fourth gear.
2. Explain how overdrive is achieved in the transmission.
3. Identify the various parts of a disassembled four-speed transmission with overdrive.
4. Reassemble a four-speed manual transmission with overdrive.

 9-1 Servicing the four-speed–overdrive transmission This chapter describes in detail the disassembly and reassembly of a four-speed manual transmission with overdrive. The transmission provides overdrive in fourth gear.

Overdrive is discussed in ✿ 4-11. The gearing and its actions are described in ✿ 4-12 and illustrated in Figs. 4-35 to 4-40. Removal of the transmission from the vehicle is discussed in ✿ 6-2.

✿ **9-2 Disassembling the four-speed transmission with overdrive** Actually, this transmission is not much different from the standard four-speed transmission discussed in Chap. 8. The major difference is in the main-shaft and countergear assemblies.

Figure 9-1 shows the four-speed overdrive transmission completely disassembled. This illustration will be your guide as we describe the disassembly and reassembly of the transmission.

1. Bearing retainer	29. Mainshaft yoke bushing
2. Bearing retainer gasket	30. Oil seal
3. Bearing retainer oil seal	31. Main drive gear
4. Snap ring, bearing (inner)	33. Needle bearing rollers
5. Snap ring, bearing (outer)	34. Snap ring
6. Pinion bearing	35. Stop ring
7. Transmission case	36. Snap ring
8. Filler plug	37. Shift strut spring
9. Gear, second speed	38. Clutch gear
10. Snap ring	39. Shift strut spring
11. Shift strut springs	40. Clutch sleeve
12. Clutch gear	41. Stop ring
13. Shift struts (3)	42. OD gear
14. Shift strut spring	43. Mainshaft (output)
15. Snap ring	44. Shift struts (3)
16. First and second clutch sleeve gear	45. Woodruff key
17. Stop ring	46. Countershaft
18. First-speed gear	47. Thrustwasher, gear (1)
19. Bearing retainer ring	48. Spacer ring needle roller bearing
20. Rear bearing	49. Needle bearing rollers
21. Snap ring	50. Bearing spacer
24. Baffle	51. Countershaft gear (cluster)
25. Gasket, case to extension housing	52. Needle bearing rollers
26. Lockwasher	53. Spacer ring needle roller bearing
27. Bolt	54. Thrustwasher, gear (1)
28. Extension housing	60. Spring, reverse detent ball

61. Ball, reverse detent
62. Woodruff key
63. Reverse idler gear shaft
64. Bushing, reverse idler gear
65. Gear, reverse idler
67. Reverse lever
68. Oil seal, reverse lever shaft
69. Reverse operating lever
72. Nut, lever
73. Gearshift control housing
74. First and second operating lever
77. Nut, lever
80. Third and OD operating lever
83. Interlock lever (2)
84. E-ring
85. Spring
86. Oil seal (2)
87. Third and OD lever
88. First and second lever
89. Third and OD speed fork
90. First and second speed fork
91. Drain plug
92. Gasket, shift control housing
93. Expansion plug

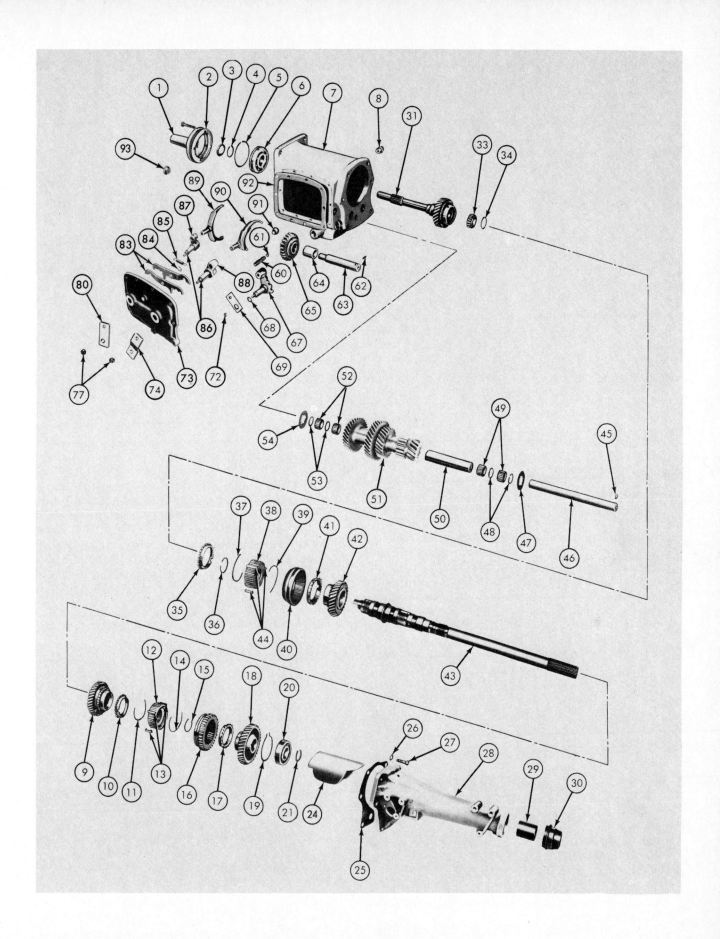

Fig. 9-1 Disassembled four-speed transmission with overdrive. *(Chrysler Corporation)*

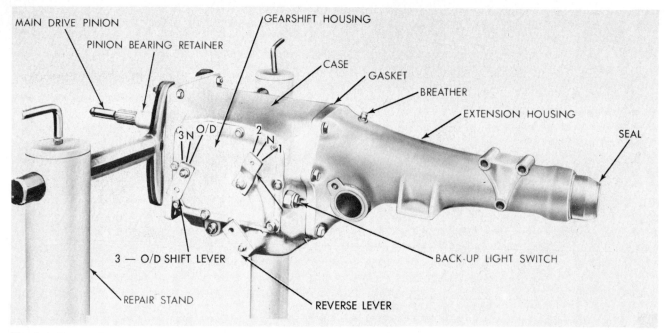

Fig. 9-2 Transmission mounted on repair stand. *(Chrysler Corporation)*

With the transmission on the bench, or mounted on a repair stand (Fig. 9-2), proceed as follows:

1. Shift the reverse-shift lever to neutral and remove the reverse-shift operating lever from the shaft. Remove the bolts that attach the gearshift housing to the transmission case (Figs. 9-2 and 9-3).
2. With the operating levers in the neutral detent position, pull the housing away from the case. The two forks may remain in place in the grooves in the synchronizers. If they do, work them out. If there are signs of oil leakage around the gearshift-lever shafts, or if the interlock levers are cracked or damaged, disassemble the gearshift housing further, as follows:

   a. Remove the nuts that attach the shift operating levers to the shafts. Disengage the levers from the flats on the shafts and remove them.

   b. Before removing the shafts, make sure they are free of burrs. If there are burrs, remove them with a fine file. Burrs could score the bores and cause leaks after reassembly.

   c. With the lever shafts out, remove the O-ring retainers and rings.

   d. Remove the E ring from the interlock-lever pivot pin and take out the interlock levers and spring.

   e. Remove the reverse detent spring and ball from the bore in the side of the case.
3. Remove the bolts that attach the extension housing to the transmission case.
4. Rotate the extension on the main shaft to expose the rear of the countershaft (Fig. 9-4). Put a bolt into the flange on the extension to hold the flange up out of the way.
5. Use a center punch or drill to make a hole in the countershaft expansion plug at the front of the case.

6. Reach through this hole to push the countershaft to the rear until the woodruff key is exposed. Remove the key.
7. Push the countershaft forward against the expansion plug. Use a brass drift punch to tap the countershaft forward far enough to drive the expansion plug out of the case.
8. Use the countershaft arbor, or short dummy shaft, to push the countershaft out of the rear of the case. Do not allow the countershaft washers to fall out of position. Lower the countergear assembly to the bottom of the case.

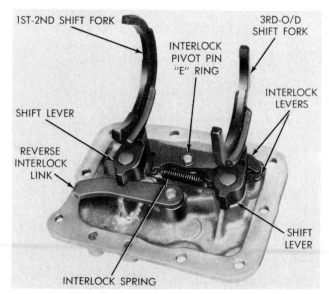

Fig. 9-3 Gearshift-housing assembly. *(Chrysler Corporation)*

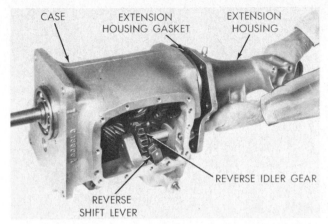

Fig. 9-4 Removing the extension-housing-and-main-shaft assembly. *(Chrysler Corporation)*

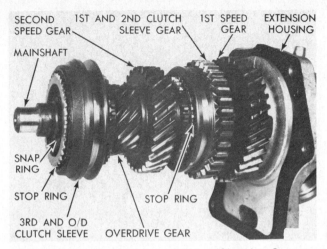

Fig. 9-5 Main-shaft-gear assembly. *(Chrysler Corporation)*

9. Rotate the extension back to the normal position.
10. Remove the bolts that attach the drive-gear-bearing retainer. Slide the retainer and gasket from the drive-gear shaft. Pry the seal from the retainer. Do not scratch the metal surfaces in the retainer bore or leakage could result on reassembly.
11. Use a brass drift punch to tap the drive-gear-and-bearing assembly forward and out of the case.

NOTE: If the purpose of the service job is to replace the drive gear or bearing, then no further disassembly is required. However, if additional work is required, proceed as follows.

12. Slide the third-and-overdrive synchronizer sleeve slightly forward. Slide the reverse idler gear to the center of its shaft. Then use a soft-faced hammer to tap on the extension housing (in a rearward direction). Slide the extension-and-main-shaft assembly out and away from the case (Fig. 9-4).
13. Study the main shaft and Fig. 9-5 before disassembling the main shaft so you know where every part is located and the relationship of that part to all other parts.
14. Remove the snap ring (Fig. 9-6) that retains the third-and-overdrive synchronizer to the main shaft. Then slide the synchronizer off the main shaft.
15. Slide the overdrive gear and stop ring off the main shaft. Mark and separate the synchronizer parts for cleaning and inspection.
16. Use long-nose pliers to compress the snap ring that retains the main-shaft ball bearing in the extension housing (Fig. 9-7). Hold the snap ring compressed so that you can pull the main-shaft assembly and bearing out of the extension housing (Fig. 9-8).
17. Remove the snap ring that retains the main-shaft bearing on the shaft (Fig. 9-9).
18. Remove the bearing from the main shaft by inserting steel plates on the front side of the first-speed gear. Then press or drive the main shaft through the bearing. Do not damage the gear teeth!
19. Remove the bearing retainer ring, first-speed gear, and first-speed stop ring from the main shaft.
20. Remove the snap ring that retains the first-speed-

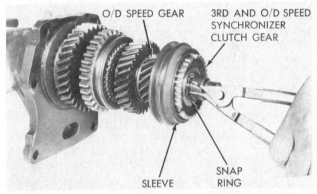

Fig. 9-6 Disassembling the main shaft by removing the snap ring. *(Chrysler Corporation)*

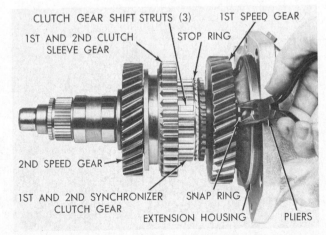

Fig. 9-7 Removing the main-shaft-bearing snap ring from the extension housing. *(Chrysler Corporation)*

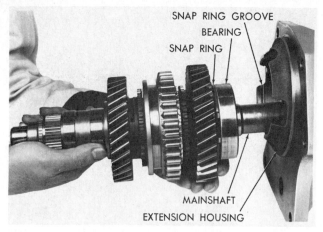

Fig. 9-8 Removing the main shaft. *(Chrysler Corporation)*

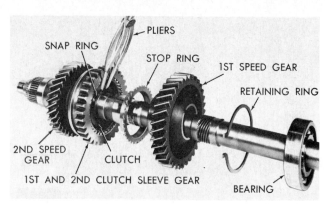

Fig. 9-10 Removing the clutch-gear snap ring. *(Chrysler Corporation)*

gear-and-sleeve assembly to the main shaft (Fig. 9-10). Slide the first-and-second-synchronizer-and-gear-assembly from the main shaft. Before cleaning, mark all parts so you will know which goes where on reassembly.

21. Figure 9-11 shows the main shaft. Inspect all bearing surfaces for wear or other damage.

22. If the drive-gear bearing is to be replaced and the old gear reused, use the shop press to remove the bearing. Remove the snap ring and 16 bearing rollers from the cavity in the drive gear.

23. Take the countergear assembly from the case and remove the arbor (short dummy shaft) and 76 needle bearings, thrust washers, and spacers.

24. The reverse-idler-gear shaft is a press fit in the case. To remove it requires a special tool, as shown in Fig. 9-12. Turning the nut will force the shaft out.

25. Clean and inspect all parts (✿ 7-4 to 7-6). Discard defective parts.

✿ 9-3 Assembling the four-speed transmission with overdrive

1. If the reverse shaft was removed to fix an oil leak, reinstall the parts as follows:

a. Install the reverse shift lever in the case bore, followed by the greased O ring and retainer.

b. Put the reverse-idler-gear shaft in position in the end of the case and drive it in far enough to permit you to put the reverse idler gear on the shaft. The fork slot should be to the rear (Fig. 9-12). Engage the slot with the reverse shift fork.

c. Drive the shaft into the case far enough to install the woodruff key. Then drive the shaft in flush with the end of the case.

d. If the backup-light switch has been removed, install it with the gasket.

2. Countergear. Coat the inside of the bore in each end of the countergear with grease and install roller-bearing spacers with the special arbor (dummy shaft). Install 19 rollers, 1 spacer ring, 19 more rollers, and another spacer ring in each end of the gear (items 48, 49, 52, and 53 in Fig. 9-1).

a. Then coat thrust washers with grease and stick them over the arbor with the tang side toward the case boss (Fig. 9-13).

b. Install the countergear assembly in the case, resting it on the bottom of the case. Make sure the thrust washers stay in place.

3. Drive pinion (Fig. 9-14). Press the pinion bearing on the drive-pinion shaft. Be sure the snap-ring groove is toward the front. Seat the bearing against the shoulder on the gear.

a. Install a new snap ring on the shaft to retain the

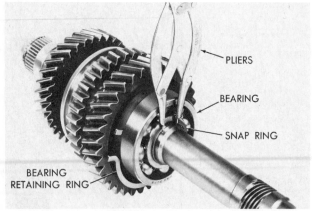

Fig. 9-9 Removing the main-shaft-bearing snap ring from the main shaft. *(Chrysler Corporation)*

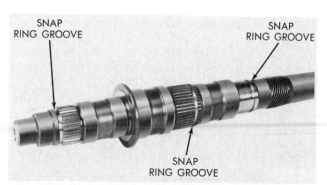

Fig. 9-11 Main-shaft-bearing surfaces. *(Chrysler Corporation)*

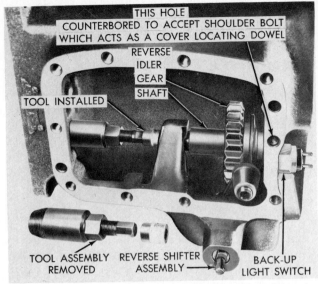

Fig. 9-12 Removing the reverse-idler-gear shaft. *(Chrysler Corporation)*

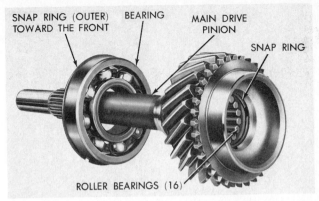

Fig. 9-14 Drive-pinion-and-bearing assembly. *(Chrysler Corporation)*

bearing. Be sure the snap ring is seated. This snap ring is a select fit to achieve minimum end play.

b. Clamp the drive-gear (or *drive-pinion*) shaft in the soft jaws of a vise and install 16 bearing rollers in the cavity of the shaft. Coat the rollers with grease to hold them in place. Install the bearing-retainer snap ring in its groove.

c. Use the special tool to install a new oil seal in the bore of the bearing retainer (item 1 in Fig. 9-1).

4. If the bushing at the rear end of the extension housing needs replacement, remove the yoke seal and drive the bushing out of the housing with the special tools required. Slide the new bushing on the end of the tool. Align the oil hole in the bushing with the oil slot in the housing and drive the bushing into place. Install a new seal by driving it into place with the special tool required.

5. Main shaft. Start the reassembly by assembling the synchronizers and then install the various parts on the main shaft as follows:

a. Refer to Figs. 9-15 and 9-16 for the proper relationship of the two synchronizer-assembly parts. To assemble a synchronizer, put a stop ring flat on the bench, followed by the gear and sleeve. Drop the struts into their slots and snap in a strut spring, placing the tang inside one strut. Turn the assembly over and install a second strut spring with the tang in a different strut.

b. Slide the second-speed gear over the main shaft (cone toward rear) and down against the shoulder on the shaft (Fig. 9-10).

c. Slide the first-and-second-synchronizer assembly (including the stop ring), with the lugs indexed in the hub slots, over the main shaft and down against the second-gear cone. Secure with a new snap ring (Fig. 9-10). Slide the next stop ring over the shaft and index the lugs in the clutch hub slots.

d. Slide the first-speed gear (cone toward gear just installed) over the main shaft and into position against the sleeve gear.

e. Install the main-shaft-bearing-retainer ring and rear bearing. Use an arbor and a suitable tool to drive or press the bearing down into position. Install a new snap ring (Fig. 9-5). This snap ring is a select fit for minimum end play.

f. Install the partly assembled main shaft into the extension housing far enough to engage the bearing

Fig. 9-13 Removing the countershaft-gear-and-arbor assembly. *(Chrysler Corporation)*

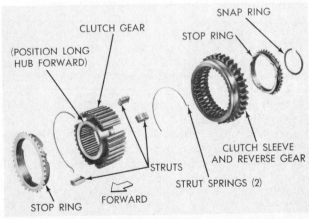

Fig. 9-15 First-and-second-speed synchronizer, disassembled. *(Chrysler Corporation)*

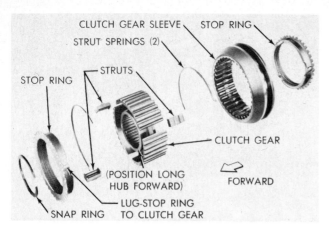

Fig. 9-16 Third-and-fourth (overdrive) synchronizer, disassembled. *(Chrysler Corporation)*

retaining ring in the slot in the extension housing (Fig. 9-8). Compress the ring with needle-nose pliers so that the main-shaft ball bearing can move in and bottom against its thrust shoulder in the extension housing. Release the ring. Make sure it seats all around in its groove in the extension housing (Fig. 9-7).

g. Slide the overdrive gear over the main shaft (cone toward the front) and install the overdrive-gear stop ring.

h. Install the third-overdrive-synchronizer-gear assembly (sleeve, shift struts, and springs) on the main shaft with the shift-fork slot toward the rear. Be sure the rear stop ring indexes with the clutch gear struts.

i. Install the retaining snap ring (Fig. 9-6). Coat the front stop ring with grease and put it over the drive gear, again indexing the ring lugs with the struts.

j. Coat a new extension-housing gasket with grease and put it in place on the extension-housing.

k. To provide clearance so that assembly will be possible, slide the reverse idler gear to the center of its shaft and move the third-overdrive synchronizer sleeve as far forward as practical. Do not lose the struts.

l. Now, slowly insert the main-shaft assembly into the case (Fig. 9-4), tilting it as required to clear the idler-gear-and-countergear assembly.

m. Place the third-overdrive synchronizer sleeve in neutral.

n. Rotate the extension on the main shaft to expose the rear of the countershaft. Install one extension bolt to hold the extension flange up out of the way and to prevent it from moving away from the case.

o. Install the drive-gear assembly, with bearing, through the front of the case and position it in the case bore. Install the outer snap ring in the bearing groove. Tap lightly on the end of the shaft. If everything is in position, the outer snap ring will bottom onto the case face without excessive effort. If it does not bottom easily, check to see if a strut, pinion roller, or stop ring is out of position.

p. Turn the transmission assembly upside down, holding the countergear assembly to prevent gear damage.

q. Lower the countergear assembly into position with the teeth meshed with the drive-gear teeth. Make sure that the thrust washers remain in position on the ends of the arbor and that the tangs are aligned with the slots in the case.

r. Start the countershaft into the bore at the rear end of the case. Push it forward until the shaft is about halfway into the case and countergear. Install the woodruff key. Push the shaft forward until its end is flush with the rear of the case face. Remove the dummy shaft (the arbor).

s. Rotate the extension into proper position and secure it with bolts tightened to the specified torque.

t. Turn the transmission right side up. Install the drive-pinion-gearing retainer and gasket. Coat threads with sealing compound, and then install the attaching bolts and tighten to the specified torque.

u. Install a new expansion plug, coated with sealing compound, in the countershaft bore at the front of the case.

6. Gearshift housing and mechanism (Fig. 9-17). Install the interlock levers on the pivot pins and fasten with E rings. Use pliers to install the spring on the interlock-lever hangers.

a. Grease the housing bores and push each shaft into its proper bore followed with a greased O ring and O-ring retainer.

b. Install the operating levers and tighten the retaining nuts to the specified torque. Be sure the third-overdrive operating lever points downward.

c. Rotate each shift-fork lever to the neutral position (straight up) and install the third-overdrive shift fork in the bore in its shift lever. It should be under the interlock levers.

d. Position both synchronizer sleeves in neutral. Put the first-second shift fork in the groove of the first-second sleeve. Slide the reverse idler gear to neutral.

e. Lay the transmission on its right side and position the gearshift-housing gasket, using grease to hold it in place.

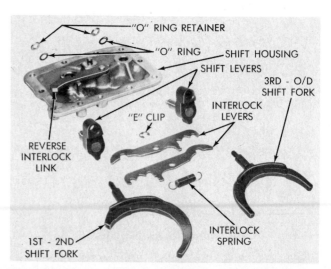

Fig. 9-17 Shift housing, disassembled. *(Chrysler Corporation)*

f. Install the reverse detent ball and spring in the bore in the side of the case.

g. Lower the gearshift housing into place, guiding the third-overdrive shift fork into its synchronizer groove. Then lead the shaft of the first-second shift fork into its bore in the first-second lever. Hold the reverse interlock link against the first-second shift lever to provide clearance as the shift cover is lowered into position.

h. To finish the installation, use a screwdriver to raise the interlock lever against its spring tension. This allows the first-second shift-fork shaft to slip under the levers. Be sure the reverse detent spring is in place in the cover bore. The housing will now seat against the gasket on the case.

i. Secure the housing with the shift-housing retaining bolts run up finger tight. Shift through all gears to make sure everything is operating properly.

j. Eight of the retaining bolts are shoulder bolts for accurately locating the shifter mechanism on the case. One bolt shoulder is longer and acts as a dowel, passing through the cover and into the transmission case at the center of the rear flange (Fig. 9-12). Two bolts are standard, located at the lower rear of the cover. Tighten all bolts to specifications.

k. Grease the reverse shaft and install the operating lever and nut. Tighten to specified torque.

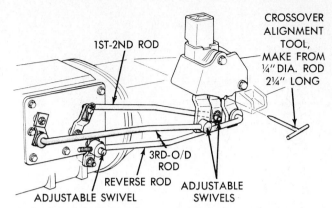

Fig. 9-18 Gearshift-linkage adjustment. *(Chrysler Corporation)*

l. Put the required amount of the specified lubricant in the transmission. After the transmission has been installed in the vehicle and the linkages properly adjusted, road test the car to make sure the transmission is working properly. Figure 9-18 shows the gearshift linkages and the alignment tool used to adjust them.

Chapter 9 review questions

Select the *one* correct, best, or most probable answer to each question. Then check your answers against the correct answers given at the end of the book.

1. When the transmission is in overdrive, the output or main shaft is turning faster than the:
   a. countergear
   b. clutch gear
   c. engine crankshaft
   d. all of the above
2. Before you remove the gearshift housing from the case, the:
   a. operating levers must be in reverse
   b. operating levers must be in the neutral detent position
   c. extension housing must be off
   d. side cover must be removed

3. The purpose of the short dummy shaft is to:
   a. keep the main-shaft roller bearings in place
   b. hold the clutch-gear thrust washers in place
   c. hold the countershaft bearing rollers in place
   d. drive the main shaft from the case
4. As a first step in disassembling the main-shaft assembly, remove the:
   a. third-speed gear
   b. third-and-overdrive synchronizer snap ring
   c. idler gear
   d. clutch gear
5. As a first step in reassembling the main-shaft assembly:
   a. assemble the synchronizers
   b. install the third-speed gear
   c. install the first-speed gear
   d. assemble the first-and-reverse-gear assembly

FIVE-SPEED OVERDRIVE MANUAL-TRANSMISSION SERVICE

After studying this chapter, and with proper instruction and equipment, you should be able to:

1. Disassemble a five-speed manual transmission which has overdrive in fifth gear.
2. Explain how overdrive is achieved in the transmission.
3. Identify the various parts of a disassembled five-speed manual transmission with overdrive.
4. Reassemble a five-speed manual transmission with overdrive.

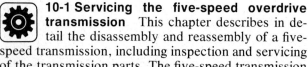

10-1 Servicing the five-speed overdrive transmission This chapter describes in detail the disassembly and reassembly of a five-speed transmission, including inspection and servicing of the transmission parts. The five-speed transmission provides overdrive in fifth speed. Otherwise, it is much like the four-speed transmission with overdrive discussed in Chap. 9. The major difference is that the five-speed transmission has extra gears which give a more gradual change in gear ratios from low to direct drive.

In the four-speed overdrive transmission, the gear ratios go from low to direct in three steps (low, second, direct). In the five-speed transmission, the gear ratios go from low to direct in four steps (low, second, third, direct). In both, the top gear position is overdrive. Figures 4-41 and 4-42 and ✿ 4-13 illustrate and discuss a five-speed transmission with overdrive. Removal of the transmission from the vehicle is discussed in ✿ 6-2.

✿ 10-2 Five-speed overdrive manual-transmission disassembly Figure 10-1 shows a completely disassembled five-speed overdrive manual transmission. Figure 10-2 shows the main gears in the transmission. Notice that, because of the extra gear, the main shaft has an extra support in the center. First gear, reverse gear, and the first-and-reverse synchronizer are on the other side of the center support (to the right in Fig. 10-2 and parts numbered 31 to 38 in Fig. 10-1).

In the sections that follow, we describe how to disassemble the transmission and also how to remove and replace bearings and seals (✿ 10-3). Proceed as follows to disassemble the transmission:

1. Remove the selector-lever pivot from the extension housing (Fig. 10-3).
2. Remove the poppet spring and plunger by removing the plug (Fig. 10-4). Use a magnet, or turn the transmission upside down, to remove the plunger.
3. Use a pin punch to drive the roll pin from the shifter head and shift rail.
4. Remove the six bolts that fasten the extension housing and the transmission case to the center support (item 50 in Fig. 10-1).
5. Slide the transmission case forward and remove the thrust bearing and races from the drive-gear shaft or case.
6. Remove the extension housing by sliding it rearward. The shifter head, shift rail, and selector lever are not fastened to the housing and should not be permitted to drop out. Remove them from the extension housing.
7. The selector lever may be removed from the shift rail by removing the retaining ring and pin. The needle rollers are not always retained in the needle race. Catch loose needles if they fall out so that you can put them back into the race on assembly.
8. Remove the rail-selector pin and remove the rail selector (Fig. 10-5).
9. Remove the snap ring from behind the speedometer gear and remove the gear and ball.
10. To remove the first-speed gear and blocker ring, remove the snap ring and thrust washer (Fig. 10-6).

Careful: All the snap rings are not the same. Be sure to use the correct snap ring in the correct groove when reassembling the transmission.

11. Remove the snap ring from behind the synchronizer hub.

12. Turn the shift rail to position the interlock pawl in the inboard position to permit removal of the first-and-reverse shift link.

13. Slide the first-and-reverse synchronizer, shift fork, shift link, and shift rail rearward from the transmission (Fig. 10-7).

14. Remove the reverse idler gear from the reverse idler shaft. Then slide the reverse gear rearward from the output shaft (Fig. 10-8).

15. Position the interlock pawl in the center position so that the second-and-third-speed shift link and shift fork can be removed. Remove these parts (Fig. 10-9).

16. Position the interlock pawl in the outboard position to permit removal of the fourth-and-fifth-speed shift link and shift fork. Remove these parts (Fig. 10-10).

17. Remove the center support from the output shaft (Fig. 10-11). Be careful; some of the bearing rollers could drop out of the center support.

18. Remove the needle thrust bearing from the output shaft or center support (Fig. 10-12).

19. Remove the countergear from the gears on the output shaft.

20. Slip the input shaft (drive gear) from the output shaft. Note the thrust bearing in the end of the input shaft. It may remain on the synchronizer hub. Remove it.

21. Remove the fourth-and-fifth synchronizer and blocker rings (Fig. 10-13).

22. Remove the fifth-speed gear (Fig. 10-14). Spacers and bearing needles may drop out. With the correct number of needles installed, there is space for about one more needle.

23. Remove the snap ring and thrust washer from the output shaft and slide the second gear and blocker ring off the shaft (Fig. 10-15).

24. Remove the snap ring and remove the second-and-third-gear synchronizer assembly and blocker ring from the output shaft (Fig. 10-16).

25. The third-speed gear is the only gear remaining on the shaft. Slide it off.

✿ 10-3 Cleaning and servicing transmission parts

See ✿ 7-4 to 7-6 for basic procedures for cleaning and inspecting transmission parts, including bearings. Additional servicing steps, including synchronizer key and spring replacement and the removal and installation of bearings and seals, follow:

1. When replacing synchronizer keys and springs, remember that the clutch hubs and sliding sleeves are a selective fit and must be kept together as an assembly. If keys or springs are broken, they may be replaced.

   a. Mark the hub and sleeve so that they can be matched up on reassembly.

   b. Push the hub from the sleeve and remove the keys and springs.

   c. To assemble the synchronizer (Fig. 10-17), put one spring on each side of the hub so that the springs overlap the slots in the hub and so that the openings in the springs are not opposite each other.

   d. Put the shift keys in the slots of the hub with the radius side out and the springs in the grooves of the keys.

   e. Install the sliding sleeve on the hub, aligning the marks made before disassembly. Depress the keys for ease of installation.

2. To remove the input-shaft bearing, use the special tool shown in Fig. 10-18. Insert the tool so that its fingers are beyond the bearing. Expand the tool to ensure a firm grip on the bearing. Use a slide hammer as shown to pull the bearing out. To install a new bearing, apply Loctite or a similar sealer on the bearing-to-case surface. Then use a special driver to drive the bearing into the case until the tool seats on the case.

3. The countergear bearing (next to the input-shaft bearing shown in Fig. 10-18) is removed and replaced in the same way as the input-shaft bearing.

4. To remove the input-shaft-bearing seal, first remove the bearing (item 2 above). Then turn the case over and work the pointed end of the tool shown in Fig. 10-19 to drive the seal out. To install a new seal, turn the case over. Apply Loctite or a similar sealer to the seal-to-case surface. Install the new seal with a special tool, working from the inside of the case. The seal lip and spring should be against the tool. When the tool seats on the case, the seal is in place.

5. To remove the output-shaft bearing in the center support, put the center support on two wood blocks (Fig. 10-20). Use the special tools as shown to drive out the bearing from the center support. To install the bearing, turn the center support over and use the special installation tools to press the bearing into place. When the tool seats on the center support, the bearing has reached its correct position in the support.

6. To remove the countergear bearing from the center support, put the center support on two wood blocks. Use special tools to drive out the bearing. This is the same procedure used to remove the output-shaft bearing except that a different tool is required. Turn the center support over to install the bearing, using the special tools specified. First, coat the bearing-to-case surface with Loctite or a similar sealer.

7. Install the interlock pawl and retaining plate, tightening the screws to the specified torque.

8. To replace the countergear bearing in the extension housing, use the special tool and a slide hammer (Fig. 10-21). This procedure is the same as for removing and replacing the countergear bearing in the center support (see item 6 above). Coat the bearing-to-housing surface with Loctite or a similar sealer before installing the bearing.

9. To replace the oil seal in the extension housing, pry the old seal out. Coat the new seal bore in the extension housing with Permatex or a similar sealer. Use special tools to press the new seal into place.

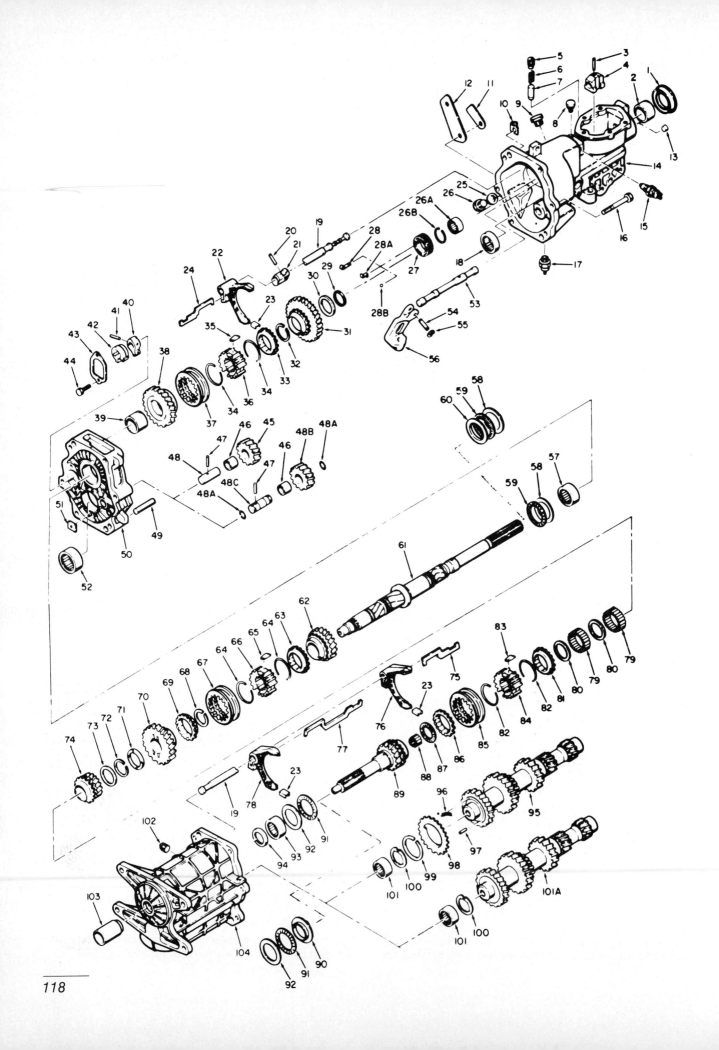

1. Oil seal	36. Clutch hub	69. Synchronizer blocking ring
2. Bushing	37. Clutch sleeve	70. Second speed gear
3. Pin	38. Reverse gear and bushing	71. Thrust washer
4. Shifter head	39. Bushing	72. Snap ring
5. Threaded plug	40. Selector arm-optional	73. Spacer
6. Poppet spring	41. Spring pin	74. Fifth speed gear
7. Mesh lock plunger	42. Interlock pawl	75. Second and third shift link
8. Breather	43. Selector arm retaining plate	76. Second and third shift fork
9. Selector lever pivot	44. Self-tapping screw	77. Forth and fifth shift link
10. Wiring harness clip*	45. Reverse idler gear and bushing-vin V & F	78. Fourth and fifth shift fork
11. Name plate	46. Bushing	79. Needle rollers
12. Back-up light bracket*	47. Spring pin	80. Spacer
13. Cup plug	48. Reverse idler shaft	81. Synchronizer blocking ring
14. Extension housing with bushing	48A. O-ring	82. Synchronizer spring
16. Bolt	48B. Reverse idler gear-vin A & P	83. Shift plate
17. Switch	48C. Reverse idler shaft	84. Clutch hub
18. Needle bearing	49. Dowel Pin	85. Clutch sleeve
19. Shift rail	50. Center support	86. Synchronizer blocking ring
20. Spring pin	51. Magnet	87. Needle thrust bearing
21. Rail selector end	52. Needle bearing	88. Bearing rollers
22. First and reverse shift fork	53. Shift rail	89. Input drive gear (clutch gear)
23. Shift fork pad	54. Pin	91. Needle thrust bearing
24. First and reverse shift link	55. Retaining clip	92. Thrust washer
25. Gasket	56. Selector lever	93. Needle bearing
26. Plug	57. Needle bearing	94. Oil seal
27. Speedometer gear	58. Thrust washer	95. Cluster gear-vin F and P
28. Speedometer gear retaining clip or ball and snap ring	59. Needle thrust bearing	96. Spring*
	60. Needle thrust race**	97. Spring pin*
29. Snap ring	61. Output shaft	98. Gear damper*
30. Thrust washer	62. Third speed gear	99. Snap ring*
31. First speed gear	63. Blocking ring	100. Thrust washer
32. Snap ring	64. Synchronizer spring	101. Needle bearing
33. Blocking ring	65. Synchronizer shift plate	101A. Cluster gear-vin V and A
34. Synchronizer spring	66. Clutch hub	102. Pipe plug
35. Shift plate	67. Clutch sleeve	103. Trans case sleeve
	68. Snap ring	104. Trans case

*Starfire only
**Mixed production

Fig. 10-1 Disassembled five-speed overdrive transmission. (*Oldsmobile Division of General Motors Corporation*)

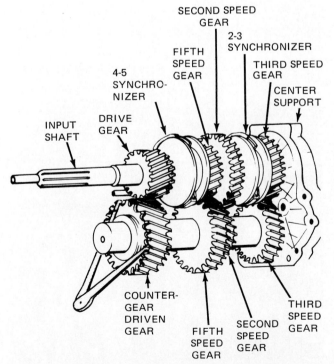

Fig. 10-2 Gears in the five-speed overdrive transmission. (*Oldsmobile Division of General Motors Corporation*)

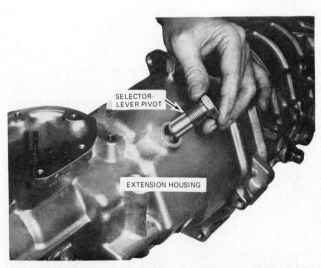

Fig. 10-3 Removing the selector-lever pivot. (*Oldsmobile Division of General Motors Corporation*)

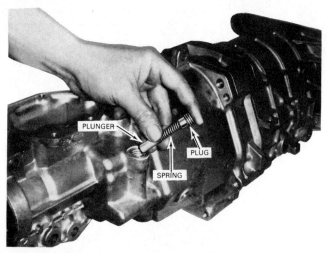

Fig. 10-4 Removing the poppet spring and plunger. (*Oldsmobile Division of General Motors Corporation*)

Fig. 10-5 Removing the rail selector. (*Oldsmobile Division of General Motors Corporation*)

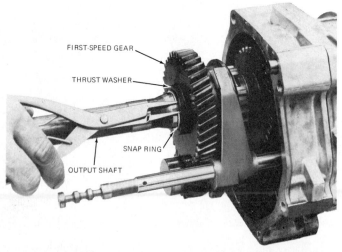

Fig. 10-6 Removing the first-gear snap ring. (*Oldsmobile Division of General Motors Corporation*)

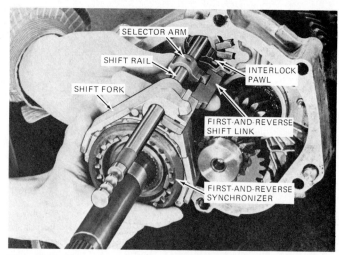

Fig. 10-7 Removing the first-and-reverse shift rail. (*Oldsmobile Division of General Motors Corporation*)

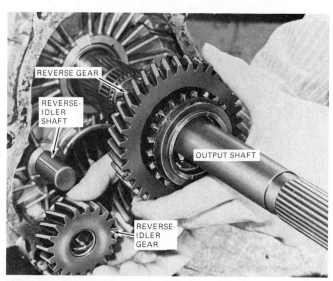

Fig. 10-8 Removing the reverse idler gear and reverse gear. (*Oldsmobile Division of General Motors Corporation*)

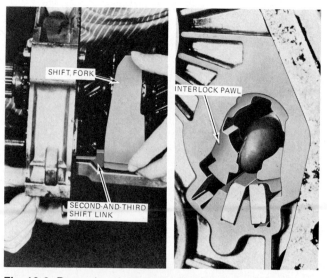

Fig. 10-9 Removing the second-and-third-speed shift rail. (*Oldsmobile Division of General Motors Corporation*)

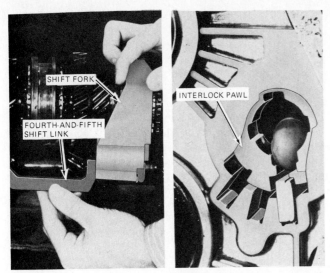

Fig. 10-10 Removing the fourth-and-fifth-speed shift rail. (Oldsmobile Division of General Motors Corporation)

Fig. 10-11 Removing the center support. (Oldsmobile Division of General Motors Corporation)

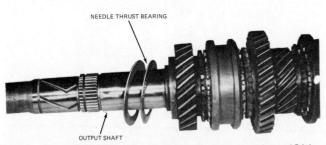

Fig. 10-12 Removing the needle thrust bearing. (Oldsmobile Division of General Motors Corporation)

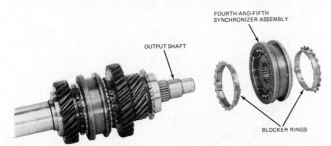

Fig. 10-13 Removing the fourth-and-fifth-speed synchronizer. (Oldsmobile Division of General Motors Corporation)

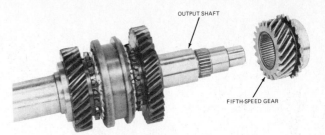

Fig. 10-14 Removing the fifth-speed gear. (Oldsmobile Division of General Motors Corporation)

Fig. 10-15 Removing the second-speed gear. (Oldsmobile Division of General Motors Corporation)

Fig. 10-16 Removing the second-and-third-speed synchronizer. (Oldsmobile Division of General Motors Corporation)

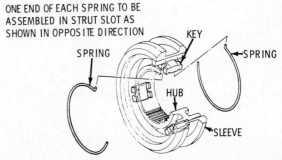

Fig. 10-17 Synchronizer assembly. (Oldsmobile Division of General Motors Corporation)

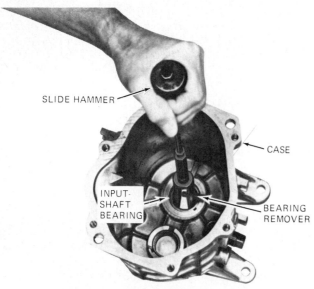

Fig. 10-18 Removing the input-shaft bearing from the case. (*Oldsmobile Division of General Motors Corporation*)

10. To replace the bushing in the extension housing, first note how deep the bushing is installed in the bore. Use the special tool and driver to push the bushing out from the rear of the housing (pushing it into the housing). Then drive a new bushing into place, to the same depth, with a bushing-installing tool. Drive the bushing in from the rear of the extension housing.

✿ 10-4 Reassembling the transmission Essentially, reassembling the transmission is just the reverse of disassembly. You start with the main shaft (output

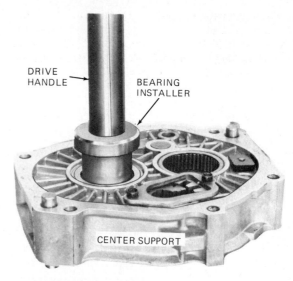

Fig. 10-20 Removing the output-shaft bearing from the center support. (*Oldsmobile Division of General Motors Corporation*)

shaft) and add the gears, blocker rings, synchronizers, and snap rings, putting them onto the shaft in the reverse order from which you removed them (Fig. 10-2). Proceed as follows:

1. Install the third-speed gear on the output shaft with the coned end toward the front.
2. Install the second-and-third-gear synchronizer. Synchronizer assemblies are similar except for different hub splines. The hub and sleeve are assembled with a selective fit that provides no more than 0.002 inch [0.052 mm] of backlash but slides easily.

Fig. 10-19 Removing the seal from the case. (*Oldsmobile Division of General Motors Corporation*)

Fig. 10-21 Removing the countergear bearing from the extension housing. (*Oldsmobile Division of General Motors Corporation*)

Mated parts should be kept together to ensure correct fit.

a. Assemble blocker rings with the slots aligned with the shift keys of the synchronizer assembly.

b. Install the synchronizer and blocker rings with the chamfer on the sleeve toward the front of the shaft. Position the synchronizer on the face of the third-speed gear.

c. Install the correct snap ring in the shaft groove (Fig. 10-16).

3. Install the second-speed gear, coned end into the blocker ring. Put the thrust washer on the face of the gear and install the snap ring (Fig. 10-15).

4. Install the fifth-speed gear (Fig. 10-14) as follows: Install 2 rows of 46 bearing needles inside the gear with a spacer between each end of the bearings. Retain with heavy grease.

Careful: Be sure you use the correct needles. Other bearings in this transmission have needles of the same length but of different diameter. Install the fifth-speed gear on the output shaft with the coned end toward the front of the shaft (Fig. 10-14). Do not let the needles or spacers drop out.

5. To install the fourth-and-fifth-speed synchronizer, first put the blocker rings into place with the slots aligned with the shift keys (Fig. 10-13). Then install the synchronizer on the shaft with the chamfered edge of the sleeve pointing toward the front.

6. Install the thrust bearing on the end of the fourth-and-fifth speed synchronizer.

7. Install 19 bearing needles in the bore of the input shaft, retaining them with heavy grease. Then install the input shaft over the end of the output shaft, being careful to avoid having needles drop out.

8. Install thrust bearings and races on the output shaft (Fig. 10-12). The inner race lip should face the rear.

9. Install the output shaft and countergear assembly in the center support. Use the special tool as shown in Fig. 10-22 to back up the gear damper one tooth so that all teeth match. Then install the assemblies in the center support (Fig. 10-23). Do not let any bearing needles fall out.

10. Install the reverse gear and reverse idler gear with their bushings or bearings (Fig. 10-8).

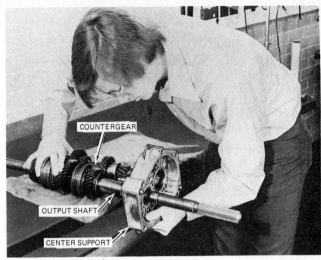

Fig. 10-23 Installing the output shaft and countergear on the center support. (*Oldsmobile Division of General Motors Corporation*)

11. Figure 10-24 shows the shift forks and links used in the transmission. Before installing them, replace the shift pads if necessary. Install the fourth-and-fifth-speed shift fork on the fourth-and-fifth synchronizer sleeve. Position the interlock pawl so that the shift link can be installed through the outboard slot of the center support. Engage the shift link with the shift fork.

12. Install the second-and-third-speed shift fork and link and the first-and-reverse shift fork and link in a similar manner (Fig. 10-7).

13. If the selector arm was removed during disassembly, install it over the shift rail. Align the hole in the arm with the hole near the middle of the rail. Drive the roll pin into the arm and rail.

14. Install the synchronizer-hub snap ring.

15. Install the blocker ring and first-speed gear over the output shaft behind the first-and-reverse synchronizer assembly. Align the notches on the blocker ring with the notches in the synchronizer hub. Install a thrust washer and snap ring behind the first-speed gear (Fig. 10-6).

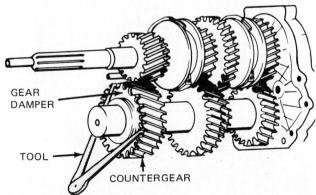

Fig. 10-22 Using a special tool to back up the gear damper so that the teeth match. (*Oldsmobile Division of General Motors Corporation*)

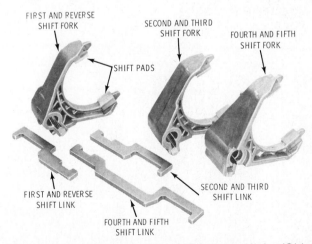

Fig. 10-24 Proper layout of the shift forks and links. (*Oldsmobile Division of General Motors Corporation*)

123

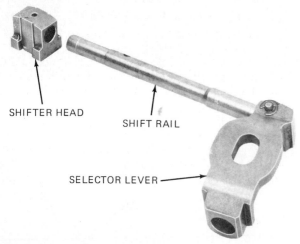

Fig. 10-25 Installing the shifter lever and shift rail. (*Oldsmobile Division of General Motors Corporation*)

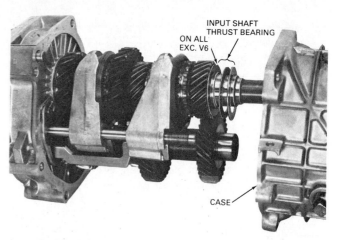

Fig. 10-27 Installing the input-shaft thrust bearing and case. (*Oldsmobile Division of General Motors Corporation*)

16. Install the speedometer-gear retainer in the hole in the output shaft with the retainer loop forward. Slide the speedometer gear over the output shaft and retainer until the retainer snaps up to lock the gear.

17. Slide the rail selector with the hole in the selector to the rear of the transmission and with the ball facing inboard (Fig. 10-5). Secure with the roll pin.

18. Install the selector lever and shift rail in the hole in the extension housing. Install the shifter head on the rail as the rail becomes exposed in the housing opening. Do not install the roll pin at this time (Fig. 10-25).

19. Apply a continuous $^{1}/_{32}$-inch [0.8-mm] bead of RTV (Room-Temperature Vulcanizing) sealer to the transmission case and the extension housing. Install the extension housing over the output shaft while guiding the selector lever so that it engages the rail selector (Fig. 10-26).

20. Install the lipped thrust race (not used on cars with V-6 engines) with the lip toward the front. Then

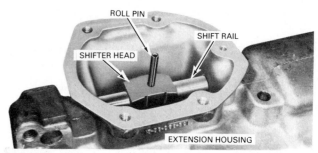

Fig. 10-28 Installing the shifter-head roll pin. (*Oldsmobile Division of General Motors Corporation*)

Fig. 10-26 Installing the extension housing. (*Oldsmobile Division of General Motors Corporation*)

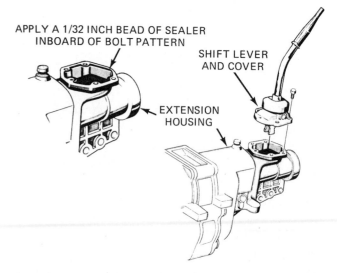

Fig. 10-29 Installing the transmission cover and shift lever. (*Oldsmobile Division of General Motors Corporation*)

install the needle thrust bearing and flat thrust washer over the input shaft (Fig. 10-27). Install the case on the front side of the center support and retain it with six bolts torqued to specifications. If the case and center support do not go together easily, check to see if the fifth-speed-gear bearing spacer has fallen down between the fifth gear and synchronizer.

21. Drive the roll pin into the shifter head and shift rail (Fig. 10-28).
22. Install the plunger, poppet, and plug (Fig. 10-4), first coating the plug threads with Loctite or an equivalent sealer. Tighten to specifications.
23. Coat the pivot threads with Loctite or an equivalent sealer and install the selector-lever pivot (Fig. 10-3). Align the hole in the selector lever with the hole in the extension housing before installing the pivot. Tighten to specifications.
24. After installing the transmission on the vehicle, fill it with the specified amount and type of lubricant.
25. Install the transmission cover and shift lever as follows (Fig. 10-29):

    a. Apply a continuous bead of RTV to the extension housing inboard of the bolt pattern.

    b. Put the transmission and shift lever in neutral and install the shift lever and cover.

    c. Install four screws and tighten to the specified torque.

    d. Tuck in the carpeting and install the boot around the shifter.

Chapter 10 review questions

Select the *one* correct, best, or most probable answer to each question. Then check your answers against the correct answers given at the end of the book.

1. Because of the extra gears required in the five-speed transmission, the main shaft:
    a. is more compact
    b. has a center support
    c. is not splined
    d. cannot be disassembled
2. When the transmission is in overdrive, the main shaft is turning faster than the:
    a. engine crankshaft
    b. clutch gear
    c. countergear
    d. all of these
3. In this transmission:
    a. all the snap rings are identical
    b. no snap rings are used
    c. all the snap rings are not the same
    d. only one snap ring is used
4. The last gear to be taken off the main shaft is the:
    a. ring gear
    b. output gear
    c. first-speed gear
    d. third-speed gear
5. The front end of the output or main shaft is supported by a:
    a. transmission case
    b. center support
    c. clutch gear or input shaft
    d. extension housing

MANUAL-TRANSAXLE TROUBLE DIAGNOSIS

After studying this chapter, and with proper instruction and equipment, you should be able to:

1. List and describe the 16 possible kinds of transaxle noise that might cause a customer to complain, and explain what might cause each.
2. Explain what might cause the transaxle to slip out of gear and how to correct the trouble.
3. Discuss what might cause difficulty shifting into gear and how to correct the trouble.
4. Explain what might cause the transaxle to stick in gear and how to correct the trouble.
5. Discuss what might cause the gears to clash in shifting and how to correct the trouble.
6. Explain where the transaxle might leak lubricant and how to stop the leaks.

11-1 Diagnosing manual-transaxle trouble This chapter discusses the trouble diagnosis of manual transaxles that are used in front-drive cars with transversely mounted engines. The purpose and operation of the transaxle are covered in ☼ 4-14 and 4-15, and the transaxle is illustrated in Figs. 4-43 to 4-54. Later chapters describe the general procedure for removing and installing transaxles and how to disassemble, service, and reassemble them.

☼11-2 Transaxle trouble-diagnosis chart Before attempting any repair of the clutch, transaxle, or operating linkages for any reason except an obvious failure, try to identify the problem and its possible cause. Many clutch and transaxle problems show up as shifting difficulties. These include excessive effort needed to shift, gear clashing and grinding, and the inability to shift into some gears.

In addition, there may be noise problems. These vary with vehicle size, type and size of engine, and amount of body insulation used. The fact that the entire drive train is located at the front of the car almost under the feet of the driver makes any drive-train noise more audible to the driver (Fig. 11-1). But noises that you might believe are coming from the drive train could be coming from the tires, road surfaces, wheel bearings, engine, or exhaust system. For this reason, a thorough and careful check should be made to locate the cause of the noise before removing and disassembling the transaxle.

NOTE: See ☼ 3-1 for a discussion of diagnosis theory.

The chart on pages 127 and 128 lists various transaxle troubles and their possible causes and corrections. For example, if a knock at low speeds is caused by drive-axle joints, the remedy is to service the joints. If vibration is being caused by rough wheel bearings, the correction is to replace the bearings.

NOTE: The troubles and possible causes are not listed in the order of frequency of occurrence. Item 1, or item a, does not necessarily occur more often than item 2, or item b. Also, note that the chart applies especially to the General Motors transaxle used in their smaller, front-drive cars. However, the chart can be used as a guide when investigating trouble in any transaxle.

☼11-3 Noises Transaxle gears, like any mechanical device, are not absolutely quiet and will produce some noise. If the noise is annoying, try the following steps to see if you can determine whether or not the noise is excessive, and, if it is excessive, what could be causing it.

1. Drive on a smooth, level asphalt road, which will reduce tire and road noise to a minimum.

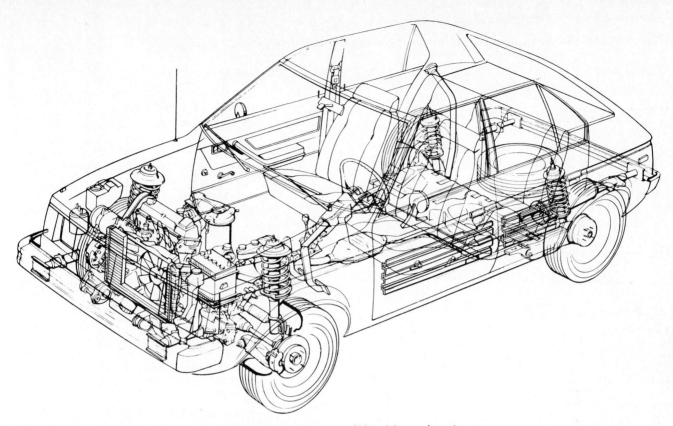

Fig. 11-1 With the entire drive train almost under the feet of the driver, almost any drive-train noise is more audible to the driver. *(Chrysler Corporation)*

Transaxle Trouble-Diagnosis Chart

(See ✹ 11-3 to 11-9 for explanations of the trouble causes listed below and their corrections.)

COMPLAINT	POSSIBLE CAUSE
1. Noise is the same in drive or coast (✹ 11-3)	a. Road noise
	b. Tire noise
	c. Front-wheel bearing noise
	d. Incorrect drive-axle angle (standing height)
2. Noise changes on different types of road (✹ 11-3)	a. Road noise
	b. Tire noise
3. Noise tone lowers as car speed is lowered (✹ 11-3)	Tire noise
4. Noise is produced with engine running, whether vehicle is stopped or moving (✹ 11-3)	a. Engine noise
	b. Transaxle noise
	c. Exhaust noise
5. Knock at low speeds (✹ 11-3)	a. Worn drive-axle joints
	b. Worn side-gear-hub counterbore
6. Noise most pronounced on turns (✹ 11-3)	Differential-gear noise
7. Clunk on acceleration or deceleration (✹ 11-3)	a. Loose engine or transaxle mounts
	b. Worn differential pinion shaft in case, or side-gear-hub counterbore in case worn oversize
	c. Worn or damaged drive-axle inboard joints
8. Clicking noise in turns (✹ 11-3)	Worn or damaged outboard joint
9. Vibration (✹ 11-3)	a. Rough wheel bearing
	b. Damaged drive-axle shaft
	c. Out-of-round tires
	d. Tire unbalance
	e. Worn joint in drive-axle shaft
	f. Incorrect drive-axle angle
10. Noisy in neutral with engine running (✹ 11-3)	Damaged input-gear bearings

Transaxle Trouble-Diagnosis Chart *(Continued)*

(See ✿11-3 to 11-9 for explanations of the trouble causes listed below and their corrections.)

COMPLAINT	POSSIBLE CAUSE
11. Noisy in first only (✿11-3)	a. Damaged or worn first-speed constant-mesh gears b. Damaged or worn 1-2 synchronizer
12. Noisy in second only (✿11-3)	a. Damaged or worn second-speed constant-mesh gears b. Damaged or worn 1-2 synchronizer
13. Noisy in third only (✿11-3)	a. Damaged or worn third-speed constant-mesh gears b. Damaged or worn 3-4 synchronizer
14. Noisy in high gear only (✿11-3)	a. Damaged or worn 3-4 synchronizer b. Damaged fourth-speed gear or output gear
15. Noisy in reverse only (✿11-3)	a. Worn or damaged reverse idler gear or idler bushing b. Worn or damaged 1-2 synchronizer sleeve
16. Noisy in all gears (✿11-3)	a. Insufficient lubricant b. Damaged or worn bearings c. Worn or damaged input gear (shaft) and/or output gear (shaft)
17. Transaxle slips out of gear (✿11-5)	a. Worn or improperly adjusted linkage b. Transmission loose on engine housing c. Shift linkage does not work freely; binds d. Bent or damaged cables e. Input-gear-bearing retainer broken or loose f. Dirt between clutch cover and engine housing g. Stiff shift-lever seal
18. Hard shifting into gear (✿11-6)	a. Gearshift linkage out of adjustment or needs lubricant b. Clutch not disengaging c. Clutch linkage needs adjustment d. Internal trouble in transaxle
19. Transaxle sticks in gear (✿11-7)	a. Gearshift linkage out of adjustment, disconnected, or needs lubricant b. Clutch not disengaging c. Internal trouble in transaxle
20. Gears clash in shifting (✿11-8)	a. Incorrect gearshift-linkage adjustment b. Clutch not disengaging c. Clutch linkage needs adjustment d. Internal trouble in transaxle
21. Lubricant leaks out (✿11-9)	a. Axle-shaft seals faulty b. Excessive amount of lubricant in transmission c. Loose or broken input-gear (shaft)-bearing retainer d. Input-gear-bearing-retainer O ring and/or lip seal damaged e. Lack of sealant between case and clutch cover or loose clutch cover f. Shift-lever seal leaks

2. Drive the vehicle long enough to warm up all lubricant.
3. Note the speed at which the noise occurs and in which gear range.
4. Stop the vehicle and see if the noise is still present with transaxle in neutral. Then listen with the transaxle in gear with clutch pedal depressed.
5. Determine under which of the following driving conditions the noise is most noticeable:
   a. Driving—light acceleration or heavy pull.
   b. Float—constant vehicle speed with light throttle on a level road.
   c. Coast—partly or fully closed throttle with transaxle in gear.
   d. All of the above.
6. After road testing the vehicle, consider the following:
   a. If the noise is the same in drive or coast, it could be due to excessive drive-axle angle. The front suspension may be binding, or the springs may be weak.

This could cause the drive-axle universal jo[...]
driving through an excessive angle.

b. A knock at low speed could be caused [...]
drive-axle universal joints or by worn coun[...]
in the side-gear hubs.

c. A clunk on acceleration or deceleration c[...]
caused by loose engine or transaxle moun[...]
11-2) or by items 7b and 7c in the trouble-di[...]
chart (✹ 11-2).

d. Refer to the chart on page 127 for other p[...]
causes of noise.

e. See the following section (✹ 11-4), whic[...]
cusses bearing noises.

NOTE: Chapters 17 to 20 cover drive axles and d[...]
entials.

✹ 11-4 Bearing noise   Bad bearings usually pro[...]
a rough growl or grating noise rather than the w[...]
that is typical of gear noise. If bearing noise is suspe[...]
(see items 1, 9, 10, 15, and 16 in the chart on pages[...]
and 128), it will be necessary to remove the transa[...]
and disassemble it so that the bearings can be [...]
spected.

NOTE: See ✹ 7-4 to 7-6 for the cleaning and inspecti[...]
of bearings.

Clean the bearing assembly in solvent and allow it [...]
dry. If the bearing has become magnetized, norm[...]
cleaning methods will not remove any metal particles
attached to the bearing. The bearing must be demagnetized first. Bearings fail by lapping, spalling, or locking (Fig. 7-10).

Lapping is caused by fine particles of abrasive material, such as scale, sand, or emery. These particles
circulate with the oil and cause wear of the roller and
race surfaces. Bearings which are worn loose but appear smooth without spalling or pitting have been running with dirty oil.

Spalling is caused by overloading or faulty assembly.
Bearings that fail by spalling have either flaked or pitted

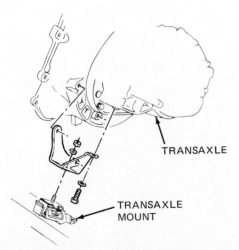

Fig. 11-2 Manual transaxle mount. *(Pontiac Motor Division of General Motors Corporation)*

TRANSAXLE

TRANSAXLE
MOUNT

✹ 11-9 Lubricant leaks   If the transaxle is overfilled [...] the [...]
with lubricant, some will leak out. If this is not the [...]
cause, then one or more of the seals are faulty. (S[...]
item 21 in the transaxle trouble-diagnosis chart.)

✹ 11-10 Broken transaxle mounts   Raise [...]
a lift. Push up and pull down on the tr[...]
while watching the mounts. If the [...]
from the metal plate of the mount o[...]
up but not down (mount is botto[...]
mount (Fig. 11-2). If there i[...]
metal plate of the mount and[...]
the screws or nuts that a[...]
cross member.

✹ 11-11 Low lu[...]
location of th[...]
the level o[...]
plug. Th[...]
mm][...]
is [...]

[...]ough highway on a ve-
[...]carrier. The repeated pounding of the race by the
roller or ball because of the rough road causes the race
to be indented.

To check for brinnelling, spin the wheel by hand
while listening at the hub for brinnelling or rough-bearing noise. Wheel bearings are not serviceable as a separate item. They must be replaced as an integral part
of the hub and spindle.

✹ 11-5 Transaxle slips out of gear   This could be
caused by a worn or out-of-adjustment linkage, a transaxle that is loose on the engine housing, a binding shift
lever, bent or damaged shift cables, a stiff shift-lever
seal, or trouble inside the transaxle (see item 17 in the
chart on page 128).

✹ 11-6 Hard shifting into gear   This can be caused
by the linkage being out of adjustment or needing lubrication, by the clutch not disengaging, by the clutch
linkage being out of adjustment, or by trouble inside
the transaxle.

✹ 11-7 Transaxle sticks in gear   This could be
caused by the gearshift linkage being out of adjustment
or disconnected or needing lubrication. It could also be
caused by the clutch not disengaging or by trouble
inside the transaxle.

✹ 11-8 Gears clash in shifting   When gears clash
during a shift, it could be that the gearshift linkage is
not properly adjusted. Also, it could be that the clutch
is not disengaging or the clutch linkage is out of adjustment. If not due to any of these conditions, the
trouble is probably inside the transaxle.

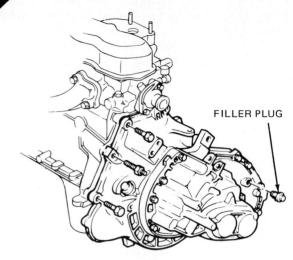

Fig. 11-3 Filler-plug position and position of the transaxle-attaching bolts. *(Pontiac Motor Division of General Motors Corporation)*

the car on
transaxle case
rubber separates
if the case moves
moved out), replace the
movement between a
its attaching point, tighten
attach the mount to the case or

bricant level Figure 11-3 shows the
filler plug for the transaxle. To check
the lubricant in the transaxle, remove the
lubricant level should be within ½ inch [13
of the lower edge of the filler opening. If lubricant
needed, add Dexron®-II, or the equivalent, to bring
the lubricant up to the proper level. Install the plug and
torque it to specifications.

Chapter 11 review questions

Select the *one* correct, best, or most probable answer to each question. Then check your answers against the correct answers given at the end of the book.

1. If the noise is the same in drive or coast, it could be:
   a. road noise
   b. tire noise
   c. front-wheel-bearing noise
   d. any of the above
2. If the noise is a knock at low speed, it probably is due to:
   a. worn drive-axle joints
   b. a worn side-gear-hub counterbore
   c. defective tires
   d. a and b
3. If the transaxle is noisy in one gear position, it probably is due to:
   a. worn constant-mesh gears or synchronizer
   b. worn main bearings
   c. a defective clutch
   d. a damaged clutch gear
4. If the transaxle slips out of gear, it could be due to:
   a. an improperly adjusted linkage
   b. transmission loose on engine
   c. dirt between the clutch cover and engine
   d. any of the above

5. Hard shifting into gear can be caused by failure of the clutch to disengage, linkage out of adjustment, or:
   a. worn pilot bearing in engine crankshaft
   b. internal transaxle troubles
   c. worn wheel bearings
   d. loose extension housing
6. Gear clash when shifting could be caused by an incorrect linkage adjustment, internal transaxle problems, or:
   a. a damaged drive axle
   b. internal differential problems
   c. the clutch not disengaging
   d. worn constant-mesh gears
7. A clunk on acceleration or deceleration could be caused by:
   a. loose engine or transaxle mounts
   b. worn parts in the differential
   c. worn joints in drive axles
   d. any of above
8. Bad bearings produce a:
   a. whine
   b. thump
   c. growl or grating noise
   d. constant noise regardless of car speed

CHAPTER 12

MANUAL-TRANSAXLE REMOVAL AND INSTALLATION

After studying this chapter, and with proper instruction and equipment, you should be able to:

1. Remove and replace a manual transaxle in a car.
2. Service and adjust transaxle gearshift linkages.

 12-1 Replacing manual transaxles This chapter covers the removal and installation of two manual transaxles, a General Motors unit (Fig. 1-2) and a Chrysler Corporation unit. It also covers typical gearshift-linkage adjustments.

Chapter 11 describes various possible transaxle troubles that might require transaxle removal for service. The gearing and operation of transaxles are covered in ✿ 4-14 and 4-15. Following chapters describe the disassembly and reassembly of typical manual transaxles.

✿ **12-2 General Motors manual-transaxle removal** The General Motors transaxle used on their smaller front-drive cars is removed as follows:

1. Disconnect the battery-ground cable from the transaxle case and attach the cable to the upper radiator hose with wire or tape.
2. Some vehicles have two transaxle-strut-bracket bolts on the left side of the engine compartment. Where present, remove the bolts.
3. Remove the top four engine-to-transaxle bolts and the one bolt toward the rear of the car near the cowl. The bolt nearer the cowl is installed from the engine side (Fig. 12-1).
4. Break loose the engine-to-transaxle bolt near the starting motor (at the front of the car) but do not remove the bolt yet.
5. Disconnect the speedometer cable from the transaxle. On cars with cruise control, remove the transaxle speedometer cable at the cruise-control transducer.
6. Remove the retaining clip and washer from the transaxle-shift linkage at the transaxle. Remove the clips securing the shift cables to the mounting bosses on the transaxle case.
7. Install an engine-holding fixture so that one end is supported on the cowl tray over the wiper motor and the other end rests on the radiator support

(Figs. 12-2 and 12-3). Attach a fixture hook to the engine lift ring and raise the engine just enough to take the weight off the engine mounts.

CAUTION: The engine-support fixture must be located in the center of the cowl and the fasteners must be properly torqued before supporting the engine. The fixture is not intended to support the entire weight of the engine and transaxle. You could be hurt if the fixture is not properly installed because it could then slip and let the engine drop down.

8. Unlock the steering column and raise the car on a lift. Place a transmission jack under the transaxle in readiness to support it.
9. Remove the two nuts attaching the stabilizer bar to the driver-side lower control arm (Fig. 12-4).

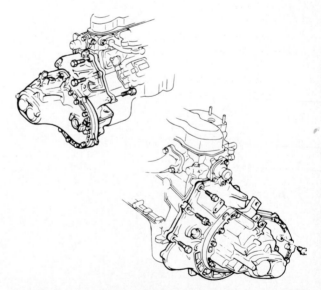

Fig. 12-1 Transaxle-to-engine attachment. *(Pontiac Motor Division of General Motors Corporation)*

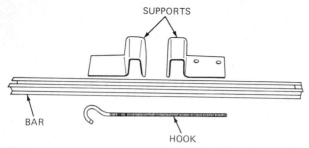

Fig. 12-2 Engine-support tools. *(Pontiac Motor Division of General Motors Corporation)*

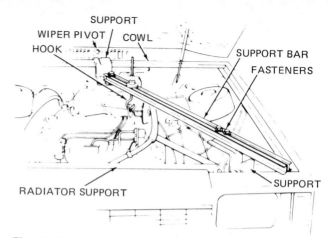

Fig. 12-3 Using support tools to support the engine. *(Pontiac Motor Division of General Motors Corporation)*

10. Remove the four bolts that attach the retaining plate that holds the driver-side stabilizer bar to the cradle (Fig. 12-4).
11. Loosen the four bolts holding the stabilizer bracket on the passenger side of the cradle (Fig. 12-4).
12. If the exhaust pipe is in the way, disconnect and remove it.
13. Pull the stabilizer bar down on the driver's side.
14. Remove four nuts and disconnect the front and rear transaxle mounts at the cradle (Fig. 12-4).
15. Remove the two rear center cross-member bolts (Fig. 12-4).
16. Remove from the passenger side three front-cradle-attaching bolts. The nuts can be reached by pulling back the splash shield next to the frame rail (Fig. 12-4).

17. If so equipped, remove the top bolt from the lower front transaxle damper shock absorber.
18. Remove the left front wheel. Remove the front cradle-to-body bolts on the driver's side of the cradle. Remove the rear cradle-to-body bolts (Fig. 12-4).
19. Use the special tool to pull the driver-side (left) driveshaft from the transaxle assembly. The passenger-side (right) driveshaft is then disconnected from the case. When the transaxle assembly is removed, the right shaft can be swung out of the way.

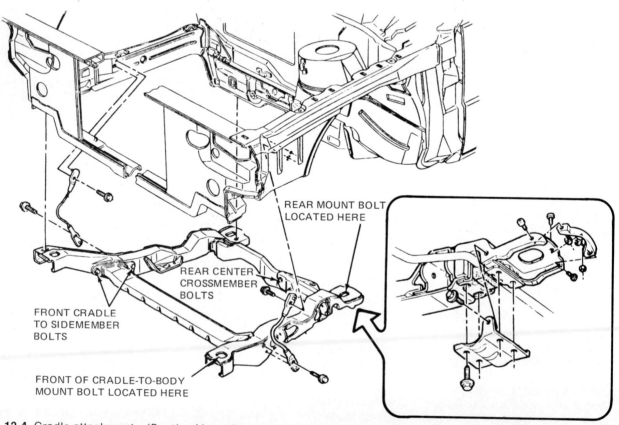

Fig. 12-4 Cradle attachments. *(Pontiac Motor Division of General Motors Corporation)*

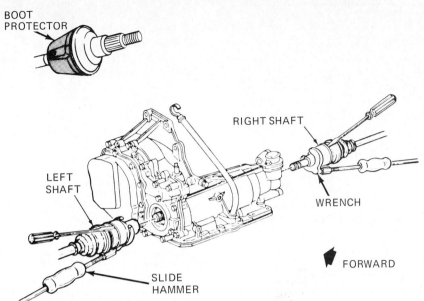

Fig. 12-5 Removing driveshafts from the transaxle. *(Pontiac Motor Division of General Motors Corporation)*

Be sure to use the boot protector when disconnecting the driveshafts (Fig. 12-5).

20. Swing the partial cradle to the driver's side and wire it securely outboard of the fender well.
21. Remove the bolts from the flywheel shield and the starting-motor shield and remove the shields.
22. If so equipped, remove the two transaxle-extension bolts from the engine-to-transaxle bracket.
23. Securely attach the transaxle case to the transmission jack.
24. Remove the last transaxle-to-engine bolt.
25. Remove the transaxle by sliding it to the driver's side, away from the engine. Lower the jack and move the transaxle to the bench for service.

✿ 12-3 Manual-transaxle installation The installation of the General Motors transaxle is basically the reverse of removal. These special points should be watched:

1. When installing the transaxle, position the right drive-axle shaft in its bore as the transaxle is being moved into place. This shaft cannot be installed after the transaxle is connected to the engine.
2. When the transaxle is fastened to the engine, swing the cradle into position and immediately install the cradle-to-body bolts. After this step, installation can be completed by reversing the removal procedure.
3. When moving the cradle back into the installed position, be sure to guide the left drive axle into the case bore.

✿ 12-4 Adjusting the shift cables After installation of the transaxle is complete, connect and adjust the shift cables as follows (Fig. 12-6):

1. Shift the transaxle into first gear. At the control assembly, with the shifter boot and retainer off, install two pins (⁵/₃₂-inch [3.8-mm] or No. 22 drill bits)

in the alignment holes in the control assembly. This will keep the transaxle in first gear. Then attach the two shift cables to the control assembly with studs and pin retainers. Be sure the cables are routed correctly and operate freely.
2. Make sure the transaxle is still in first gear by pushing the rail selector shaft inward (down) until you feel the resistance of the inhibitor spring. Then rotate the shift lever fully counterclockwise.
3. Install the stud (with cable A attached) in the slotted area in shift lever F. Install the stud (with cable B attached) in the slotted area of selector lever D, while lightly pulling on lever D to remove lash. Remove the two drill bits or pins and road test the car. Make sure that the shift lever feels free in neutral during shifting. This is described as *having a good neutral-gate feel.* Make any slight additional adjustment necessary.

✿ 12-5 Chrysler Corporation transaxle removal The transaxle made by Chrysler for some of its front-drive cars is discussed in ✿ 4-15. To remove this transaxle, first work inside the engine compartment as follows:

1. Disconnect the negative (ground) cable from the battery.
2. Remove the following parts from the transaxle:
   a. Clutch cable
   b. Speedometer cable
   c. Backup-light-switch harness
   d. Starting motor
   e. Front roll rod
   f. Four top mounting bolts

Then, from outside the engine compartment, proceed as follows:

1. Raise the car on a lift and remove the front wheels (Fig. 12-7).

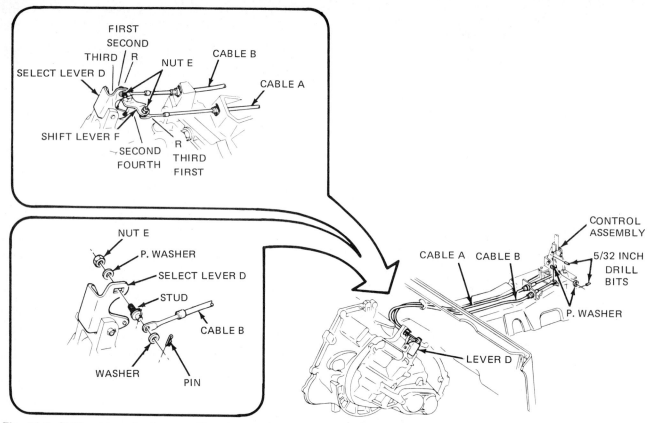

Fig. 12-6 Shift-cable adjustments. *(Pontiac Motor Division of General Motors Corporation)*

2. Remove the undercover and disconnect the shift rod and extension (Fig. 12-8).
3. Drain the lubricant from the transaxle.
4. Remove the right and left driveshafts from the transaxle case. This procedure is covered in Chap. 18.

Careful: The driveshaft retainer ring should be replaced with a new ring on reassembly. Do not damage the drive boots.

5. Disconnect the range-selector cable.
6. Remove the engine rear cover.
7. Remove the coupling bolt at each end of the front roll rod.
8. Loosen the engine side-roll-rod bracket.
9. Position a transmission jack under the transaxle or use a chainfall as recommended by Chrysler.
10. Remove the transaxle-mount-insulator-to-transaxle-mount bracket nuts. Loosen the transaxle-mount bracket.
11. Remove and lower the transaxle.

✿ 12-6 Chrysler Corporation transaxle installation Basically, the installation of the transaxle is the reverse of removal. During installation, the coupling bolts at each end of the front roll should be temporarily tightened. After the transaxle is installed, the bolts should be tightened to specifications.

Make sure that the shift-rod-to-transaxle set-screw lock wire is installed.

Fill the transaxle with the specified amount and type of lubricant. Adjust the clutch-pedal linkage.

✿ 12-7 Gearshift-assembly service The gearshift assembly is shown disassembled at the top in Fig. 12-9. The shift patterns for both the range selector and the gearshift selector are shown in Fig. 4-54.

Fig. 12-7 To remove the transaxle, raise the car on a lift and remove the front wheels. *(Chrysler Corporation)*

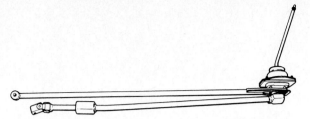

Fig. 12-8 Main gearshift-lever assembly. *(Chrysler Corporation)*

1. Removal To remove the gearshift assembly, proceed as follows:

1. Remove the knobs from the two levers.
2. Loosen the console-box mounting screws and disconnect the meter-harness connectors. Remove the console box.
3. Remove the range-selector bracket (Fig. 12-10).
4. Raise the car and, from underneath, remove the undercover and the transaxle extension (Fig. 12-11).
5. Remove the lock wire from the shift-rod set screw. Loosen the set screw and remove the transaxle shift rod (Fig. 12-12).
6. Remove the heat protector (Fig. 12-13).
7. Remove the gearshift assembly (Fig. 12-8).

2. Disassembly To disassemble the gearshift assembly (Fig. 12-9):

1. Remove the E ring and coil spring.
2. Remove the extension mounting screws to detach the extension (Fig. 12-14).
3. Use a punch and hammer to drive out the pin attaching the fulcrum ball to the main gearshift lever (Fig. 12-15). Remove the ball.

NOTE: The main gearshift lever and shift rod are held together by staking and so cannot be disassembled.

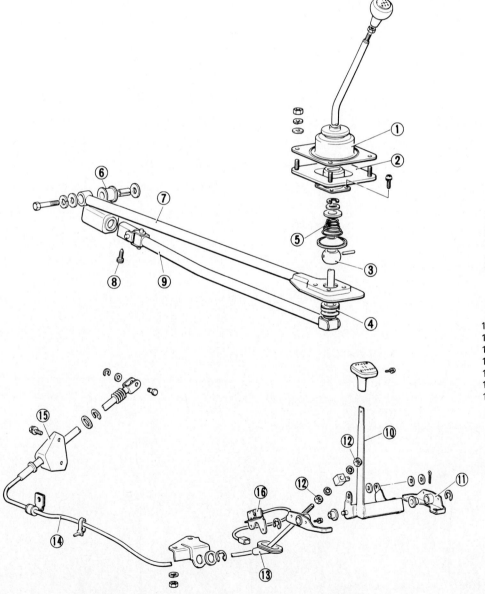

1. Shift lever cover
2. Insulator
3. Fulcrum ball
4. Cover
5. Spring
6. Bushing
7. Extension
8. Set screw
9. Shift rod
10. Range selector lever
11. Selector lever mounting bracket
12. Selector cable adjusting nut
13. Dust cover
14. Selector cable
15. Cable bracket
16. Selector switch

Fig. 12-9 Transaxle control components. *(Chrysler Corporation)*

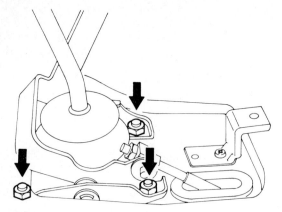

Fig. 12-10 Removing bracket. *(Chrysler Corporation)*

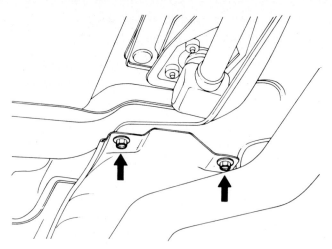

Fig. 12-13 Removing heat protector. *(Chrysler Corporation)*

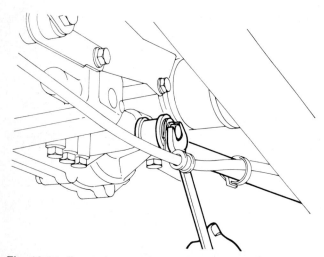

Fig. 12-11 Removing extension. *(Chrysler Corporation)*

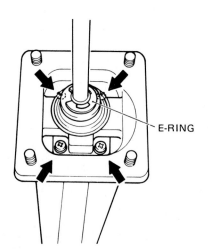

E-RING

Fig. 12-14 Detaching extension from shift rod. *(Chrysler Corporation)*

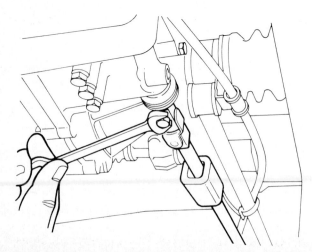

Fig. 12-12 Removing shift rod. *(Chrysler Corporation)*

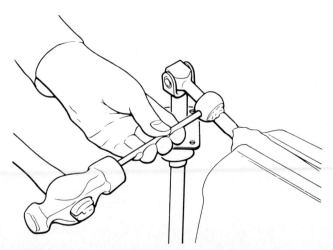

Fig. 12-15 Removing fulcrum ball. *(Chrysler Corporation)*

3. **Inspection** Inspect for bent rods, loose shift-rod joints, damaged insulator or dust cover, worn fulcrum ball, and weak spring.

4. **Reassembly and installation** To reassemble, apply the specified grease to the fulcrum ball surface, spring seat, and insulator sliding surface. Also apply grease inside the dust covers of the joints (Fig. 12-16).

Attach the extension to the insulator after putting the parts together. Install the assembly, heat protector, and extension. Attach the shift rod. Be sure to install a new lock wire on the shift-rod set screw.

✿ 12-8 Chrysler range-selector control service On Chrysler Corporation cars that have a dual-range manual transaxle, the cables may become broken or damaged. To inspect or replace a cable, proceed as follows:

1. **Removal** To remove the cable:

1. Remove the knobs of the main gearshift lever and range-selector lever.
2. After loosening the console-box mounting screws, disconnect the meter-harness connectors and remove the console box.
3. Remove the adjusting nut to detach the cable from the lever (Fig. 12-17).
4. Lift up the car and remove the undercover.
5. Remove the cable clips from the cross member and extension.
6. Remove the ring that couples the lever with the cable to detach the lever from the cable. Remove the cable bracket from the transaxle (Fig. 12-18).

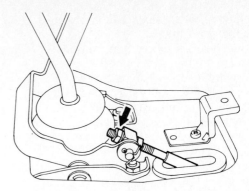

Fig. 12-17 Detaching cable from lever. *(Chrysler Corporation)*

7. After removing the cable mounting bracket under the floor, remove the cable.

2. **Inspection** Check for:

1. Broken or damaged cable.
2. Binding or excessive sliding resistance between inner and outer cables.

3. **Installation** Proceed as follows:

1. Adjust the cable length with the lever at the ECONOMY position (with the back of lever kept in contact with the floor).
2. Move the lever from ECONOMY position to POWER position and then back again to make sure that the lever is operating properly.

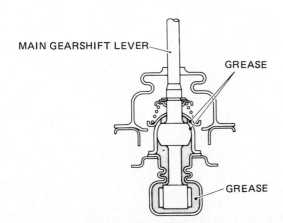

Fig. 12-16 Grease points for main gearshift lever. *(Chrysler Corporation)*

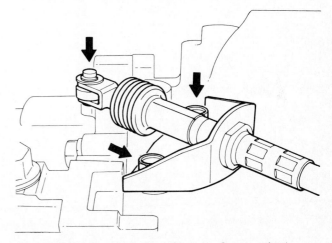

Fig. 12-18 Removing cable. *(Chrysler Corporation)*

Chapter 12 review questions

Select the *one* correct, best, or most probable answer to each question. Then check your answers against the correct answers given at the end of the book.

1. The first step in transaxle removal is to:
   a. disconnect the engine-to-transaxle bolts
   b. disconnect the battery-ground cable

c. install the engine support
d. remove the front wheels

2. The purpose of the transmission jack is to:
   a. raise the engine to take the weight off the mounts
   b. raise the car to get under and remove the transaxle

c. support the transaxle as it is detached
d. detach the clutch
3. The two levers used on some transaxles in Chrysler-built cars are the:
  a. forward lever and reverse lever
  b. gearshift lever and range-selector lever
  c. clutch lever and gearshift lever
  d. power lever and economy lever
4. The purpose of the engine-support fixture recommended by General Motors is to:
  a. support the engine weight while the transaxle is removed
  b. lift the engine from the vehicle
  c. lift the transaxle from the vehicle
  d. lift the engine just enough to take the weight off the engine mounts
5. When installing the General Motors transaxle:
  a. slip the right drive-axle shaft into the transaxle after it is installed
  b. install the right drive-axle shaft while assembling the transaxle
  c. slip the right drive axle into the transaxle as the transaxle is moved into place
  d. install both drive axles before installing the transaxle

CHAPTER 13

MANUAL-TRANSAXLE SERVICE

After studying this chapter, and with proper instruction and equipment, you should be able to:

1. Disassemble, inspect, service, and reassemble a General Motors transaxle.
2. Identify the parts of a disassembled transaxle.

13-1 Servicing manual transaxles This chapter describes the disassembly and reassembly of the transaxle used in many General Motors front-drive cars. Also included is the inspection and servicing of the transaxle parts. This transaxle is described in ✿ 4-14. It has four forward speeds and reverse. The fourth-gear position is overdrive.

The Chrysler Corporation transaxle is covered in Chap. 14. It also has four forward speeds and reverse, with the fourth-gear position being overdrive. In addition, some Chrysler models have extra gearing that gives the transaxle two speed ranges.

Chapter 11 covers transaxle trouble diagnosis. Chapter 12 covers the removal and installation of transaxles as well as adjustment of the shift linkages.

✿ **13-2 Transaxle disassembly** Figure 13-1 shows a completely disassembled transaxle. This figure will be your guide for the disassembly and reassembly of the transaxle. Proceed as follows:

1. Put the transaxle on a suitable stand, such as the one shown in Fig. 13-2. This figure shows the clutch cover removed and laid on its side behind the transaxle case.
2. To remove the clutch cover, remove the 15 bolts attaching the cover to the transaxle case.
3. Use a plastic hammer and carefully tap the clutch cover loose from the case. A special plastic sealant is used between the cover and case instead of a gasket.
4. Remove the differential assembly and ring gear as shown in Fig. 13-3. These are the parts numbered 76 to 89 in the lower left corner of Fig. 13-1. Disassembly of the differential is described later.
5. Position the shifter shaft in the neutral position so that the shifter can move freely and is not engaged in any drive gear.

6. Bend back the tab on the lock and remove the bolt from the shifter shaft. Remove the shifter shaft and the shift-fork shaft from the synchronizer forks as shown in Fig. 13-4.
7. Remove the reverse-shift fork shown in Fig. 13-5 by disengaging it from the guide pin and interlock bracket.
8. Remove the lock bolt securing the reverse-idler-gear shaft. Remove the reverse idler gear and shaft (Fig. 13-6).
9. Remove the detent shift lever and interlock assembly (Fig. 13-7). Leave the shift forks temporarily engaged with the synchronizers.
10. Grasp the input and output shafts and lift them out as an assembly from the case. Note the position of the shift forks (Fig. 13-8) so that you can put them back in the same places on reassembly.

You are now ready to disassemble the subassemblies you have removed or detached from the transaxle.

✿ **13-3 Disassembling the input shaft** Figure 13-9 shows the input- and output-shaft assemblies. We discuss the disassembly of the input shaft first (to the left in Fig. 13-9). In the following discussions and illustrations, the letters "RH" and "LH" are used. These refer to the positions of the parts in the transaxle when they are installed in the car. The LH refers to the end farthest from the clutch, and the RH refers to the end nearest the clutch. The end of the input shaft pointing up in Fig. 13-9 would be labeled the RH end because it is nearest the clutch. The shaft splines at the top carry the friction disk of the clutch (Figs. 2-26 and 2-27).

To disassemble the input shaft, several parts that have been press-fitted on the shaft must be pressed off in a shop press. The basic rule in pressing a gear or bearing on or off a shaft is that the force must be applied at the correct points. For example, force through a ball or roller bearing can ruin it. Special supports must be

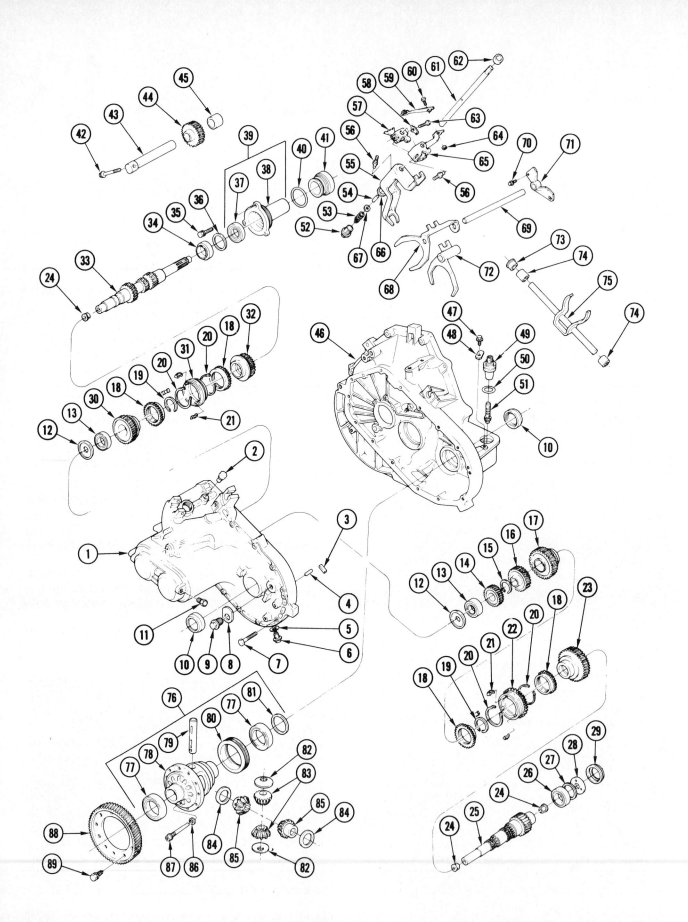

1. Case assembly
2. Vent assembly
3. Magnet
4. Pin
5. Washer, drain screw
6. Screw, drain
7. Bolt
8. Washer, fill plug
9. Plug, fill
10. Seal assembly, axle shaft
11. Plug
12. Shield, oil
13. Bearing assembly
14. Gear, fourth speed output
15. Ring, third speed output gear retaining
16. Gear, third speed output
17. Gear, second speed output
18. Ring, synchronizer blocking
19. Ring, synchronizer retaining
20. Spring, synchronizer key retaining
21. Key, synchronizer
22. Synchronizer assembly
23. Gear, first speed output
24. Sleeve, oil shield
25. Gear, output
26. Bearing assembly, output
27. Shim, output gear bearing adjustment
28. Shield, output bearing oil
29. Retainer, output gear bearing oil shield
30. Gear, fourth speed input

31. Synchronizer assembly
32. Gear, third speed input
33. Gear, input cluster
34. Bearing assembly, input
35. Screw
36. Shim, input gear bearing adjustment
37. Seal assembly, input gear
38. Retainer, input gear
39. Retainer assembly, input gear bearing
40. Seal, input gear bearing retainer
41. Bearing assembly, clutch release
42. Screw and washer, reverse idler
43. Shaft, reverse idler
44. Gear assembly, reverse idler
45. Spacer, reverse idler shaft
46. Housing assembly, clutch and differential
47. Screw
48. Retainer, speedo gear fitting
49. Sleeve, speedo driven gear
50. Seal, speedo gear sleeve
51. Gear, speedo driven
52. Seat, reverse inhibitor spring
53. Spring, reverse inhibitor
54. Pin
55. Lever, reverse shift
56. Stud, reverse lever locating
57. Lever assembly, detent
58. Washer, lock detent lever
59. Spring, detent
60. Bolt

61. Shaft, shift
62. Seal assembly, shift shaft
63. Bolt
64. Nut
65. Interlock, shift
66. Shim, shift shaft
67. Washer, reverse inhibitor spring
68. Fork, third and fourth shift
69. Shaft, shift fork
70. Screw
71. Guide, oil
72. Fork, first and second shift
73. Seal assembly, clutch fork shaft
74. Bearing, clutch fork shaft
75. Shaft assembly, clutch fork
76. Differential assembly
77. Bearing assembly, differential
78. Case, differential
79. Shaft, differential pinion
80. Gear, speedo drive
81. Shim, differential bearing adjustment
82. Washer, pinion thrust
83. Gear, differential pinion
84. Washer, side gear thrust
85. Gear, differential side
86. Lockwasher
87. Screw, pinion shaft
88. Gear, differential ring
89. Bolt

Fig. 13-1 Manual transaxle completely disassembled. *(Chevrolet Motor Division of General Motors Corporation)*

used that will fit around the part while force is being applied to the shaft. Proceed as follows:

1. Use the input shaft LH bearing remover shown in Fig. 13-10 to press the fourth-speed gear and LH bearing from the shaft.
2. Remove the brass blocker ring and the snap ring from the 3-4 synchronizer.
3. Position support plates behind the third-speed gear as shown in Fig. 13-11 and press the third-speed gear and 3-4 synchronizer from the input shaft.
4. Using the special tool shown in Fig. 13-12, press the RH bearing from the shaft.

✿ 13-4 Disassembling the output shaft Figures 13-13 to 13-16 show the removal of the various gears,

synchronizers, and bearings from the output shaft. Each operation requires a special support positioned under the gear or the bearing. Proceed as follows:

1. Use support plates behind the fourth-speed gear and press the gear and LH bearing off the shaft (Fig. 13-13).
2. Remove the snap ring holding the third-speed gear in place.
3. Slide the 1-2 synchronizer assembly into first gear to allow the support plates to support the second-speed gear (Fig. 13-14). Press the second-speed gear and third-speed gear from the output shaft.
4. Remove the snap ring holding the 1-2 synchronizer in position.

Fig. 13-2 Clutch cover removed. *(Pontiac Motor Division of General Motors Corporation)*

Fig. 13-3 Removing the ring gear and differential assembly. *(Pontiac Motor Division of General Motors Corporation)*

Fig. 13-4 Removing the shifter shaft. *(Pontiac Motor Division of General Motors Corporation)*

Fig. 13-6 Removing the reverse idler gear. *(Pontiac Motor Division of General Motors Corporation)*

5. Support the first-speed gear as shown in Fig. 13-15 and press this gear and the 1-2 synchronizer from the output shaft.
6. Install the special tool (Fig. 13-16) and press the RH bearing off the shaft.

❋ 13-5 Synchronizer service Figure 13-17 shows the parts in a synchronizer. To disassemble it, carefully pry the springs from the synchronizer. Then mark the hub and sleeve so that you can put them back together in the same relative positions. Separate the parts, noting the positions of the keys. Clean and inspect the parts. Replace defective parts.

NOTE: The hub and sleeve are a matched set, so if one is replaced the other should be replaced with it. Do not mix the parts between the two synchronizers.

To reassemble the synchronizer, put the hub in the sleeve, with the lip on the hub away from the shift-fork groove in the sleeve, aligning the previously made marks.

Carefully install one retainer spring. Then carefully pry the spring back and insert the keys one at a time. Be sure to position the spring so that it is captured by the keys.

Install the spring on the opposite side, with the open segment of the ring in a different position from the open

Fig. 13-5 Removing the reverse-shift fork. *(Pontiac Motor Division of General Motors Corporation)*

Fig. 13-7 Removing the interlock assembly. *(Pontiac Motor Division of General Motors Corporation)*

Fig. 13-8 Shaft-and-shift-forks assembly. *(Pontiac Motor Division of General Motors Corporation)*

segment of the first spring. The openings should not be opposite each other.

✿13-6 Assembling the input shaft Lubricate all parts before starting the reassembly. Refer to Fig. 13-9, which shows the proper relationship of the parts that go onto the shaft. Proceed as follows:

Fig. 13-9 Input- and output-shaft assemblies. *(Pontiac Motor Division of General Motors Corporation)*

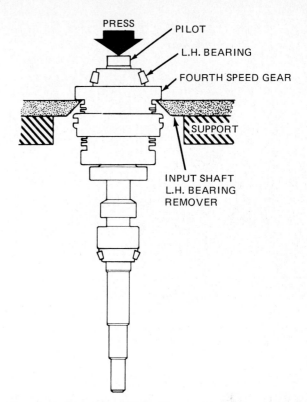

Fig. 13-10 Removing LH bearing and fourth-speed input gear. *(Pontiac Motor Division of General Motors Corporation)*

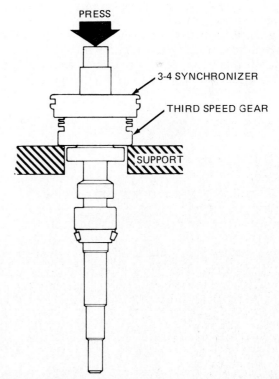

Fig. 13-11 Removing third-fourth synchronizer and third-speed input gear. *(Pontiac Motor Division of General Motors Corporation)*

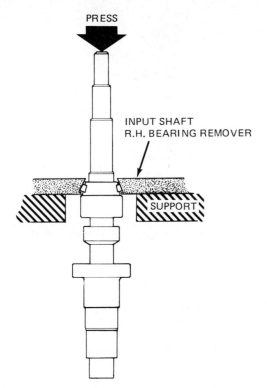

Fig. 13-12 Removing RH bearing. *(Pontiac Motor Division of General Motors Corporation)*

1. Press the RH bearing onto the shaft with the special bearing installer shown in Fig. 13-18. The tool is shown partly cut away so that you can see that the force is being applied to the inner race of the tapered-roller bearing.
2. Put the third-speed gear on the shaft, with the groove toward the synchronizer (Fig. 13-19).
3. Install the brass blocker ring onto the gear cone; then install the 3-4 synchronizer. Use the proper tool

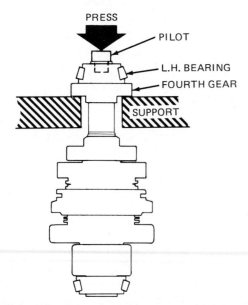

Fig. 13-13 Removing LH bearing and fourth-speed output gear. *(Pontiac Motor Division of General Motors Corporation)*

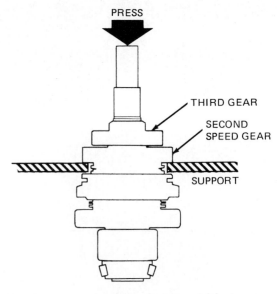

Fig. 13-14 Removing third-speed and second-speed output gears. *(Pontiac Motor Division of General Motors Corporation)*

(Fig. 13-19) to press against the hub and not against the sleeve.
4. Install a snap ring to retain the 3-4 synchronizer. Be sure to position the beveled or pointed edges of the gap away from the synchronizer (Fig. 13-20) so that it can be easily removed later if necessary.
5. Install the brass blocker ring.
6. Install the fourth-speed gear on the shaft, in the position shown in Fig. 13-21. Then install the LH bearing on the shaft using the inner-race installer as shown (Fig. 13-21).

⚙ **13-7 Assembling the output shaft** Lubricate all parts before starting the reassembly. Refer to Fig.

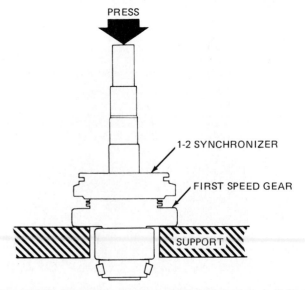

Fig. 13-15 Removing first-second synchronizer and first-speed output gear. *(Pontiac Motor Division of General Motors Corporation)*

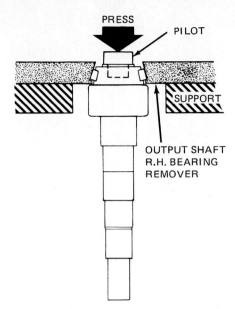

Fig. 13-16 Removing RH bearing. *(Pontiac Motor Division of General Motors Corporation)*

13-9, which shows the proper relationship of the parts that go onto the shaft. Proceed as follows:

1. Install the RH bearing on the shaft with the bearing installer as shown in Fig. 13-22.
2. Place the first-speed gear on the shaft in the position shown in Fig. 13-23. Put a brass blocker ring onto the gear cone, and install the 1-2 synchronizer with a special tool that presses on the hub. Do not press on the sleeve.
3. Install the snap ring to retain the 1-2 synchronizer. See Fig. 13-20 for proper positioning of the ring on the shaft.
4. Put a brass blocker ring onto the synchronizer.
5. Put the second-speed gear onto the shaft in the position shown in Fig. 13-24. Then press the third-speed gear onto the shaft with the hub toward the fourth-speed gear (Fig. 13-24).
6. Install the snap ring to hold the third-speed gear in position (see Fig. 13-20).
7. Press the fourth-speed gear onto the shaft with its hub toward the third-speed gear. Use the support plate and special tools as shown in Fig. 13-25.

Fig. 13-17 Synchronizer. *(Pontiac Motor Division of General Motors Corporation)*

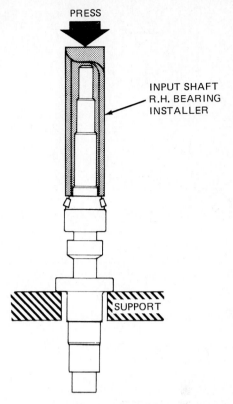

Fig. 13-18 Installing RH bearing on the input shaft. *(Pontiac Motor Division of General Motors Corporation)*

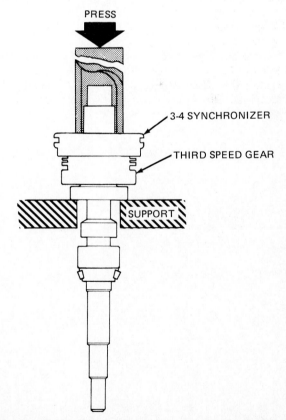

Fig. 13-19 Installing third-speed gear and third-fourth synchronizer on the input shaft. *(Pontiac Motor Division of General Motors Corporation)*

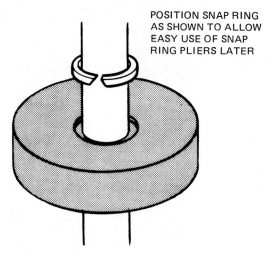

Fig. 13-20 Correct position of snap ring upon installation. *(Pontiac Motor Division of General Motors Corporation)*

✿13-8 Transaxle-case overhaul Do not use the transaxle stand (Fig. 13-2) to support the case when pressing bearing races into the case. The stand is not intended to support heavy loads. To overhaul the transaxle case, proceed as follows:

1. Remove the reverse-inhibitor fitting from outside the case (Fig. 13-26). From inside the case, remove the spring and pilot spacer.
2. If the bearing races need replacement, use the special bearing-cup remover which uses force to exert

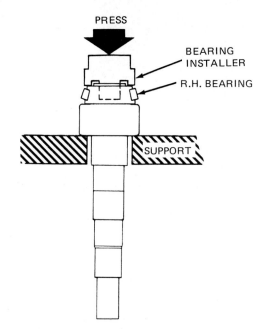

Fig. 13-22 Installing RH bearing on the output shaft. *(Pontiac Motor Division of General Motors Corporation)*

the pull. To install the tool, turn the set screw counterclockwise to retract the tool so that it will pass the bearing race. Then turn the set screw clockwise to expand the tool so that it grasps the race (Fig. 13-27). Use the special bearing-cup installer to install the races.

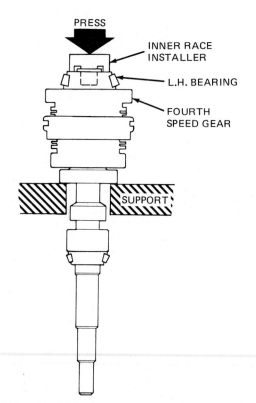

Fig. 13-21 Installing fourth-speed gear and LH bearing on the input shaft. *(Pontiac Motor Division of General Motors Corporation)*

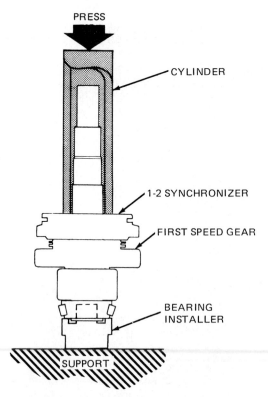

Fig. 13-23 Installing first-speed gear and first-second synchronizer on the output shaft. *(Pontiac Motor Division of General Motors Corporation)*

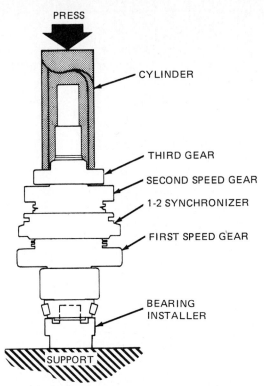

Fig. 13-24 Installing the second-speed and third-speed gears on the output shaft. *(Pontiac Motor Division of General Motors Corporation)*

3. Remove the oil slingers (Fig. 13-28).
4. Remove the side-bearing race with a special tool, if the race needs replacement. Use the special tool to install a new race.

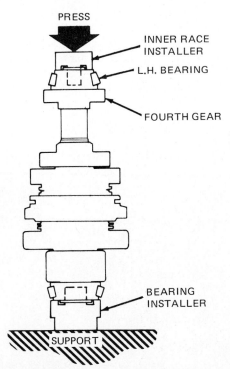

Fig. 13-25 Installing the fourth-speed gear and LH bearing on the output shaft. *(Pontiac Motor Division of General Motors Corporation)*

Fig. 13-26 Removing the reverse-inhibitor fitting. *(Pontiac Motor Division of General Motors Corporation)*

5. Check the two guide pins for the interlock bracket and reverse-shift fork. Check the magnet and clean it if necessary. Remove the sealant from the mating surface with a special tool. Do not gouge or damage the aluminum surface because any scratch or nick could cause oil leaks.

⚙ **13-9 Clutch-cover overhaul** Chapter 3 covers servicing of the clutch itself. However, there are several items to check in the clutch cover, as follows:

1. Remove the differential side-bearing race and shim if the race requires replacement.

Fig. 13-27 Removing bearing cups. *(Pontiac Motor Division of General Motors Corporation)*

Fig. 13-28 Removing oil slingers. *(Pontiac Motor Division of General Motors Corporation)*

2. Remove the other bearing races (input-shaft and output-shaft RH bearings) if they require replacement.
3. Remove the shim from in back of the input bearing race. Remove the oil shield, shim, and retainer from in back of the output-shaft-bearing race.
4. Remove the three bolts securing the input-gear-bearing retainer (the release-bearing sleeve—item 38 in Fig. 13-1). Remove the sleeve. Tap it carefully if necessary to loosen it (Fig. 13-29).
5. Remove the external oil ring and internal oil seal from the sleeve (Fig. 13-30).
6. Remove the plastic oil scoop (Fig. 13-31).
7. If it is necessary to replace the clutch-fork shaft or bushing, use the special tool as shown in Fig. 13-32. Always install a new shaft seal after installation of the shaft or bushing.

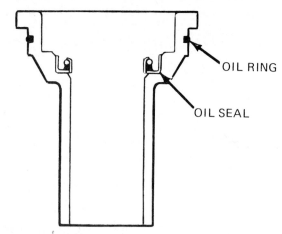

Fig. 13-30 Oil seal and oil ring. *(Pontiac Motor Division of General Motors Corporation)*

8. Remove the bead of sealant from the mating surfaces. Do not damage the surfaces, as scratches or nicks could allow leaks.

✿ **13-10 Clutch-cover reassembly** After cleaning and inspecting all parts, proceed as follows:

1. Install the plastic oil scoop.
2. Install the external square-cut oil ring on the sleeve. Install the input-bearing retainer and tighten the three bolts to specifications.
3. Install the internal oil seal with the special tool.

✿ **13-11 Differential-case overhaul** See the lower left of Fig. 13-1 (items 76 to 89), which shows the parts that go into the differential. To overhaul the differential, proceed as follows:

1. Remove the ring gear from the differential case.
2. Remove the lock screw (item 87) and slide the pinion shaft (79) out of the differential case (78). Then roll the gears and thrust washers out through the openings in the case. Observe the positions of the parts

Fig. 13-29 Removing release-bearing sleeve. *(Pontiac Motor Division of General Motors Corporation)*

Fig. 13-31 Removing oil scoop. *(Pontiac Motor Division of General Motors Corporation)*

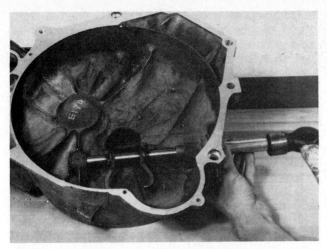
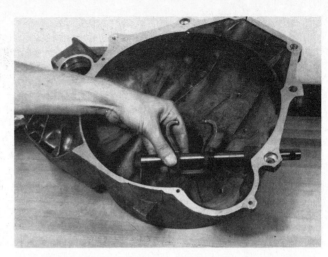

Fig. 13-32 Replacing clutch-fork shaft and bushings. *(Pontiac Motor Division of General Motors Corporation)*

before removing them so that you can put them back in the same positions on reassembly.

3. If the differential side bearings require replacement, use the special tools required to remove and replace them.
4. Clean and inspect all parts. Replace parts as required.
5. Install gears and thrust washers in the case. Install the pinion shaft and secure it with the lock screw, tightened to specifications.
6. Attach the ring gear to the differential case.

✿ **13-12 Shim selection** The proper shims must be used when the transaxle is reassembled in order to correctly preload the bearings. Shim selection starts when the input- and output-shaft assemblies and the differential assembly are ready for reinstallation in the transaxle case.

1. Put the transaxle case in the holding fixture and install the assemblies in the case (Fig. 13-2). Put the three RH bearing races onto their bearings.
2. Position the three gauges on the bearings as shown in Fig. 13-33. Be sure the bearing races fit smoothly into the bores of the gauges.
3. On the output-shaft gauge, install the metal oil-shield retainer in the bore on top of the gauge.
4. Carefully assemble the clutch cover over the three

Fig. 13-33 Gauges in position to check for shim selection. *(Pontiac Motor Division of General Motors Corporation)*

INPUT GAUGE

OIL SHIELD RETAINER

OUTPUT GAUGE

DIFFERENTIAL GAUGE

Fig. 13-34 Compressing the gauges. *(Pontiac Motor Division of General Motors Corporation)*

Fig. 13-35 Measuring gap, or clearance. *(Pontiac Motor Division of General Motors Corporation)*

gauges and on the case, using seven spacers placed evenly around the perimeter (Fig. 13-34). Retain with the bolts provided.

5. Draw the cover to the case by gradually tightening the bolts, making the circuit of the bolts several times. This compresses all three gauge sleeves.

6. Rotate each gauge to seat the bearings. Rotate the differential case through three revolutions in each direction.

7. With the gauges compressed, the gap between the outer sleeve and the base pad is the correct thickness for the preload shim at each location. Carefully compare the gap with the available shims. The largest shim that can be placed in the gap and drawn through without binding is the correct shim for reassembly (Fig. 13-35).

8. After selecting the correct shims for each location, remove the clutch cover, seven spacers, and three gauges.

9. Place the shims in their proper bores in the clutch cover, add the metal shield, and install the bearing races, using the special tools.

✹ 13-13 Transaxle-case reassembly Proceed as follows:

1. Put the two shaft assemblies on the bench and install the shift forks (Fig. 13-8).

2. Pick up the two shafts as an assembly and carefully lower them into the transaxle case. Do not nick the gears.

3. Put the interlock bracket onto the special guide pin. Be sure the bracket engages the fingers on the shift forks. Put the detent shift lever in the interlock (Fig. 13-7).

4. Install the shifter shaft through the interlock bracket and detent shift lever. Do not push it in any farther at this time.

5. Install the reverse-shift fork on the guide pin. Be sure the reverse-shift fork engages the interlock bracket.

6. Install the reverse idler gear and shaft. Be sure that the long end of the shaft points up and that the large chamfered ends of the gear teeth face up. Install the spacer on the shaft (Fig. 13-6). The flat on the reverse idler shaft faces the input shaft.

7. Fully install the shifter shaft through the reverse-shift fork until the shaft pilots into the inhibitor-spring spacer. Remove the guide pin. With the shaft in the neutral position, install the bolt and lock through the detent shift lever. Bend a tab of the lock over the bolt head so that the bolt will not come loose.

8. Install the fork shaft through the synchronizer forks and into the bore of the case.

9. Carefully install the ring-gear-and-differential-case assembly.

10. Install the magnet if it has been removed.

11. Apply a thin bead of the special sealer which is used instead of a gasket. The bead should be applied to the clutch cover. Then install the cover on the transaxle case. Use dowel pins to guide the cover into position. Tap the cover gently with a soft hammer to ensure that the parts are seated.

12. Install the 15 attaching bolts and torque to specifications.

13. Torque the idler-shaft retaining bolt to specifications.

14. Shift through the gear ranges to make sure the internal parts move freely.

15. After installing the transaxle on the vehicle (Chap. 12) and adjusting the linkages (✹ 12-4), road test the car to make sure the transaxle is working properly.

Chapter 13 review questions

Select the *one* correct, best, or most probable answer to each question. Then check your answers against the correct answers given at the end of the book.

1. The first step in disassembling the transaxle is to:
   a. remove the extension
   b. shift the transaxle into neutral
   c. remove the clutch cover
   d. none of the above

2. The input and output shafts are:
   a. removed separately from the case
   b. removed as an assembly from the case
   c. disassembled piece by piece while in the case
   d. should not be removed from the case

3. The letters LH and RH in the transaxle discussions and illustrations refer to the:
   a. positions of the axles
   b. positions of the parts when the transaxle is on the car
   c. left-hand and right-hand steering
   d. left wheel and right wheel

4. The letters LH mean:
   a. left wheel
   b. lever handle
   c. transaxle parts farthest from the clutch
   d. transaxle parts nearest the clutch

5. The first part to be removed from the output shaft is the:
   a. fourth-speed gear
   b. third-speed gear
   c. second-speed gear
   d. reverse gear

6. The first part to be removed from the input shaft is the:
   a. fourth-speed gear
   b. third-speed gear
   c. second-speed gear
   d. reverse gear

7. The first part to be removed when disassembling a synchronizer is a:
   a. gear
   b. sleeve
   c. hub
   d. spring

8. The first part to be installed on the input shaft during reassembly is the:
   a. input gear
   b. output gear
   c. RH bearing
   d. LH bearing

9. During a complete overhaul of the differential case, the first part to remove from the case is the:
   a. clutch gear
   b. ring gear
   c. pinion shaft
   d. differential pinions

10. Shim selection is required so that the:
    a. transaxle case will fit properly on the engine
    b. gears will be properly meshed
    c. bearings will be correctly preloaded
    d. differential case will fit correctly into the transaxle case

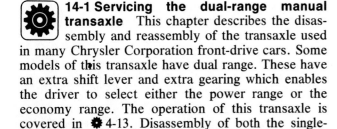

CHAPTER 14

DUAL-RANGE MANUAL-TRANSAXLE SERVICE

After studying this chapter, and with proper instruction and equipment, you should be able to:

1. Disassemble and reassemble the Chrysler dual-range transaxle, including inspection and servicing of transaxle parts.
2. Identify the parts of a disassembled dual-range transaxle.

14-1 Servicing the dual-range manual transaxle This chapter describes the disassembly and reassembly of the transaxle used in many Chrysler Corporation front-drive cars. Some models of this transaxle have dual range. These have an extra shift lever and extra gearing which enables the driver to select either the power range or the economy range. The operation of this transaxle is covered in ❈ 4-13. Disassembly of both the single-range and the dual-range transaxle is covered in this chapter.

Chapter 11 covers transaxle trouble diagnosis. Chapter 12 covers removal and installation of transaxles, as well as adjustment of the shift linkages.

❈ 14-2 Transaxle-body disassembly Figure 14-1 shows all the internal parts in the transaxle, including the parts for the two basic models. One of these is the model KM160, which is the single-range transaxle. It has one special part that is marked with a single asterisk (item 3). All parts marked with two asterisks are used only in the model KM165, which is the dual-range unit. Figure 14-2 shows the shift-control system disassembled. Figures 14-1 and 14-2 are your guides as you service the transaxle.

To disassemble the transaxle body, proceed as follows:

1. With the transaxle on the bench, remove the clutch-operating-cable bracket and the transaxle mounting bracket.
2. Remove the backup-light switch. Take out the steel balls from inside the transaxle case.
3. Remove the rear cover.
4. Remove the two spacers from the rear of the tapered-roller bearing.
5. Remove the transaxle case.

6. Put all shift rails in neutral. If any one of the shift rails is in any other position, the interlock works to lock all the shift rails.
7. Remove the three poppet plugs. Remove the three springs and three steel balls (Fig. 14-3).
8. Pull out the reverse idler shaft (Fig. 14-4). Remove the reverse idler gear. Sometimes the reverse idler shaft comes off with the transaxle case when it is removed.
9. Remove the reverse-shift-lever assembly (Fig. 14-4).
10. Pull off the reverse-shift rail (Fig. 14-4).
11. Remove the third-and-fourth-speed shift-rail spacer collar.
12. Use a flat punch to remove the spring pins from the first-and-second-speed and the third-and-fourth-speed shift forks (Fig. 14-5).
13. Pull the first-and-second-speed shift rail loose from the case (Fig. 14-6). It cannot be removed yet.
14. Remove the third-and-fourth-speed shift rail from the case. Remove the first-and-second-speed shift rail and fork together with the third-and-fourth-speed shift rail (Fig. 14-7).
15. Move the third-and-fourth-speed synchronizer sleeve to the fourth-speed side. Remove the output-shaft assembly. This is item 28 with its related parts (29 through 47) in Fig. 14-1.
16. Remove the differential assembly (items 48 through 57 in Fig. 14-1).
17. On the dual-range transaxle, remove the plug, poppet spring, and ball (Fig. 14-8).
18. Remove the input-shaft-bearing retainer (Fig. 14-9).
19. Remove the input-shaft assembly and shift rail and fork (on the dual-range transaxle) together with the intermediate-shaft assembly (Fig. 14-10).
20. Remove the shift-shaft spring retainer (Fig. 14-11).

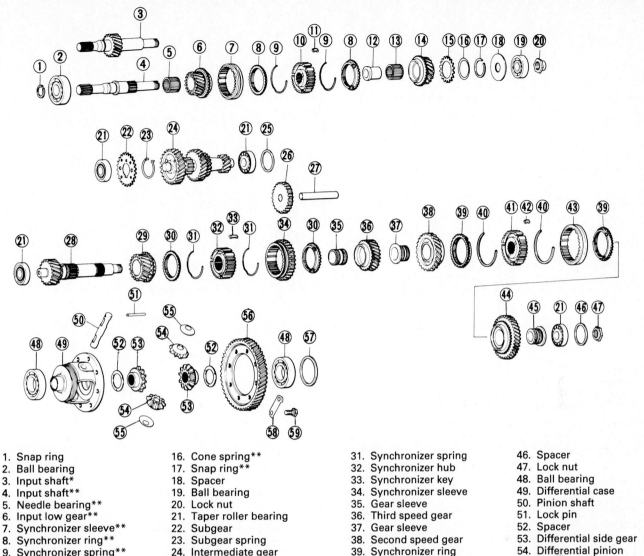

1. Snap ring
2. Ball bearing
3. Input shaft*
4. Input shaft**
5. Needle bearing**
6. Input low gear**
7. Synchronizer sleeve**
8. Synchronizer ring**
9. Synchronizer spring**
10. Synchronizer hub**
11. Synchronizer key**
12. Gear sleeve**
13. Needle bearing**
14. Input high gear**
15. Subgear**

16. Cone spring**
17. Snap ring**
18. Spacer
19. Ball bearing
20. Lock nut
21. Taper roller bearing
22. Subgear
23. Subgear spring
24. Intermediate gear
25. Spacer
26. Reverse idler shaft
27. Idler gear shaft
28. Output shaft
29. Fourth speed gear
30. Synchronizer ring

31. Synchronizer spring
32. Synchronizer hub
33. Synchronizer key
34. Synchronizer sleeve
35. Gear sleeve
36. Third speed gear
37. Gear sleeve
38. Second speed gear
39. Synchronizer ring
40. Synchronizer spring
41. Synchronizer hub
42. Synchronizer key
43. Synchronizer sleeve
44. First speed gear
45. Gear sleeve

46. Spacer
47. Lock nut
48. Ball bearing
49. Differential case
50. Pinion shaft
51. Lock pin
52. Spacer
53. Differential side gear
54. Differential pinion
55. Washer
56. Differential drive gear
57. Spacer
58. Lock washer
59. Bolt

*Single-range only
**Dual-range only

Fig. 14-1 Disassembled power train of a transaxle. Two versions are shown. Some parts are used only in the dual-range transaxle. *(Chrysler Corporation)*

21. Pull out the shift-shaft spring pin with pliers. Then use the shift-shaft hole (B in Fig. 14-12) to push the shift shaft out, as shown. Use a small punch inserted in the hole.

22. Remove the control finger, two springs, spacer collar with poppet spring, and ball. Note that when the shift shaft is pulled out, the poppet ball will jump out of the control-finger hole (A in Fig. 14-12). To prevent this, slip your finger over the hole when pulling out the shaft.

23. Remove the selector-finger lockpin, shaft, and selector finger (dual-range only), as shown in Fig. 14-13.

24. Remove the tapered-roller-bearing outer race.

Mark the race and shaft so that the race can be reinstalled in its original position.

25. Remove the speedometer-driven-gear assembly from inside the case.

❀14-3 Disassembling the single-range input shaft Refer to Fig. 14-1, which shows this input shaft (item 3) together with the parts assembled on it. Refer also to Fig. 4-52, which shows in partial cutaway view the input shaft and its related parts.

1. Remove the front-bearing snap ring and use the bearing puller and adapter shown in Fig. 14-14 to press the shaft out of the bearing.

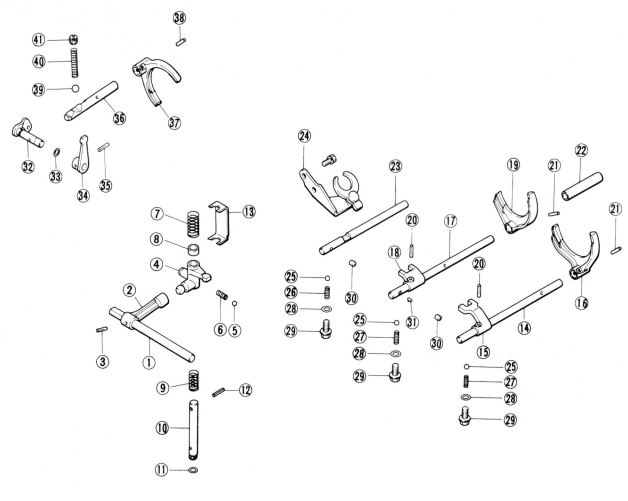

1. Control shaft
2. Control lug
3. Lock pin
4. Control finger
5. Steel ball
6. Spring
   (Length 18.9 mm [0.740 inch])
7. Neutral return spring
8. Spacer collar
9. Reverse restrict spring
10. Shift shaft

11. O-ring
12. Spring pin
13. Spring retainer
14. First and second speed shift rail
15. Shift lug
16. Shift fork
17. Third and fourth speed shift rail
18. Shift lug
19. Shift fork
20. Lock pin
21. Spring pin

22. Spacer collar
23. Reverse shift rail
24. Reverse shift lever assembly
25. Steel ball
26. Reverse spring
    (Length 16.6 mm [0.650 inch])
27. Spring
    (Length 18.9 mm [0.740 inch])
28. Gasket
29. Plug
30. Interlock plunger A

31. Interlock plunger B
32. Selector shaft*
33. O-ring*
34. Selector finger*
35. Lock pin*
36. Shift rail*
37. Shift fork*
38. Lock pin*
39. Steel ball*
40. Spring*
41. Plug*

*Dual-range only

Fig. 14-2 Disassembled control system for a transaxle that may have either a single range or a dual range. *(Chrysler Corporation)*

2. Straighten the locknut lock at the rear end of the input shaft. Remove the locknut. Hold the input shaft in the soft jaws of a vise as you loosen the nut. Do not damage the splines, gear, or ground surfaces of the shaft.

3. To remove the rear bearing, use a spacer under the inner race of the bearing as shown in Fig. 14-15 and press out the shaft on a shop press. Do not let the shaft drop as it is pressed out.

❋ **14-4 Disassembling the dual-range input shaft** Refer to Fig. 14-1, which shows this input shaft (item 4) with its related parts (items 5 through 20). Refer

also to Fig. 4-53, which shows the dual-range input shaft and its related parts in partial cutaway view.

1. After removing the front-bearing snap ring, pull the front bearing from the shaft as shown in Fig. 14-14.

2. Straighten the locknut lock at the rear of the input shaft and remove the locknut.

3. Support the input low-speed gear on a press base (Fig. 14-16) and press down on the rear end of the input shaft. This removes the input high-speed gear, gear sleeve, synchronizer assembly, input low-speed gear, and rear bearing.

Careful: When pressing down on the input shaft, be careful not to set the supporting plate on the input-low-

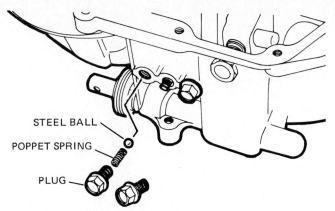

STEEL BALL

POPPET SPRING

PLUG

Fig. 14-3 Removing poppet and related parts. *(Chrysler Corporation)*

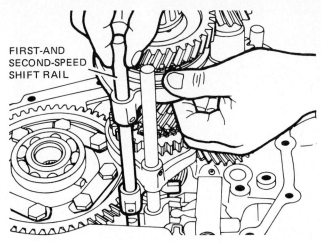

FIRST-AND
SECOND-SPEED
SHIFT RAIL

Fig. 14-6 Removing first-and-second-speed shift rail. *(Chrysler Corporation)*

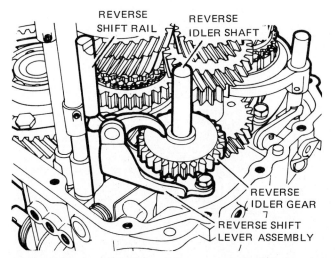

REVERSE
SHIFT RAIL

REVERSE
IDLER SHAFT

REVERSE
IDLER GEAR

REVERSE SHIFT
LEVER ASSEMBLY

Fig. 14-4 Removing reverse idler gear and related parts. *(Chrysler Corporation)*

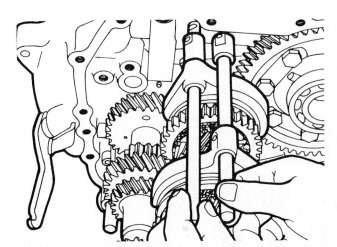

Fig. 14-7 Removing third-and-fourth-speed shift rail. *(Chrysler Corporation)*

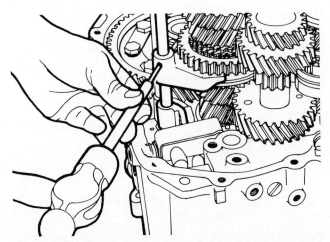

Fig. 14-5 Removing spring pin. *(Chrysler Corporation)*

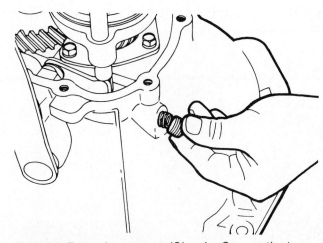

Fig. 14-8 Removing poppet. *(Chrysler Corporation)*

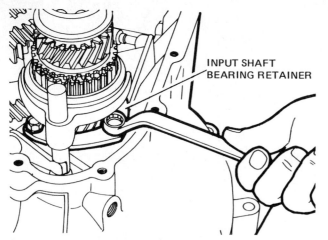

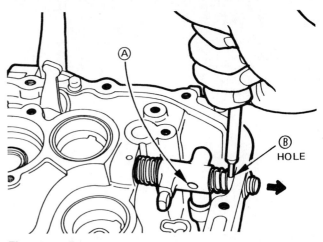

Fig. 14-9 Removing bearing retainer. *(Chrysler Corporation)*

Fig. 14-12 Removing shift shaft. *(Chrysler Corporation)*

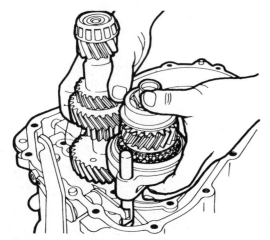

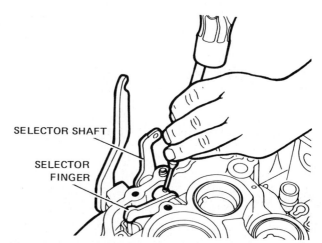

Fig. 14-10 Removing input shaft and intermediate shaft. *(Chrysler Corporation)*

Fig. 14-13 Removing selector shaft. *(Chrysler Corporation)*

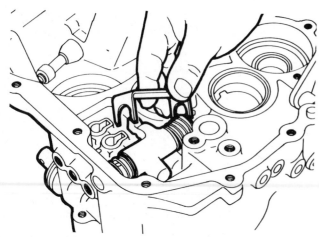

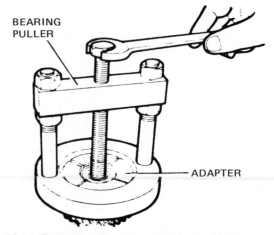

Fig. 14-11 Removing spring retainer. *(Chrysler Corporation)*

Fig. 14-14 Removing front bearing. *(Chrysler Corporation)*

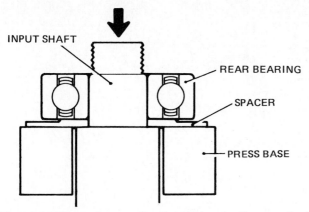

Fig. 14-15 Removing rear bearing. *(Chrysler Corporation)*

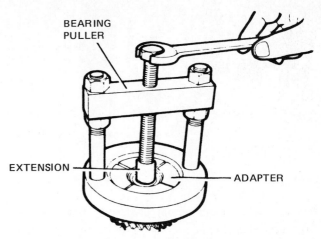

Fig. 14-17 Removing rear bearing. *(Chrysler Corporation)*

speed-gear clutch gear. This would allow the input low-speed gear and the clutch gear to become disengaged.

4. To remove only the rear bearing, use the bearing puller, extension, and rear-bearing adapter as shown in Fig. 14-17. The input high-speed gear will come off with the rear bearing.

☀ 14-5 Disassembling the intermediate gear This is the same as the countergear assembly in standard manual transmissions. Figure 14-1 numbers this gear shaft with its assembled gears item 24. Figure 4-53 lists it as item 8. The only parts that can be removed from this assembly are the bearings, the subgear, and the subgear spring.

1. To remove the front tapered-roller bearing, use the bearing puller, extension, and front-bearing adapter as shown in Fig. 14-17. Discard the old bearing. On reassembly, use a new bearing and a new outer race.
2. Remove the subgear and spring.
3. Remove the rear tapered-roller bearing in the same way you removed the front tapered-roller bearing (Fig. 14-17). Discard the old bearing and outer race.

☀ 14-6 Disassembling the output-shaft assembly The output-shaft assembly is shown in partial cut-

away view in Fig. 14-40. See also Fig. 14-1, where the output shaft and related parts are numbered 28 through 47. When removing the parts, start at the end of the shaft and take off the parts, one by one, using the appropriate tools, as follows:

1. Unlock the output-shaft rear locknut and remove it (item 14 in Fig. 14-40). When loosening the locknut, clamp the shaft in the soft jaws of a vise. Do not damage the shaft!
2. Use the bearing puller, extension, and proper adapter as shown in Fig. 14-17 to remove the front and rear tapered-roller bearings.
3. Use the puller as shown in Fig. 14-18 to remove the first-speed gear, gear sleeve, first-and-second-speed synchronizer assembly, and second-speed gear.
4. Use the gear puller shown in Fig. 14-19 to remove the second-speed-gear sleeve, third-speed gear and gear sleeve, third-and-fourth-speed synchronizer assembly, and fourth-speed gear.

☀ 14-7 Disassembling the differential The differential is shown disassembled as items 48 to 59 in the lower left-hand corner of Fig. 14-1. Figures 4-52 and 4-53 show the differential in cutaway view.

1. Use the bearing puller, extension, and adapter as shown in Fig. 14-20 to remove the ball bearings.

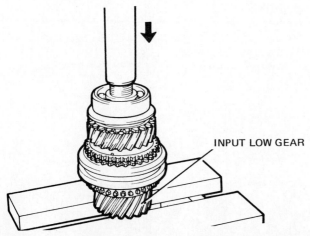

Fig. 14-16 Disassembling input shaft. *(Chrysler Corporation)*

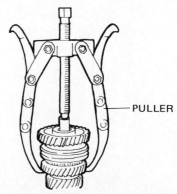

Fig. 14-18 Disassembling output shaft, step one. *(Chrysler Corporation)*

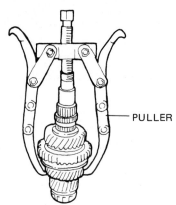

Fig. 14-19 Disassembling output shaft, step two. *(Chrysler Corporation)*

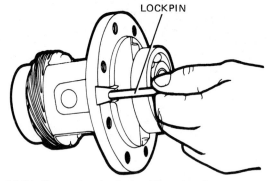

Fig. 14-21 Removing lockpin. *(Chrysler Corporation)*

2. Observe the relationship of the gears and other parts in the differential case so that you can put them back together properly.
3. Straighten the lockwashers and remove the drive-gear attaching bolts, working in a criss-cross pattern.
4. Remove the differential drive gear.
5. Pull out the lockpin (Fig. 14-21).
6. Pull out the pinion shaft and remove the differential pinions and washers.
7. Remove the differential side gears and spacers. Mark them so that you can put them back in the case in the same positions.

✿ 14-8 Inspecting the parts General inspection procedures, including the cleaning and inspection of bearings, are covered in ✿ 7-4 to 7-6. Detailed inspection of the transaxle parts is as follows:

1. Input shaft (Fig. 14-22) Check the splines ① for damage and wear. Also check the fit of the clutch-friction-disk hub on the splines to make sure it is not too loose or too tight. Check the oil-seal area ② for damage and wear. Check the shaft outside diameter ③ for damage and wear.

2. Intermediate gear and subgear Check the gear teeth for damage and wear.

3. Output shaft (Fig. 14-23) Check the output gear ①, the fourth-speed-gear contacting area ②, and the fourth-speed-bearing area ③ for damage or wear.

4. Gear and gear sleeve (Fig. 14-24) Check the gear at points ① through ⑤ for damage or wear. Check the sleeve for damage or wear at ⑥. Check each speed gear for ease of rotation or looseness when assembled on the shaft.

5. Synchronizer ring (Fig. 14-25) Check the synchronizer rings as follows: inspect the tooth surface ①, groove ②, and thread ③ for damage or wear. Put the ring in place on the gear as shown in Fig. 14-26 and check the clearance A. If the clearance is less than 0.0315 inch [0.8 mm], or if the internal screw threads of the ring are worn or damaged, replace the ring.

6. Synchronizer hub, sleeve, and shift fork (Fig. 14-26) The synchronizer hub and sleeve are a matched set and should be replaced as a set if one or the other is damaged. Check the hub external splines ①, key groove ②, and gear-contacting surfaces ③ for damage or wear.

Check the sleeve internal splines ④ and the shift-fork grooves ⑤ for wear or damage. The original width of the groove is 0.2756 inch [7 mm]. The original clearance between the groove and the shift fork is 0.0039 to 0.0118 inch [0.1 to 0.3 mm]. Check the clearance. If it is greater than 0.020 inch [0.5 mm], replace the worn part or parts.

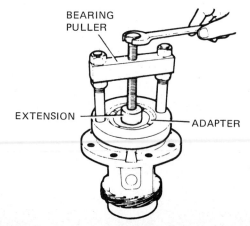

Fig. 14-20 Removing ball bearing. *(Chrysler Corporation)*

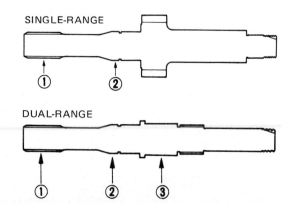

Fig. 14-22 Checkpoints of the input shaft. *(Chrysler Corporation)*

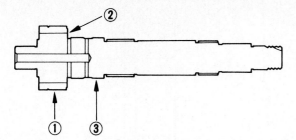

Fig. 14-23 Checkpoints of the output shaft. *(Chrysler Corporation)*

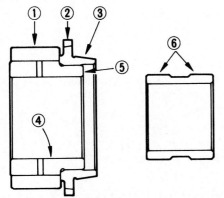

Fig. 14-24 Checkpoints of the gear and sleeve. *(Chrysler Corporation)*

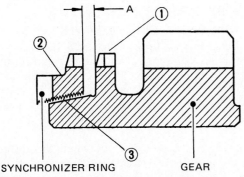

SYNCHRONIZER RING   GEAR

Fig. 14-25 Checkpoints of the synchronizer ring. *(Chrysler Corporation)*

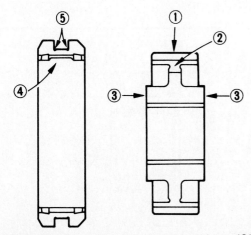

Fig. 14-26 Checkpoints of the synchronizer. *(Chrysler Corporation)*

Check the hub and sleeve to make sure they slide easily, but without excessive looseness, in the direction of rotation.

7. Synchronizer key and spring (Fig. 14-27) Check the arrowed areas for wear or damage.

8. Reverse idler gear, shift lever, and shaft (Fig. 14-28) Check the reverse idler gear at points ①, ②, and ③ for damage or wear. Check the gear-to-shaft clearance. The standard outside diameter of the shaft is 0.6299 inch [16 mm]. The original clearance of the gear to the shaft is 0.0009 to 0.0022 inch [0.024 to 0.056 mm]. If the clearance is excessive, replace the shaft and gear.

Check the shift-lever ends ④ and ⑤ for damage and wear. Check the clearance between the shift lever ⑤ and the reverse-shift rail. Dimension ⑤ is originally 0.5906 inch [15 mm], and the original clearance is 0.002 to 0.0083 inch [0.05 to 0.21 mm].

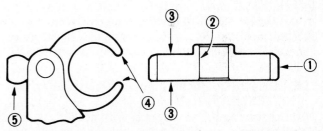

Fig. 14-27 Checkpoints of the synchronizer spring and key. *(Chrysler Corporation)*

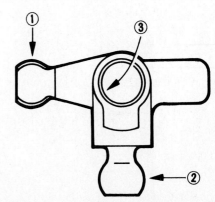

Fig. 14-28 Checkpoints of the reverse gear. *(Chrysler Corporation)*

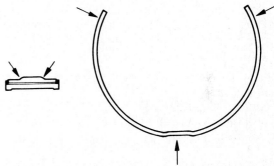

Fig. 14-29 Checkpoints of the control finger. *(Chrysler Corporation)*

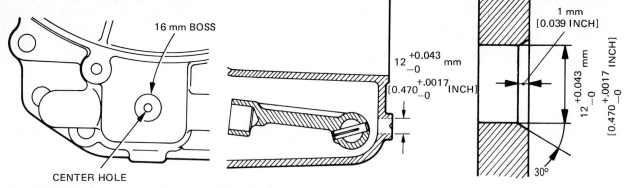

Fig. 14-30 Replacing control shaft and lug. *(Chrysler Corporation)*

9. **Control lug and finger, and shift lug (Fig. 14-29)** Check the control fingers ① and ② for damage or wear. Check the clearance between each control finger and its mating shift lug. The original diameter of the control finger is 0.5906 inch [15 mm], and the original clearance is 0.002 to 0.0083 inch [0.05 to 0.21 mm] for finger ① and 0.0008 to 0.0071 inch [0.02 to 0.18 mm] for finger ②.

Check the control finger inside the diameter ③ for damage and wear and for control-finger-to-shaft clearance. The original diameter of the shaft is 0.5512 inch [14 mm], and the clearance is 0.0002 to 0.002 inch [0.006 to 0.051 mm].

Check the shift-shaft poppet-ball slot for damage and wear.

If the control shaft or control lug requires replacement, proceed as follows: Bore a hole of the size shown in the location indicated in Fig. 14-30. Pull out the lockpin and remove the control shaft and lug. Then, install the new control shaft and lug. Drive in a new lockpin. Seal the hole with a sealing cap as shown in Fig. 14-31. Apply sealant to the cap before driving it in.

10. **Selector finger and shift rail (dual-range) (Fig. 14-32)** Check ① for damage or wear and for clearance with the high-low shift rail. Specifications are the same as for the control fingers and shift lugs (item 9 above).

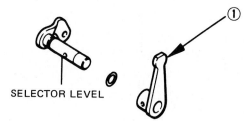

SELECTOR LEVEL

Fig. 14-32 Selector finger and lever. *(Chrysler Corporation)*

11. **Shift-rail poppet ball, spring, and interlock** Check the parts for wear or damage.

12. **Differential** Check gear teeth for damage or wear (Fig. 14-33). Check differential side gears (items 53 in Fig. 14-1). Splines ① should be in good condition, and the clearance between the gear outside diameter ② and the differential case should not be excessive. Original outside diameter is 1.378 inch [35 mm]. The clearance is originally 0.001 to 0.003 [0.025 to 0.075 mm].

Check for excessive lash between the differential side gears and the driveshaft splines in the driving direction. The lash should not exceed 0.0079 inch [0.2 mm].

Check the differential pinion and shaft for damage and proper clearance.

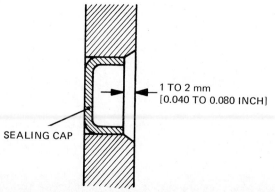

SEALING CAP

1 TO 2 mm [0.040 TO 0.080 INCH]

Fig. 14-31 Installing sealing cap. *(Chrysler Corporation)*

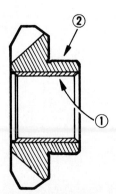

Fig. 14-33 Checkpoints for the differential gears. *(Chrysler Corporation)*

13. **Oil seal** Check oil-seal lips for damage and wear. If oil leaks at a seal, replace the seal. Also check the contacting surface on the shaft for wear.

❋ 14-9 Reassembly As a first step in reassembling the transaxle, the subassemblies—differential, output shaft, and input shaft—are assembled. Then they are installed in the transaxle case. Apply the specified lubricant to all rotating and sliding surfaces during reassembly. Following sections cover reassembly of the subassemblies and then reassembly of the complete unit.

❋ 14-10 Differential reassembly Press the ball bearings onto both ends of the differential case (Fig. 14-34). Apply the force to the inner race of the bearing. Applying the force on the outer race can ruin the bearing.

1. Install spacers on the backs of the differential side gears and install the gears in the differential case. If reusing these parts, be sure to install them in the original positions. If using new parts, select spacers of medium thickness. The range of gear spacers is shown in Fig. 14-35.
2. Install washers on the backs of the differential pinions. Engage both pinions at the same time with the side gears and rotate the pinions to slip them into place in the differential case.
3. Insert the pinion shaft.
4. Measure the backlash between the differential side gear and pinion with a dial indicator set up as shown in Fig. 14-36. If the backlash is not correct, select spacers of the proper thickness (Fig. 14-35), disassemble and reassemble the gears, and recheck. Repeat if necessary. Backlash should be 0 to 0.003 inch [0.076 mm].
5. Align the hole in the pinion shaft with the hole in the case, and insert the lockpin.
6. Install the differential drive gear on the case, using new lockwashers, as shown in Fig. 14-37. Note that one of the lockwashers also locks the pinion-shaft lockpin. Apply lubricant to the bolt threads before

Part No.	Thickness of Spacer mm (inch)
MA 180876	$1.0 \begin{array}{c} +0.16 \\ +0.09 \end{array}$ (.0429 to .0457)
MA 180875	$1.0 \begin{array}{c} +0.08 \\ +0.01 \end{array}$ (.0398 to .0425)
MA 180860	$1.0 \begin{array}{c} 0 \\ -0.07 \end{array}$ (.0366 to .0394)
MA 180861	$1.0 \begin{array}{c} -0.08 \\ -0.17 \end{array}$ (.0327 to .0362)
MA 180862	$1.0 \begin{array}{c} -0.18 \\ -0.25 \end{array}$ (.0295 to .0323)

Fig. 14-35 Part numbers of differential gear spacers of different thickness. *(Chrysler Corporation)*

Fig. 14-36 Measuring backlash between the differential side gear and pinion. *(Chrysler Corporation)*

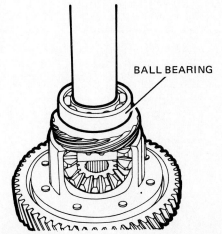

Fig. 14-34 Installing differential bearing. *(Chrysler Corporation)*

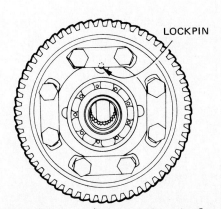

Fig. 14-37 Installing lockwashers. *(Chrysler Corporation)*

installing them. Tighten the bolts in a criss-cross pattern to the specified torque (Fig. 14-38).

7. Bend the lockwasher tangs against the inside flats of the bolt heads. If any tang breaks off, install another lockwasher.

❋ 14-11 Output-shaft reassembly Figure 14-39 identifies the different synchronizers (two for the single-range model and three for the dual-range model). Figure 14-40 shows the output shaft assembled. Proceed as follows to assemble it:

1. Assemble the third-and-fourth-speed synchronizer (to left in Fig. 14-39) by putting the hub in the sleeve with the slot in the oil groove facing in the direction shown.
2. Install the synchronizer keys (Fig. 14-41).
3. Install the synchronizer springs with their stepped parts positioned on the synchronizer keys (Fig. 14-42). Be sure that the stepped parts of the two springs are not positioned on the same key.
4. Assemble the first-and-second-speed synchronizer in the same manner.
5. Put the fourth-speed gear (2 in Fig. 14-40) on the output shaft in the position shown.
6. Install one synchronizer ring (3 in Fig. 14-40).
7. Use the special bearing installer as shown in Fig. 14-43 to press the third-and-fourth-speed synchronizer (4 in Fig. 14-40) onto the output shaft. Make sure the synchronizer-ring keyway is aligned with the synchronizer ring. After the synchronizer is pressed on, make sure the fourth-speed gear rotates easily.

Careful: The splines are sometimes chipped by the synchronizer when it is pressed into place. Remove all chips completely with compressed air.

8. Install the other synchronizer ring with its keyways correctly aligned with the keys.
9. Use the same tool you used to press on the third-and-fourth-speed synchronizer (Fig. 14-43) to press on the third-speed-gear sleeve (6 in Fig. 14-40). Install the third-speed gear (5 in Fig. 14-40).

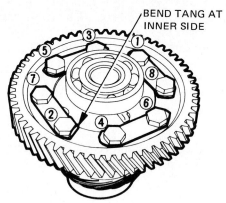

Fig. 14-38 Installing differential gear. The numbers indicate the bolt-tightening sequence. *(Chrysler Corporation)*

10. Use the same tool you used to press on the third-and-fourth-speed synchronizer (Fig. 14-43) to press on the second-speed-gear sleeve (8 in Fig. 14-40). Make sure the third-speed gear turns smoothly.
11. Install the second-speed gear (7 in Fig. 14-40).
12. Install one first-and-second-speed-synchronizer ring (9 in Fig. 14-40).
13. Use the same tool you used to press on the third-and-fourth-speed synchronizer (Fig. 14-43) to press on the first-and-second-speed synchronizer (10 in Fig. 14-40). Make sure the second-speed gear rotates smoothly.
14. Install the other first-and-second-speed-synchronizer ring (9 in Fig. 14-40). Be sure its keyways align with the keys.
15. Assemble the first-speed gear to the gear sleeve (11 and 12 in Fig. 14-40). Use the same tool as in Fig. 14-43 to press the sleeve onto the output shaft. Make sure the first-speed gear rotates smoothly.
16. Use the special bearing installer shown in Fig. 14-44 to press the tapered-roller bearings onto the front and rear ends of the output shaft. Be sure that you push only on the inner bearing race.

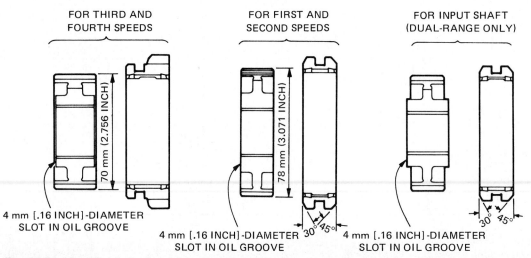

Fig. 14-39 Identification of synchronizer hubs and sleeves. *(Chrysler Corporation)*

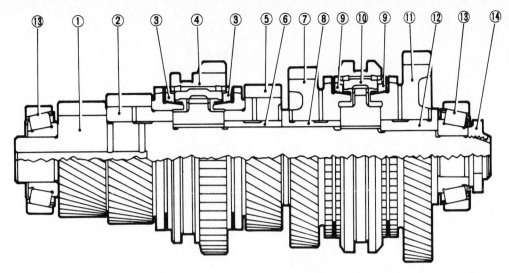

1. Output shaft
2. Fourth-speed gear
3. Synchronizer ring
4. Third-and-fourth-speed synchronizer assembly
5. Third-speed gear
6. Gear sleeve
7. Second-speed gear
8. Gear sleeve
9. Synchronizer ring
10. First-and-second-speed synchronizer assembly
11. First-speed gear
12. Gear sleeve
13. Tapered-roller bearing
14. Locknut

Fig. 14-40 Output-shaft assembly. *(Chrysler Corporation)*

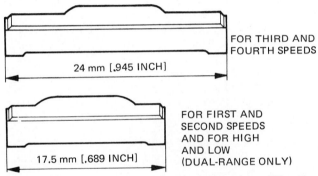

FOR THIRD AND FOURTH SPEEDS

24 mm [.945 INCH]

FOR FIRST AND SECOND SPEEDS AND FOR HIGH AND LOW (DUAL-RANGE ONLY)

17.5 mm [.689 INCH]

Fig. 14-41 Identification of synchronizer key. *(Chrysler Corporation)*

17. Install the locknut on the rear end of the output shaft. Tighten it to the specified torque and stake the nut securely (Fig. 14-45).

⚙ **14-12 Intermediate-gear reassembly** The intermediate gear is item 24 in Fig. 14-1. Install the subgear spring on the shaft (Fig. 14-46) in the exact position shown. Install the subgear (Fig. 14-46). Fit the spring end into the smallest hole.

Use the special bearing installer to press new tapered-roller bearings onto the shaft (Fig. 14-47).

Careful: Do not reuse bearings that have been removed from the shaft. Discard them.

TEETH NOT PROVIDED HERE

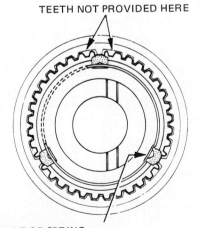

STEPPED PART OF SPRING
(FIX THE STEPPED PART OF THE OPPOSITE SIDE SPRING TO OTHER KEY.)

Fig. 14-42 Installing synchronizer springs. *(Chrysler Corporation)*

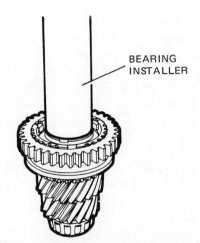

BEARING INSTALLER

Fig. 14-43 Installing third-and-fourth-speed synchronizer. *(Chrysler Corporation)*

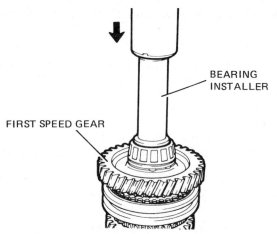

Fig. 14-44 Installing tapered-roller bearing. *(Chrysler Corporation)*

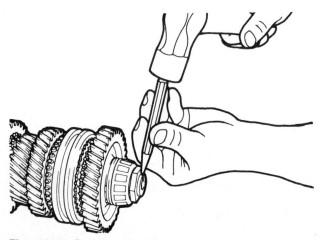

Fig. 14-45 Staking the locknut on the output shaft. *(Chrysler Corporation)*

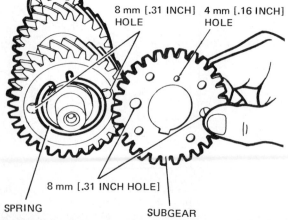

Fig. 14-46 Installing subgear spring. *(Chrysler Corporation)*

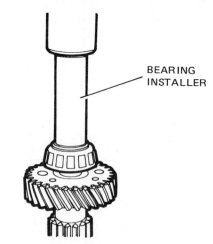

Fig. 14-47 Installing tapered-roller bearing. *(Chrysler Corporation)*

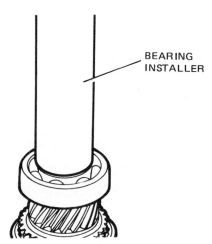

Fig. 14-48 Installing input-shaft front bearing. *(Chrysler Corporation)*

Part No.	Thickness mm (inch)	Identification color
MD 706497	2.24 ± 0.03 (.0870 to .0894)	None
MD 706498	2.31 ± 0.03 (.898 to .0921)	Blue
MD 706499	2.38 ± 0.03 (.0925 to .0949)	Brown

Fig. 14-49 Thickness range of snap rings with identifying color and part number. *(Chrysler Corporation)*

Fig. 14-50 Selecting snap ring thickness. *(Chrysler Corporation)*

⚙ 14-13 Single-range input-shaft reassembly

This is item 3 in Fig. 14-1 and item 4 in Fig. 4-52. Install the front bearing first.

1. Use the bearing installer to install the front bearing (Fig. 14-48). Select the thickest snap ring that will fit into the snap-ring groove (Figs. 14-49 and 14-50). This must be a new snap ring. Do not reuse an old snap ring.
2. Use snap-ring pliers to install the snap ring. Be careful not to damage the surface of the input shaft around which the oil seal fits.
3. Install a spacer at the rear of the input shaft with the stepped side facing the rear bearing. Then install the rear bearing (Fig. 14-51).
4. Install and tighten the rear locknut and stake it (Fig. 14-52).

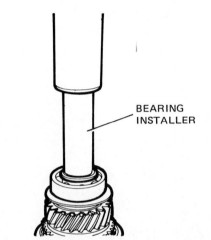

BEARING INSTALLER

Fig. 14-51 Installing rear bearing. *(Chrysler Corporation)*

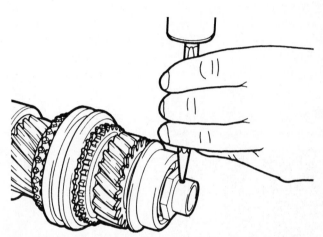

Fig. 14-52 Staking the locknut on the input shaft. *(Chrysler Corporation)*

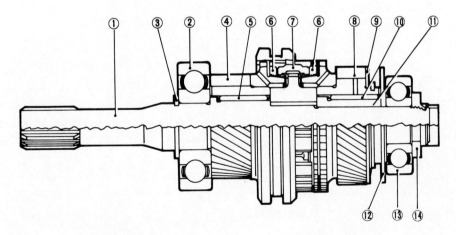

1. Input shaft
2. Front bearing
3. Snap ring
4. Input low gear
5. Needle bearing
6. Synchronizer ring
7. Synchronizer assembly
8. Input high gear
9. Subgear
10. Needle bearing
11. Gear sleeve
12. Spacer
13. Rear bearing
14. Locknut

Fig. 14-53 Input-shaft assembly for dual-range transaxle. *(Chrysler Corporation)*

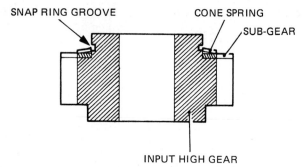

Fig. 14-54 Installing subgear and cone spring. *(Chrysler Corporation)*

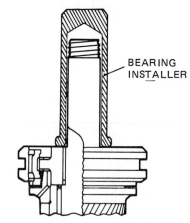

Fig. 14-55 Installing synchronizer. *(Chrysler Corporation)*

Careful: Stake the locknut only at the notch in the shaft. If the shaft is deformed by staking, it will break the breather.

☼ 14-14 Dual-range input shaft reassembly This assembly includes items 4 through 20 in Fig. 14-1. Figure 14-53 shows the input shaft assembled. Proceed as follows:

1. Install the front bearing on the input shaft and secure it with the proper new snap ring as described in items 1 and 2 of ☼ 14-13.
2. Assemble the input synchronizer in the same manner in which the output shaft (☼ 14-11 and Fig. 14-39) was assembled.
3. Install the subgear on the input high-speed gear (9 and 8 in Fig. 14-53).
4. Install the cone spring in the direction shown in Fig. 14-54.
5. Install a new snap ring. Make sure the inner side of the cone spring is not in the snap-ring groove.
6. Install the input low gear and the needle bearing (4 and 5 in Fig. 14-53).
7. Install a synchronizer ring (6 in Fig. 14-53).
8. Use the bearing installer as shown in Fig. 14-55 to press the synchronizer assembly onto the input shaft. The synchronizer key must align correctly with the ring keyway (Fig. 14-53). Do not install the synchronizer assembly backwards!
9. After installing the synchronizer, make sure the input low gear rotates smoothly.
10. Use the special installing tool to install the input high-gear sleeve (11 in Fig. 14-53).
11. Install the second synchronizer ring.

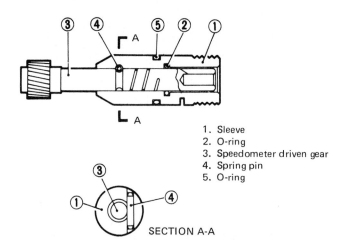

1. Sleeve
2. O-ring
3. Speedometer driven gear
4. Spring pin
5. O-ring

Fig. 14-56 Speedometer-driven gear. *(Chrysler Corporation)*

12. Install the input high gear and needle bearing.
13. Install the spacer with the stepped side facing the rear-bearing side.
14. Use the special tool to install the rear bearing on the input shaft.
15. Install the locknut and tighten to the specified torque. Then stake it in the same way you staked the locknut on the single-range input shaft (item 3 in ☼ 14-13).

☼ 14-15 Speedometer-driven-gear installation
Figure 14-56 is a cutaway view of the speedometer

1. Control shaft	9. Reverse restrict spring	18. Shift lug	31. Interlock plunger B
2. Control lug	10. Shift shaft	19. Shift fork	32. Selector shaft*
4. Control finger	13. Spring retainer	23. Reverse shift rail	34. Selector finger*
5. Steel ball	14. First-and-second-speed	25. Steel ball	36. Shift rail*
6. Spring	shift rail	27. Spring	37. Shift fork*
(Length 18.9 mm	15. Shift lug	(Length 18.9 mm	39. Steel ball*
[0.740 INCH])	16. Shift fork	[0.740 INCH])	40. Spring*
7. Neutral return spring	17. Third-and-fourth-speed	29. Plug	41. Plug*
8. Spacer collar	shift rail	30. Interlock plunger A	

*Dual-range only

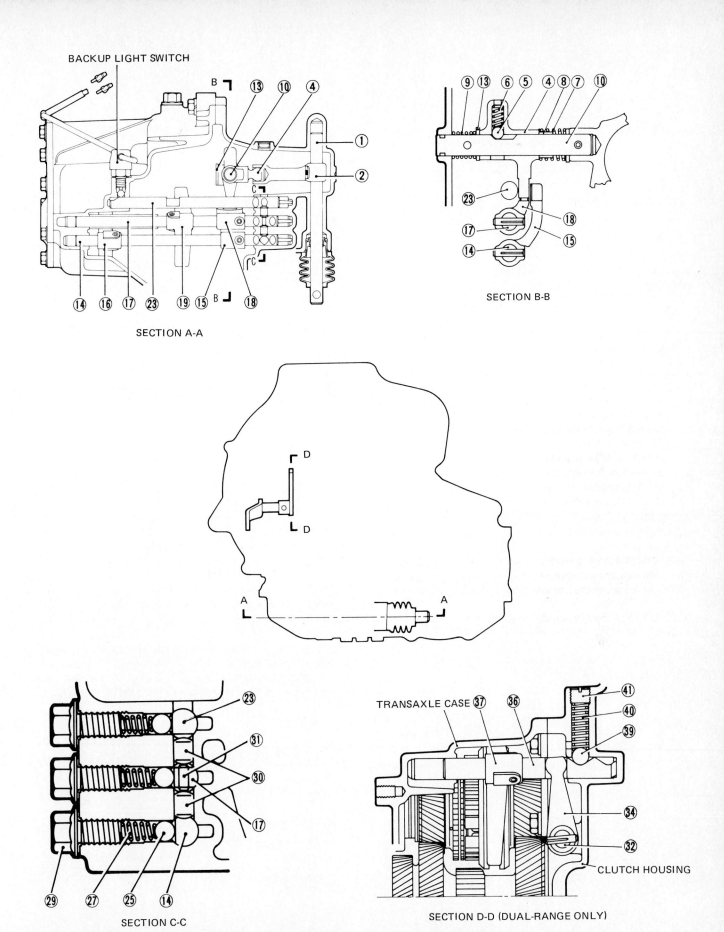

BACKUP LIGHT SWITCH

SECTION A-A

SECTION B-B

SECTION C-C

SECTION D-D (DUAL-RANGE ONLY)

TRANSAXLE CASE

CLUTCH HOUSING

Fig. 14-57 Sectional views of transaxle control system. The parts are shown disassembled in Fig. 14-2. *(Chrysler Corporation)*

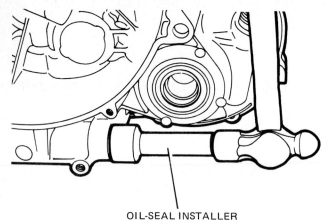

Fig. 14-58 Installing control-shaft oil seal. (Chrysler Corporation)

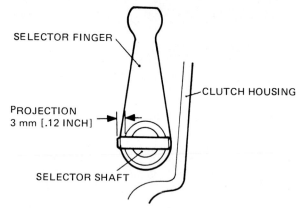

Fig. 14-60 Installing selector finger and shaft. (Chrysler Corporation)

driven gear. If it has been disassembled, assemble it as follows:

1. Install a new O ring (2 in Fig. 14-56) in the inside diameter of the sleeve (1).
2. Apply a small amount of lubricant to the gear shaft (3) and insert it into the sleeve.
3. Align the sleeve pin hole with the pin slot in the shaft and insert the pin (4), making sure the spring-pin slot will not come in contact with the shaft.
4. Install a new O ring (5) on the outside groove of the sleeve.

⚙ **14-16 Reassembling the transaxle** With all the subassemblies ready for installation in the case, install the control-system parts in the clutch housing as follows. Figure 14-2 shows these control parts for both the single-range and the dual-range transaxle. Figure 14-57 shows sectional views of the control system.

1. Apply the recommended sealant to the outside surface of a new control-shaft oil seal. Install the seal with the special oil-seal installer as shown in Fig. 14-58.

2. Install the input-shaft front oil seal with the special tool (Fig. 14-59).
3. Apply a small amount of lubricant to the speedometer sleeve and insert the speedometer-driven-gear assembly into the clutch housing. Install the locking plate snugly in the groove cut in the sleeve.
4. Install a new O ring on the selector shaft (dual-range) and lubricate the ring lightly. Insert the selector shaft into the clutch housing and install the selector finger (Fig. 14-60). With the finger and shaft-lockpin holes aligned, drive in the lockpin. Do not drive it in too deeply.
5. Install the poppet spring and steel ball in the control finger. Push the ball down into its retracted position with the poppet-ball guide (Fig. 14-61). This is done by rotating the guide 180° as shown in Fig. 14-61.
6. Install a new O ring on the shift shaft (2 in Fig. 14-62). Insert the shaft into the clutch housing and then into the reverse-restrict spring (4) and control finger (5). Press it in until it pushes the poppet-ball guide out.
7. Install the spacer collar (6) and neutral-return spring (7). Then press the shift shaft on in. Align the hole in the shift shaft with the hole in the housing and drive the spring pin in with its slit on the shift-shaft center line (upper right in Fig. 14-62).

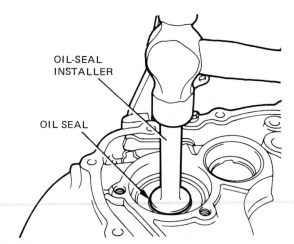

Fig. 14-59 Installing input-shaft front oil seal. (Chrysler Corporation)

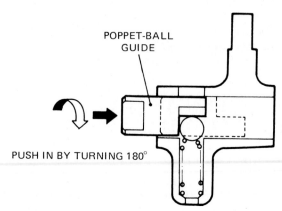

Fig. 14-61 Installing poppet with special tool. (Chrysler Corporation)

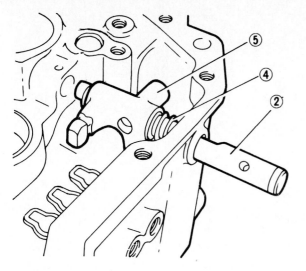

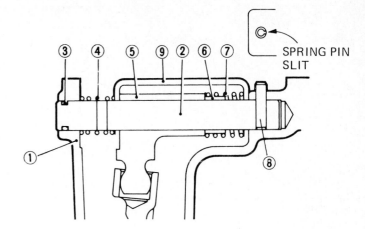

SPRING PIN SLIT

1. Clutch housing
2. Shift shaft
3. O-ring
4. Reverse restrict spring
5. Control finger
6. Spacer collar
7. Neutral return spring
8. Spring pin
9. Spring retainer

Fig. 14-62 Installing control finger and related parts. *(Chrysler Corporation)*

8. Install the spring retainer (9 in Fig. 14-62).
9. Install the differential assembly in the clutch housing. Check and adjust the differential-case end play as follows:

   a. Lay two short pieces of "fuse" about 0.800 inch [20 mm] long on the outer race of the ball bearing (Fig. 14-63). The fuses are plastic or soft-metal round strips. When the assembly is completed, they are crushed down. Their flatness is then measured to determine the clearance in the assembly. This determines the thickness of the spacer that must be installed, as explained below.

   b. Install the transaxle case and gasket. Secure temporarily with the six bolts around the differential, tightened to the specified torque. See Figs. 4-52 and 4-53, which show one of the bolts at the bottom of the pictures. The bolt is unnumbered.

   c. Remove the transaxle case and use a micrometer as shown in Fig. 14-64 to measure the amount the fuses have been flattened. Then select a spacer of the proper thickness so that the differential will have the correct end play (Fig. 14-65).

10. Turn the subgear in the direction of the arrow to align its proper hole with the hole in the end gear on the intermediate-gear assembly (Fig. 14-66). These holes are 0.310 inch [8 mm] in diameter. Insert a short bar or bolt into the holes to hold the subgear in position.

11. Install the input-shaft assembly and the intermediate-gear assembly in the clutch housing together. In the dual-range model, install the selector shift rail and shift-fork assembly together (Figs. 14-2 and 14-57).

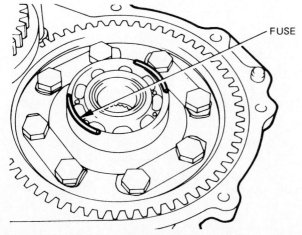

FUSE

Fig. 14-63 Measuring differential-case end play. *(Chrysler Corporation)*

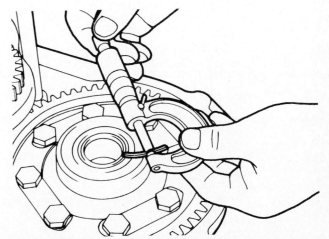

Fig. 14-64 Measuring the amount the fuses have flattened to determine the differential-case end play. *(Chrysler Corporation)*

Part No.	Thickness mm (inch)	Identification mark
MD 706574	1.31 (.0516)	E
MD 706573	1.40 (.0551)	None
MD 706572	1.49 (.0587)	C
MD 706571	1.58 (.0622)	B
MD 706570	1.67 (.0657)	A
MD 706575	1.76 (.0693)	F

Fig. 14-65 Range of spacers of different thickness with identifying marks and part numbers. *(Chrysler Corporation)*

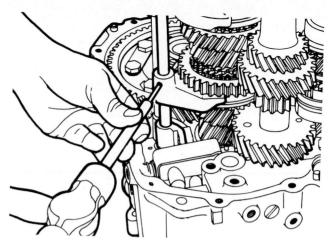

Fig. 14-68 Driving in the spring pin. *(Chrysler Corporation)*

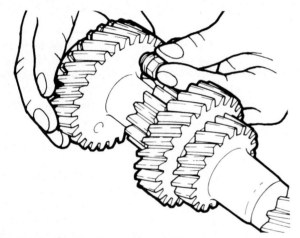

Fig. 14-66 Setting the subgear. *(Chrysler Corporation)*

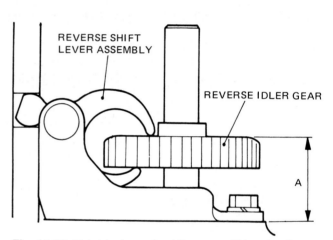

REVERSE SHIFT LEVER ASSEMBLY

REVERSE IDLER GEAR

A

Fig. 14-69 Height of reverse idler gear. *(Chrysler Corporation)*

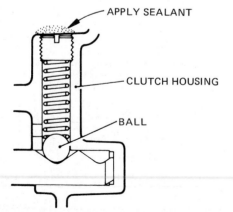

APPLY SEALANT

CLUTCH HOUSING

BALL

Fig. 14-67 Installing selector-shaft poppet. *(Chrysler Corporation)*

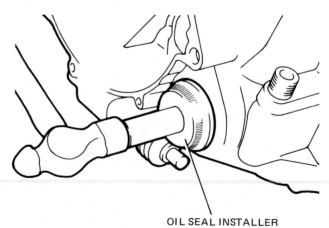

OIL SEAL INSTALLER

Fig. 14-70 Installing oil seal. *(Chrysler Corporation)*

OIL SEAL INSTALLER

Fig. 14-71 Chamfering the oil-seal hole. *(Chrysler Corporation)*

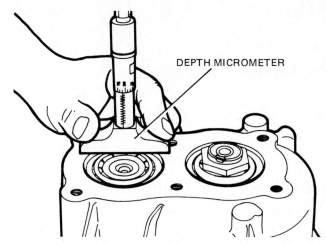

DEPTH MICROMETER

Fig. 14-72 Checking end play. *(Chrysler Corporation)*

12. Install the selector-shaft poppet ball, spring, and plug (Fig. 14-67). The plug must be turned in until its end is flush with the housing surface. Then apply sealant on top of the plug, as shown in Fig. 14-67.
13. Install the input-shaft-bearing retainer.
14. Pull out the subgear bar or bolt holding the subgear in place.
15. Install the output-shaft assembly.
16. Insert two interlock plungers into the housing. Insert an interlock plunger into the third-and-fourth-speed shift rail.
17. Install the first-and-second-speed and the third-and-fourth-speed shift rails and forks (Figs. 14-2 and 14-57).
18. With the pin holes in the shift fork and the shift rail aligned, install a spring pin with a flat punch (Fig. 14-68). Be sure the slit in the spring pin lines up with the center line of the shift rail. Check that the pin is centered in the rail.
19. Insert the reverse-shift rail.
20. Install three poppet balls, three springs, and three plugs. The poppet spring with a white-paint identification mark goes into the poppet hole of the reverse-shift rail. The other two holes take springs 0.740 inch [19 mm] in length. Install the springs with the small-diameter end toward the steel balls.
21. Install the reverse-shift-lever assembly.
22. Install the reverse idler gear and shaft (Fig. 14-69). Measure dimension A in Fig. 14-69. If it is less than standard, replace the reverse-shift-lever assembly.
23. Apply the specified sealant to the clutch-housing side of the transaxle-case gasket and lay the gasket on the clutch housing, lining up the bolt holes. Apply sealant to the transaxle side of the gasket.
24. Install the correct spacer (item 9, above).
25. Put the transaxle case in place and install the attaching bolts, tightened to the specified torque.
26. Install the intermediate-shaft and output-shaft rear tapered-roller-bearing races, pressing them in firmly by hand until they seat.
27. Use the special oil-seal installer as shown in Fig. 14-70 to press in the oil seal. If the oil-seal hole has not been chamfered (see Fig. 14-30), chamfer it with the knurls on the back of the oil-seal-installing tool (Fig. 14-71).

Careful: If the hole is not chamfered, the oil-seal outer surface will be damaged and oil leaks will result. Do not allow any chips to enter the housing. Put some grease on the knurls and clean them repeatedly to keep the chips from falling into the housing.

28. Select the spacer for the outer race of the rear bearing as follows. Make sure the race has been pressed in properly. Use a depth micrometer as shown in Fig. 14-72 to measure how far below the surface the race is. Select a spacer 0.004 inch [0.1 mm] thicker than the measured distance and install it.
    a. Install the rear cover and tighten the bolts to the specified torque.
    b. Use the oil-seal installer as shown in Fig. 14-73 to turn the control shaft. Shift to all speeds.
    c. Turn the input shaft several times. It may turn hard because of the spacer that was installed. Temporarily install the clutch friction disk to make it easier to turn the shaft.
    d. Remove the cover and the spacer. Remeasure the depth of the race with a depth micrometer (Fig. 14-72). Using the thicker spacer in the manner ex-

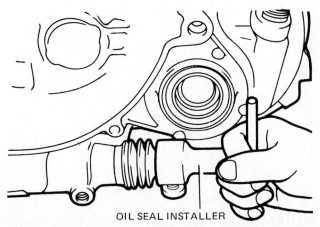

OIL SEAL INSTALLER

Fig. 14-73 Shifting the control shaft to check shifting action of transaxle. *(Chrysler Corporation)*

plained above settles the race into its final position. The depth may now be greater. Select a spacer of the proper thickness so that the specified end play will be achieved.

29. Apply sealant to the threads of the rear cover bolts. Install the bolts and tighten them to the specified torque.

30. Install the steel ball and backup-light switch. Be sure the washer is in the proper position.

31. Use the oil-seal installer (Fig. 14-73) to turn the control shaft and make sure that the shifts are made smoothly.

32. After installation of the transaxle on the vehicle and adjustment of the linkages (Chap. 12), road test the vehicle to make sure the transaxle is working properly.

Chapter 14 review questions

Select the *one* correct, best, or most probable answer to each question. Then check your answers against the correct answers given at the end of the book.

1. When disassembling the transaxle, the first parts to be removed are the:
   a. clutch housing
   b. input shaft
   c. clutch-operating bracket and transaxle mounting bracket
   d. engine mounting bracket and clutch housing

2. The first parts to be removed from the single-range input shaft are the:
   a. first-and-reverse gear and bearing
   b. input and reverse drive gears
   c. output gear and bearing
   d. snap ring and front bearing

3. The only parts that can be removed from the intermediate gear (countergear assembly) are the:
   a. bearings, subgear, and spring
   b. bearings, three gears, and washers
   c. bearings and reverse gear
   d. gears and snap rings

4. The first thing to be removed from the output-shaft assembly (Fig. 14-40) is the:
   a. tapered-roller bearing
   b. locknut

   c. first-speed gear and sleeve
   d. shifter fork

5. After removing the ball bearings from the differential case, the next thing to remove during complete disassembly is the:
   a. case
   b. differential pinions
   c. drive (ring) gear
   d. side gears

6. During reassembly of the differential, check the backlash between the:
   a. ring gear and drive gear
   b. differential side gear and pinion
   c. spacers and pinions
   d. side gear and drive gear

7. The dual-range transaxle has:
   a. one synchronizer
   b. two synchronizers
   c. three synchronizers
   d. four synchronizers

8. Bearings that have been removed from any of the shafts:
   a. should be cleaned and reused if okay
   b. should be discarded
   c. should have new rollers on reinstallation
   d. should have new races on reinstallation

TRANSFER-CASE TROUBLE DIAGNOSIS, REMOVAL, AND INSTALLATION

After studying this chapter, and with proper instruction and equipment, you should be able to:

1. Describe and list the four basic complaints about transfer cases, and explain what might cause each.
2. Remove and install a transfer case.
3. Adjust transfer-case linkages.

 15-1 Preparing for transfer-case repair This chapter discusses the various troubles that a transfer case might have, and what could cause each. Then how to remove a transfer case and reinstall it is explained. Also described is the procedure of adjusting the transfer-case linkages. Chapter 16 describes transfer-case service, including how to disassemble and reassemble transfer cases.

✿ 15-2 Transfer-case trouble diagnosis The chart that follows lists transfer-case complaints, possible causes, and checks or corrections. Typical troubles with transfer cases include excessive noise, shift lever hard to move, gears slip out of engagement, and leaking lubricant.

NOTE: See ✿ 3-1 for a discussion of diagnosis theory.

✿ 15-3 Transfer-case removal The purpose and operation of the transfer case is covered in ✿ 4-17. Removal is as follows:

1. Raise the vehicle on a lift. Drain the lubricant from the transfer case (✿ 15-5). Transfer case attachment points are shown in Fig. 15-1.
2. Disconnect the speedometer cable and remove the skid plate and cross-member supports (Fig. 15-2) as necessary. On cars with automatic transmissions, remove the strut rod (Fig. 15-3).
3. Disconnect both the rear driveshaft and the front driveshaft from the transfer case. Tie them up out of the way.
4. Disconnect the shift lever rod from the shift rail link. On some vehicles, the shift levers are disconnected at the transfer case.
5. Place a transmission jack under the transfer case to

support it. Remove the bolts attaching the transfer case to the transmission adapter. Move the transfer case to the rear until the input shaft clears the adapter. Lower the transfer case from the vehicle and take it to the work bench.

✿ 15-4 Installing the transfer case Support the transfer case with a transmission jack. Raise the transfer case into position and slide it forward so that the input shaft enters properly. Move the transfer case up against the transmission adapter. Install the bolts attaching the case to the adapter and torque to specifications. Remove the jack and proceed as follows:

1. Install the connecting rod to the shift rail link or connect the shift levers to the transfer case, according to the design. On the GM Model 203 full-time transfer case, be sure the nylon spacer is in place before installing the levers.
2. Connect the front and rear driveshafts.
3. Install the cross-member support and skid plate, if removed.
4. Connect the speedometer cable.
5. Fill the transfer case to the proper level with the specified lubricant (✿ 15-5).
6. Check and adjust the shift linkage (✿ 15-6).
7. Lower the vehicle. Road test it to make sure the transfer case is operating properly.

✿ 15-5 Transfer-case lubrication The transfer-case lubricant should be changed at the intervals recommended by the manufacturer. The frequency of change depends on the type of operation. For example, one manufacturer recommends changing the oil every 24,000 miles [38,624 km] for normal off-on road work. For heavy-duty work, such as snowplowing or pulling

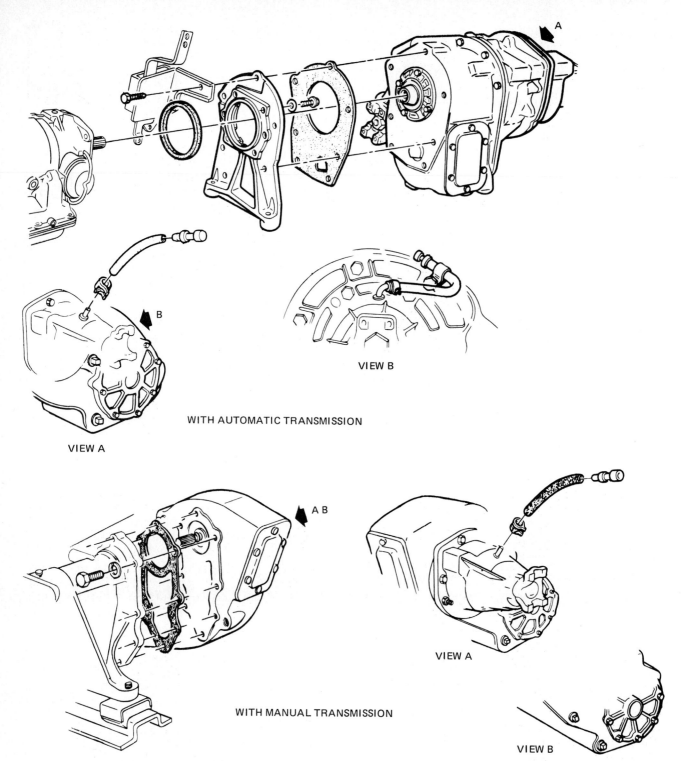

Fig. 15-1 Transfer-case attachments for both manual- and automatic-transmission installations. (*Chevrolet Motor Division of General Motors Corporation*)

a trailer, change the oil every 12,000 miles [19,312 km]. For severe use, change the oil every 1000 miles [1609 km]. If the vehicle is used in very severe work, where the transfer case is submerged, the oil should be changed every day.

Always use the type of oil or lubricant recommended by the manufacturer. Figure 15-4 shows the items to be checked. The oil change is done as follows:

1. Operate the vehicle on a rough road to agitate the lubricant so that it reaches normal operating temperature.

Transfer-Case Trouble-Diagnosis Chart

COMPLAINT	POSSIBLE CAUSE	CHECK OR CORRECTION
1. Excessive noise	a. Lubricant level low	Fill as required
	b. Worn or damaged bearings	Replace
	c. Worn or damaged chain	Replace
	d. Misalignment of driveshafts or universal joints	Align
	e. Yoke bolts loose	Torque to specifications
	f. Loose adapter bolts	Torque to specifications
2. Shift lever difficult to move	a. Dirt or contamination on linkage	Clean and lubricate
	b. Binding inside transfer case	Repair as required
3. Gears disengage from position	a. Linkage misadjusted or loose	Readjust or tighten
	b. Gears worn or damaged	Replace
	c. Shift rod bent	Replace
	d. Missing detent ball or spring	Replace
4. Lubricant leaking	a. Excessive lubricant in case	Adjust level
	b. Leaking seals or gaskets	Replace
	c. Loose bolts	Tighten
	d. Scored yoke in seal-contact area	Refinish or replace

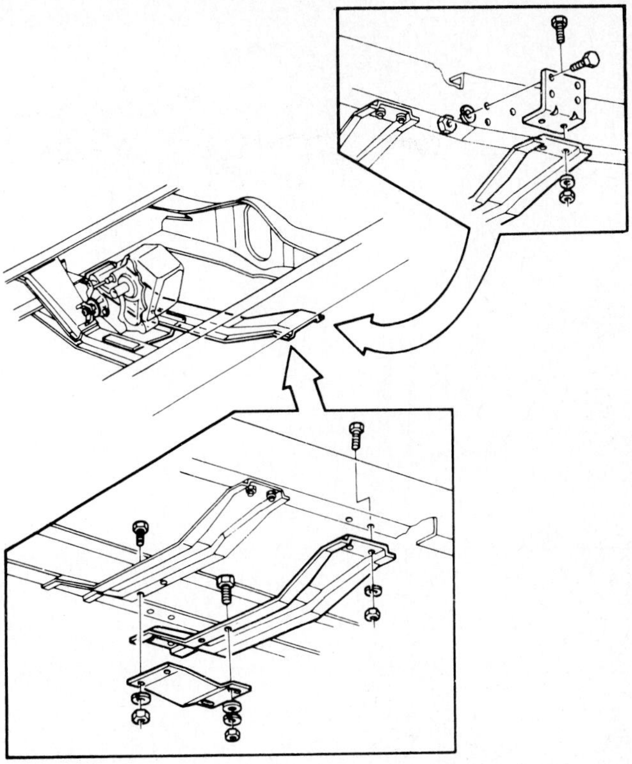

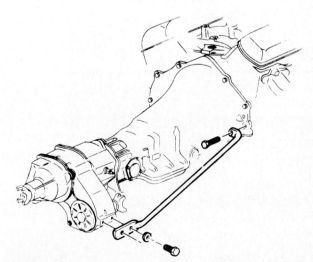

Fig. 15-2 Skid plates. (*Chevrolet Motor Division of General Motors Corporation*)

plug opening. Install the plug. Then wipe the surfaces of the case and the skid plate to remove excess oil.

9. Lower the vehicle to the floor.

✿ **15-6 Transfer-case-linkage adjustment** Figure 15-5 shows the linkage for the Model 203 full-time transfer case used in many General Motors trucks and other

2. Raise the vehicle on a lift. Remove the lubricant filler plug. Have a container in place to catch the lubricant.

3. Remove the lowest bolt from the front-output-shaft rear bearing (A in Fig. 15-4). Allow the lubricant to drain.

4. Remove the six bolts holding the power-takeoff (PTO) cover in place and remove the cover (B in Fig. 15-4).

5. Remove the speedometer driven gear (location C in Fig. 15-4).

6. Use a suction gun at locations B and C to remove as much of the lubricant as possible.

7. Install the speedometer gear, the power-takeoff cover, and the lower bolt removed in item 3, above.

8. Add about seven pints [3.3 liters] of the recommended oil through the filler-plug opening. Check the fluid level and add more oil if necessary to raise the level to about ½ inch [13 mm] below the filler-

Fig. 15-3 Strut rods. (*Chevrolet Motor Division of General Motors Corporation*)

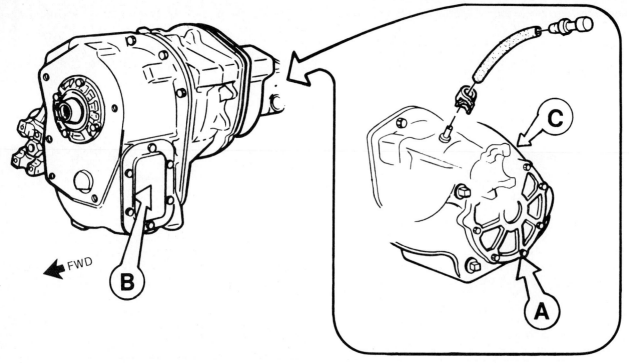

Fig. 15-4 Model 203 full-time transfer case oil-change locations. (*Chevrolet Motor Division of General Motors Corporation*)

vehicles. Refer to this illustration as we explain how to make the linkage adjustment.

1. With rods C and H removed (Fig. 15-5), align the gauge holes in levers A and B with the gauge hole in the shifter assembly and insert gauge pin J. This positions levers A and B in neutral.

2. Position arms F and G in the straight down "six-o'clock" position.
3. With swivel E and locknuts D loosely assembled in rod C, rotate the swivel until the ends of rod C will enter both lever B and arm A at the same time.
4. Lock rod C in place with retainer K.
5. Tighten locknuts D against swivel E to specified

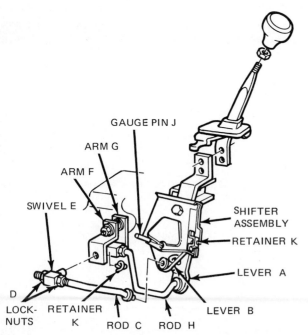

Fig. 15-5 Linkage adjustment on Model 203 full-time transfer case. (*Chevrolet Motor Division of General Motors Corporation*)

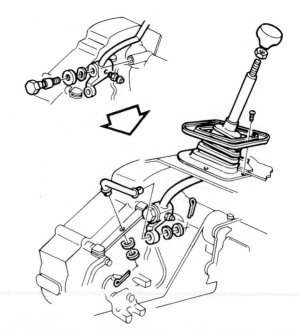

Fig. 15-6 Linkage adjustment on Model 205 part-time transfer case. (*Chevrolet Motor Division of General Motors Corporation*)

torque. Be careful not to change the position of arm F.

6. Repeat steps 3, 4, and 5 for rod H when installing the rod to lever A and arm G.
7. Remove gauge pin J.

Figure 15-6 shows the linkage for the Model 205 part-time transfer case. This linkage is simpler than the linkage for the full-time transfer case.

Chapter 15 review questions

Select the *one* correct, best, or most probable answer to each question. Then check your answers against the correct answers given at the end of the book.

1. Excessive noise from the transfer case could be caused by:
   a. low lubricant level
   b. misaligned driveshafts
   c. worn bearings
   d. all of the above
2. If the shift lever is difficult to move, it could be caused by:
   a. excessive lubricant in the case
   b. binding inside the case
   c. worn gears
   d. a defective synchronizer

3. If the shift lever will not stay in position, the trouble could be:
   a. linkage loose or out-of-adjustment
   b. gears worn or damaged
   c. missing detent ball or spring
   d. all of the above
4. The purpose of the skid plate is to:
   a. prevent the vehicle from skidding
   b. protect the transfer case from rocks or other snags it might encounter in rough areas
   c. help support the transfer case
   d. strengthen the frame
5. Leaks may be caused by:
   a. excessive lubricant in the case
   b. leaking seals or gaskets
   c. loose bolts
   d. all of the above

TRANSFER-CASE SERVICE

After studying this chapter, and with proper instruction and equipment, you should be able to:

1. Disassemble and reassemble a transfer case.
2. Identify the parts of a disassembled transfer case.

16-1 Servicing the transfer case This chapter describes the disassembly and reassembly of a full-time transfer case (discussed in ✿ 4-17). The unit has a differential so that it can be used full time without undue wear of the tires. Without a differential, the front wheels and rear wheels would have to turn at the same speed. As explained in ✿ 4-17 and shown in Fig. 4-55, the front wheels move in a wider arc than the rear wheels when a turn is made. Without a differential, the tires would skid during turns to take care of this difference in travel. With the differential, the front wheels can rotate faster than the rear wheels when a turn is made.

There are simpler transfer cases. But if you can service the full-time transfer case described in this chapter, you will be able to handle the simpler units.

✿ 16-2 Full-time transfer-case disassembly Figure 16-1 is a cutaway view of the full-time transfer case. Figure 16-2 is a completely disassembled view of the unit. These illustrations will be your guides as we describe the disassembly and reassembly of the transfer case. Figures 16-3 and 16-4 are front and rear views of the unit.

1. Position the transfer case on the work bench. If the lubricant has not been drained, remove the lower bolts from the front-output-bearing rear cover and the power-takeoff cover. Allow the lubricant to drain into a suitable container.
2. Use the companion-flange remover as shown in Fig. 16-5 and loosen the rear-output-shaft-flange retaining nut.
3. Use the same tool to remove the front-output-shaft-flange retaining nut, washer, and flange. Tap the dust shield rearward away from the bolts to get enough clearance to remove the bolts from the flange.
4. Remove the bolts holding the output-shaft front-bearing retainer (Fig. 16-6). Remove the bearing retainer and gasket from the case. Discard the gasket.

5. Use a hoist or other lifting device to position the assembly on blocks (Fig. 16-7).
6. Remove the bolts attaching the rear section of the rear output housing to the front section and separate the sections. Remove the shims and speedometer gear from the output shaft.
7. Remove the bolts attaching the front section of the rear output housing to the transfer case. Remove the housing from the case.
8. Remove the O-ring seal from the front section of the rear output housing and discard it.
9. Disengage the rear output shaft from the differential-carrier assembly. Slide the carrier from the shaft.
10. Install a water-hose clamp on the input shaft to keep from losing the bearings when the input shaft is removed from the range box (9 in Fig. 16-2).
11. Raise the shift rail and drive out the pin retaining the shift fork to the rail (Fig. 16-8).
12. Remove the shift-rail-poppet-ball plug, gasket spring, and ball from the case (Fig. 16-9).
13. Push the shift rail down, lift up on the lockout clutch, and remove the shift fork from the clutch assembly.
14. Remove the bolts attaching the front-output-shaft rear-bearing retainer to the transfer case. Tap on the front of the shaft or carefully pry the retainer away from the case. Remove the retainer from the shaft and discard the gasket. Recover any rollers which may fall from the rear cover.
15. If it is necessary to replace the rear bearing, support the cover and press the bearing from the cover. Press the new bearing into place leaving a 0.060-inch [1.5-mm] overhang.
16. From the lower side of the case remove the output-shaft front bearing. You may have to pry this off.
17. Disengage the front output shaft from the chain and remove the shaft from the transfer case (Fig. 16-10).
18. Remove the bolts attaching the intermediate-chain housing to the range box. Lift or use a chain hoist to lift the intermediate housing from the range box (Fig. 16-11).

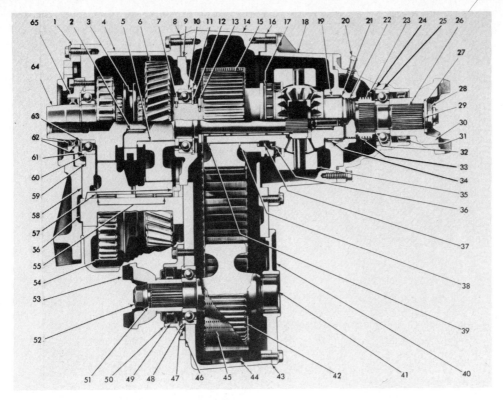

1. Adapter
2. Input Drive Gear Pilot Brgs.
3. Range Selector Sliding Clutch
4. Range Selector Housing
5. Low Speed Gear Bushing
6. Low Speed Gear
7. Thrust Washer & Locating Pin
8. Gasket
9. Input Brg. Retainer
10. Input Brg.
11. Brg. Outer Ring
12. Brg. Retaining Ring
13. Thrust Washer, Locating Pin, Lubricating Washer & Spacer
14. Intermediate (Chain) Housing
15. Drive Shaft Sprocket
16. Gasket
17. Sliding Lock Clutch
18. Rear Output Housing
19. Rear Output Front Brg.
20. Vent

21. Oil Seal
22. Oil Pump
23. Speedometer Drive Gear
24. Brg. Retainer Ring
25. Rear Outut Rear Brg.
26. Rear Output Shaft
27. Washer
28. Locknut
29. Rubber Spline Seal
30. Rear Output Yoke
31. Rear Output Seal
32. Shims
33. Input Shaft "O" Ring
34. Input Shaft Pilot Bearings
35. Differential Carrier Assembly
36. Spring Washer Cup
37. Lockout Clutch Spring
38. Snap Ring
39. Snap Ring
40. Front Output Rear Brg. Cover
41. Front Output Rear Brg.
42. Front Output Drive Sprocket
43. Gasket

44. Magnet
45. Drive Chain
46. Gasket
47. Brg. Outer Ring
48. Front Output Front Brg.
49. Front Output Shaft Seal
50. Front Output Brg. Retainer
51. Rubber Spline Seal
52. Locknut
53. Front Output Yoke
54. Countergear
55. Countergear Spacers and Brgs.
56. Countergear Shaft
57. Countergear Thrust Washer
58. Gasket
59. Brg. Retainer Gasket
60. Brg. Outer Ring
61. Input Gear Brg.
62. Input Gear Seals (2)
63. Brg. Snap Ring
64. Input Gear
65. Input Gear Brg. Retainer

Fig. 16-1 Sectional view of Model 203 full-time transfer case. *(Chevrolet Motor Division of General Motors Corporation)*

19. Remove the chain from the intermediate housing.
20. Remove the lockout clutch, drive gear, and input-shaft assembly from the range box. If a host clamp is installed on the end of the input shaft, it will prevent your losing the bearing rollers which may fall out of the clutch assembly if it is pulled off the input shaft.
21. Pull up on the shift rail and disconnect the rail from the link.
22. Remove (lift off) the input-shaft assembly from the range box. The transfer case is now completely disassembled into subassemblies. These subassemblies should be disassembled for cleaning and inspection of the parts.

☀ 16-3 Cleaning and inspection of parts The procedures for cleaning and checking the condition of bearings and other parts, such as shafts, gears, cases,

covers, and housings, are covered in detail in ☀ 7-4 to 7-6.

☀ 16-4 Differential-carrier service Figure 16-2 (bottom) shows the differential carrier completely disassembled (parts numbered 75 to 92). This includes the rear output shaft, which has a side gear that is meshed with the differential pinions. Note that the other side gear in the differential is on the end of the front output shaft (73).

This differential is different from other differentials previously described in this book (☀ 1-11). The front side gear (73) carries power into the differential. The power passes through the differential to the rear output shaft and from there to the rear wheels. Power is also flowing to the front wheels through the chain drive and sprockets.

If both the front wheels and the rear wheels are

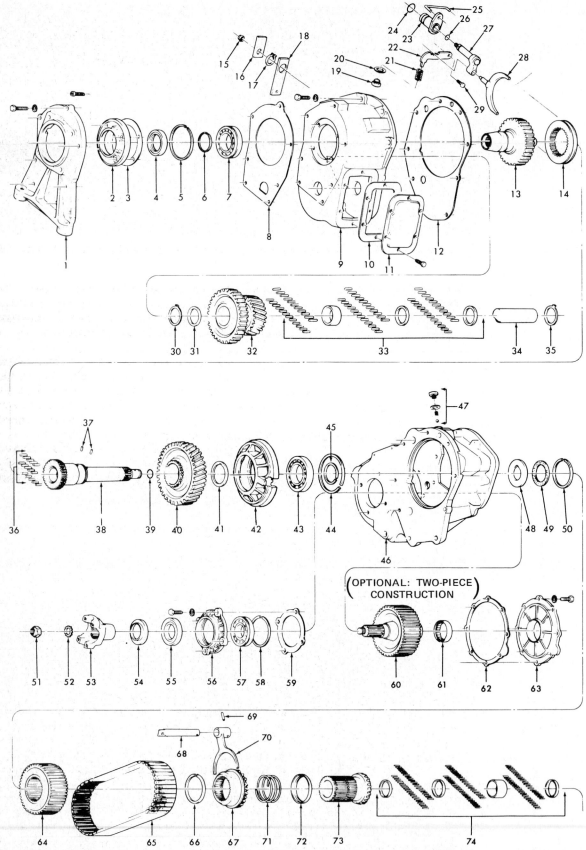

Fig. 16-2 Disassembled Model 203 full-time transfer case. *(Chevrolet Motor Division of General Motors Corporation)*

(OPTIONAL: TWO-PIECE CONSTRUCTION)

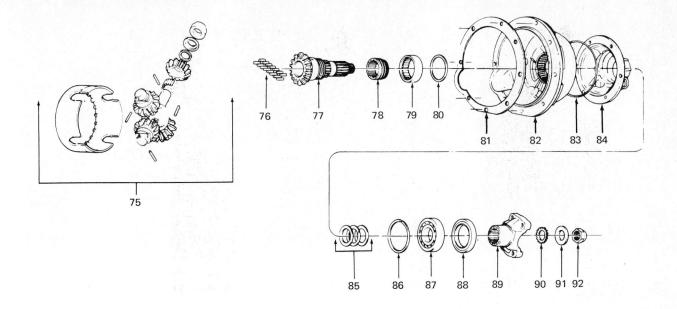

1. Adapter
2. Input gear bearing retainer
3. Input gear bearing retainer gasket
4. Input gear seals
5. Bearing outer ring
6. Bearing to shaft retaining ring
7. Input gear bearing
8. Adapter to selector housing gasket
9. Range selector housing (range box)
10. P.T.O. cover gasket
11. P.T.O. cover
12. Selector housing to chain housing gasket
13. Main drive input gear
14. Range selector sliding clutch
15. Shift lever lock nut
16. Range selector shift lever
17. Shift lever retaining ring
18. Lockout shift lever
19. Detent plate spring plug
20. Detent plate spring plug gasket
21. Detent plate spring
22. Detent plate
23. Lockout shifter shaft
24. "O" ring seal
25. Lockout shaft connector link
26. "O" ring seal
27. Range selector shifter shaft
28. Range selector shift fork
29. Detent plate pivot pin
30. Thrust washer
31. Spacer (short)
32. Range selector counter gear
33. Countergear roller bearings and spacers (72 bearings req'd.)
34. Countergear shaft
35. Thrust washer
36. Input shaft roller bearings (15 req'd.)
37. Thrust washer pins (2 req'd.)
38. Input shaft
39. O-ring seal
40. Low speed gear and bushing
41. Thrust washer
42. Input shaft bearing retainer
43. Input shaft bearing
44. Input shaft bearing retaining ring (large)
45. Input shaft bearing retaining ring
46. Chain drive housing

47. Lockout shift rail poppet plug, gasket, spring and ball.
48. Thrust washer
49. Lubricating thrust washer
50. Retaining ring
51. Flange lock nut
52. Seal
53. Front output yoke
54. Dust shield
55. Front output shaft seal
56. Front otuput shaft bearing retainer
57. Front output shaft bearing
58. Bearing outer ring
59. Bearing retainer gasket
60. Front output shaft
61. Front output shaft rear bearing
62. Front output rear bearing retainer cover gasket
63. Front output rear bearing retainer
64. Drive shaft sprocket
65. Drive chain
66. Retaining ring
67. Sliding lock clutch
68. Lockout shift rail
69. Shift fork retaining pin
70. Lockout shift fork
71. Lockout clutch spring
72. Spring washer cup
73. Front side gear
74. Front side gear bearing and spaces (123 bearings req'd.)
75. Differential carrier assembly (132 bearings req'd.)
76. Rear output shaft roller bearings (15 req'd.)
77. Rear output shaft
78. Speedometer drive gear
79. Rear output shaft front roller bearing
80. Oil pump O-ring seal
81. Rear output housing gasket
82. Rear output housing (front)
83. O-ring seal
84. Rear output housing (rear)
85. Shim pack
86. Bearing retainer
87. Rear output rear bearing
88. Rear output shaft seal
89. Rear output flange
90. Rear output shaft rubber seal
91. Washer
92. Flange nut

Fig. 16-2 *(Continued)*

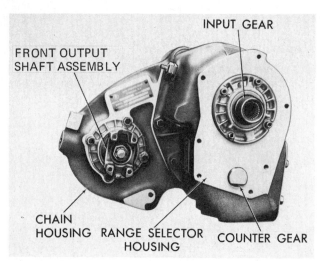

Fig. 16-3 Front view of transfer case. *(Chevrolet Motor Division of General Motors Corporation)*

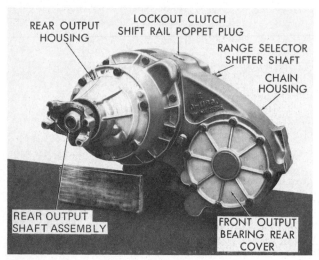

Fig. 16-4 Rear view of transfer case. *(Chevrolet Motor Division of General Motors Corporation)*

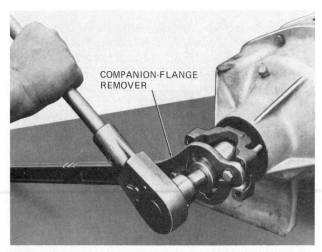

Fig. 16-5 Removing rear-output-shaft-flange nut. *(Chevrolet Motor Division of General Motors Corporation)*

Fig. 16-6 Removing front-output-shaft front-bearing retainer. *(Chevrolet Motor Division of General Motors Corporation)*

turning at the same speed (straight-ahead driving), the differential is turning as a solid unit. However, if the front wheels turn faster than the rear wheels, the differential pinions turn on their shafts to compensate. They reduce the speed of the rear output shaft (77) as compared to the front side gear (73) and front output shaft (60). Chapter 19 describes differentials in detail.

Disassemble the differential carrier as follows:

1. Remove the bolts from the carrier assembly and separate the carrier sections.
2. Lift the pinion-gear-and-spider assembly from the carrier. Observe that the undercut side of the pinion-gear spider faces toward the front side gear.
3. Remove the pinion thrust washers, pinion roller washers, pinions, and roller bearings from the spider.

Fig. 16-7 Positioning the transfer case for disassembly. *(Chevrolet Motor Division of General Motors Corporation)*

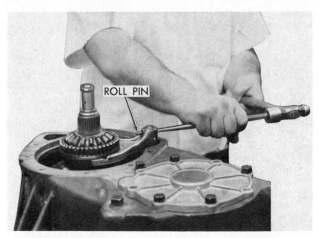

Fig. 16-8 Removing shift-fork retaining pin. *(Chevrolet Motor Division of General Motors Corporation)*

Fig. 16-9 Removing poppet-ball plug. *(Chevrolet Motor Division of General Motors Corporation)*

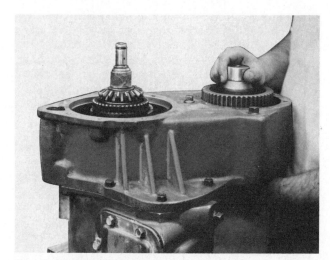

Fig. 16-10 Removing front output shaft. *(Chevrolet Motor Division of General Motors Corporation)*

Fig. 16-11 Removing intermediate (chain) housing. *(Chevrolet Motor Division of General Motors Corporation)*

To reassemble the differential carrier:

1. Clean and inspect all parts to make sure they are in good condition. Replace defective parts.
2. Use petroleum jelly to load the rollers in the pinions. Put 33 rollers in each pinion.
3. Install the pinion roller washer, pinion gear, roller washer, and thrust washer on each leg of the spider.
4. Put the spider assembly in the front half of the carrier, with the undercut surface of the spider thrust face facing downward toward the gear teeth.
5. Align the marks on the carrier sections and bring the two halves together. Install the retaining bolts and tighten to specifications.

✿ **16-5 Lockout-clutch-assembly service** The parts in this assembly are numbered 66 to 74 in Fig. 16-2. To disassemble it:

1. Remove the front side gear from the input-shaft assembly and remove the thrust washer, 123 rollers, and the spacers from the front-side-gear bore. Note the positions of the spacers so that you can put them back properly.
2. Use snap-ring pliers to remove the snap ring retaining the drive sprocket to the clutch assembly. Slide the drive sprocket from the front side gear.
3. Remove the lower snap ring.
4. Remove the sliding gear, spring, and spring cup washer from the front side gear.

After cleaning and inspecting all parts and replacing worn or damaged parts, reassemble the lockout clutch as follows:

1. Install the spring cup washer, spring, and sliding clutch gear on the front side gear.

2. Install the snap ring retaining the slide clutch to the front side gear.
3. Use petroleum jelly to load the front side gear with 123 rollers and the spacers.
4. Install the thrust washer in the gear end of the front side gear.
5. Slide the drive sprocket onto the clutch splines and install the retaining ring.

✿ 16-6 Input-shaft-assembly service The parts in this assembly are numbered 36 to 44 in Fig. 16-2. Disassemble as follows:

1. Slide the thrust washer and spacer from the shaft.
2. Use snap-ring pliers to remove the snap ring retaining the input-bearing retainer to the shaft (Fig. 16-12). Slide the bearing retainer off the shaft.
3. Support the low-speed gear and tap the shaft from the gear and thrust washer. Note the positions of the thrust-washer pins in the shaft.
4. Use a screwdriver to pry open the large snap ring in the bearing retainer (Fig. 16-13) and remove it. Then remove the bearing from the retainer.
5. Remove the 15 rollers from the end of the input shaft.
6. Remove the O ring from the end of the shaft and discard it.

After cleaning and inspecting all parts, and replacing any worn or defective parts, reassemble as follows:

1. Position the bearing in the retainer and tap or press it into place. Ball-loading slots should face the concave (dished in) side of the retainer.
2. Install the large snap ring to hold the bearing in place. Note that the snap ring is a select fit. Use the largest size you can to provide a tight fit.
3. Install the low-speed gear on the shaft with the clutch end toward the gear end of the shaft.
4. Position the thrust washer on the shaft, aligning the

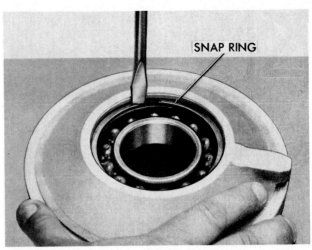

Fig. 16-13 Removing input-shaft-bearing-to-retainer snap ring. *(Chevrolet Motor Division of General Motors Corporation)*

slot in the washer with the pin in the shaft. Slide or tap the washer into place.
5. Position the input-bearing retainer on the shaft and install a snap ring to hold the bearing in place on the shaft. The snap ring is a select fit. Use the largest size to provide the tightest fit.
6. Slide a spacer and thrust washer onto the shaft. Align the spacer with a locater pin.
7. Use petroleum jelly to install the 15 rollers in the end of the shaft.
8. Install a new rubber O ring on the end of the shaft.

✿ 16-7 Range-selector-housing (range box) service The range box assembly includes the parts numbered 1 through 35 in Fig. 16-2. Proceed as follows to disassemble the range-selector housing.

1. Remove the poppet-plate spring, plug, and gasket. Discard the gasket.
2. Disengage the sliding clutch gear from the input gear and remove the clutch fork and sliding gear from the housing.
3. Remove the shift-lever-assembly retaining nut and the upper shift lever from the shifter shaft.
4. Remove the shift-lever snap ring and lower lever.
5. Push the shifter-shaft assembly downward and remove the lockout-clutch-connector link. The long end of the connector link engages the poppet plate.
6. Remove the shifter-shaft assembly from the housing and separate the inner and outer shifter shafts. Remove and discard the O rings.
7. If the poppet plate is damaged, drive the pivot shaft from the housing so that you can remove the poppet plate and spring.
8. Remove the input-gear-bearing-retainer-and-seal assembly. Discard the gasket.
9. Remove the large snap ring and tap the input gear and bearing from the housing.
10. To remove the bearing from the input gear, remove the retaining snap ring. This is a select-fit snap ring. Select the snap ring giving the tightest fit on reassembly.

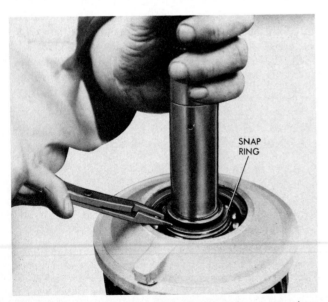

Fig. 16-12 Removing input-shaft-to-bearing snap ring. *(Chevrolet Motor Division of General Motors Corporation)*

11. Working from the intermediate-case side, use a special tool as shown in Fig. 16-14 to remove the countershaft. Then remove the cluster gear and washers from the housing. Recover the 72 rollers from the housing and shaft.

After cleaning and inspecting all parts, and replacing any worn or defective parts, reassemble as follows:

1. Use a dummy shaft and petroleum jelly to install 72 rollers and spacers in the cluster-gear bore.
2. Use petroleum jelly to hold the countershaft thrust washers in place and position them in the housing. Engage the tabs on the washers with the slots in the housing thrust surface.
3. Position the cluster-gear assembly in the housing and install the countershaft through the front face of the housing and into the gear assembly. The countershaft face with the flat should face forward and must align with the gasket.
4. Install the bearing, without the large snap ring, on the input shaft with the snap-ring groove outward. Install a new retaining ring on the shaft. Position the input gear and bearing in the housing. The retaining ring is a select fit. Use the thickest ring you can to produce a tight fit.
5. Install a snap ring on the outside diameter of the bearing.
6. Align the oil slot in the retainer with the drain hole in the housing. Install the input-gear-bearing retainer, gasket, and retaining bolts. Tighten the bolts to specifications.
7. If the poppet plate was removed, install the plate and pivot pin. Use sealant on the pin.
8. Install new O rings on the inner and outer shifter shafts. Lubricate them and assemble the inner shaft in the outer shaft.
9. Push the shafts into the housing, engaging the long end of the lockout-clutch-connector link with the outer shifter shaft before the shaft assembly bottoms.

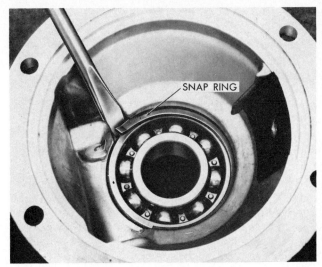

Fig. 16-15 Removing rear-output-bearing retainer ring. *(Chevrolet Motor Division of General Motors Corporation)*

10. Install the lower shift lever and retaining ring.
11. Install the upper shift lever and shifter-shaft retaining nut.
12. Install the shift fork and sliding clutch gear. Push the fork up into the shifter-shaft assembly to engage the poppet plate, sliding the clutch gear forward onto the input-shaft gear.
13. Install the poppet spring, gasket, and plug in the top of the housing. Check the spring engagement with the poppet plate.

❂ 16-8 Rear-output-shaft-housing service This part is numbered 82 in Fig. 16-2. The parts associated with it are numbers 76 to 92. The housing has two bearings to support the rear output shaft (77). These

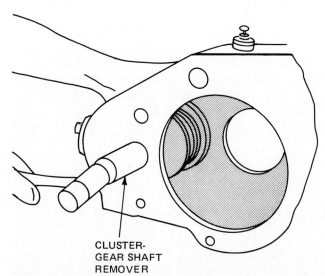

Fig. 16-14 Removing cluster-gear shaft. *(Chevrolet Motor Division of General Motors Corporation)*

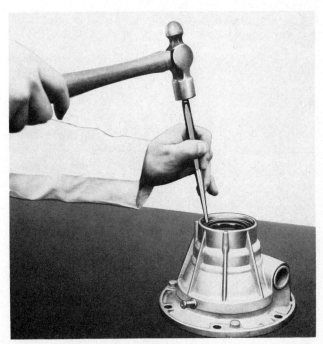

Fig. 16-16 Removing rear-output-shaft front bearing. *(Chevrolet Motor Division of General Motors Corporation)*

are the ball bearing at the rear (Fig. 16-15), and a roller bearing (Fig. 16-16). To replace the bearings, proceed as follows:

1. Remove the speedometer-driven gear.
2. Pry the old seal out with a screwdriver or other suitable tool.
3. Pry the rear-bearing snap ring out with a screwdriver (Fig. 16-15).
4. Pull or tap the bearing from the housing.
5. To remove the front bearing, use a punch as shown in Fig. 16-16. Discard the rubber seal.
6. To install the rear bearing, tap it into place and install a new snap ring. The snap ring is a select fit. Select the snap ring that provides the tightest fit.
7. Install a new seal by driving it into place with the special installer (Fig. 16-17).
8. If the vent seal has come out during disassembly, install a new seal, cementing it into place.

❁ **16-9 Front-output-shaft bearing-seal replacement** First, remove the old seal from the retainer bore. Apply sealant to the outer diameter of the new seal. Then install the new seal in the retainer with the special seal installer (Fig. 16-18).

❁ **16-10 Front-output-shaft rear-bearing replacement** To replace this bearing, remove the rear cover and gasket. Discard the gasket. Support the rear cover and press out the bearing. Put the new bearing in position. Use a piece of wood to cover it and press the bearing in until it is flush with the opening. Position a

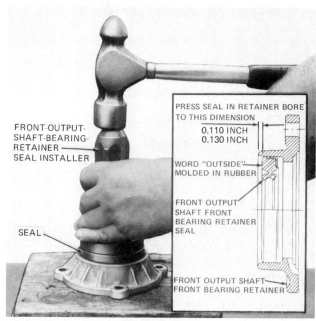

Fig. 16-18 Installing front-output-bearing-retainer seal. *(Chevrolet Motor Division of General Motors Corporation)*

new gasket and the cover and secure with retaining bolts, torqued to specifications.

❁ **16-11 Assembling the transfer case** With all subassemblies reassembled, put the range box on blocks, with the input-gear side toward the bench (Fig. 16-19).

1. Put the range-box-to-transfer-case gasket on the input housing.
2. Install the lockout-clutch-and-drive-sprocket assembly on the input-shaft assembly. Use a hose clamp on the end of the shaft to keep the bearing rollers from coming out.
3. Install the input shaft, lockout clutch, and drive-sprocket assembly in the range box. Align the tab on the bearing retainer with the notch in the gasket.

Fig. 16-17 Installing rear-output-shaft seal. *(Chevrolet Motor Division of General Motors Corporation)*

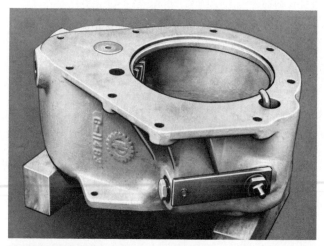

Fig. 16-19 Positioning range box for assembly. *(Chevrolet Motor Division of General Motors Corporation)*

Fig. 16-20 Drive-gear-and-lockout-clutch assembly installed. *(Chevrolet Motor Division of General Motors Corporation)*

4. Connect the lockout-clutch shift rail to the connector link and position the rail in the housing bore (Fig. 16-20). Rotate the shifter shaft while lowering the shift rail into the housing to prevent the link and rail from being disconnected.
5. Put the drive chain in the chain housing, positioning it around the outside of the housing. Install the chain housing on the range box (Fig. 16-21), engaging the shift-rail channel of the housing to the shift rail. Position the chain on the input drive sprocket.

Fig. 16-21 Installing intermediate housing on range box. *(Chevrolet Motor Division of General Motors Corporation)*

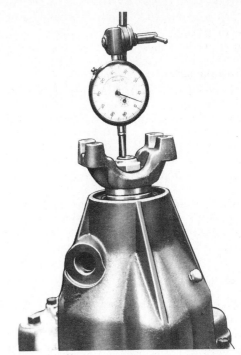

Fig. 16-22 Checking rear-output-shaft end play. *(Chevrolet Motor Division of General Motors Corporation)*

6. Install the front-output sprocket in the case, engaging it with the chain. Rotate the clutch-drive gear to assist in positioning the chain on the drive sprocket.
7 Install the shift fork on the clutch assembly and the shift rail. Then push the clutch assembly fully into the drive sprocket. Install the roll pin retaining the shift fork to the shift rail.
8. Install the front-output-shaft bearing, retainer, gasket, and retaining bolts.
9. Install the front-output-shaft flange, gasket, seal, washer, and retaining nut. Tap the dust shield back into place after installing the bolts in the flange.
10. Install the front-output-shaft rear-bearing retainer, gasket, and bolts. If the rear bearing was removed, position it on the cover and press it into the cover until it is flush with the opening.
11. Install the differential carrier assembly on the input shaft. The carrier bolts should face the rear of the shaft.
12. Position the rear-output shaft on the differential carrier and then proceed as follows to determine the thickness of the shims required (32 in Fig. 16-1):
    a. Install the rear-output housing (front), gasket, and retaining bolts.
    b. Install the speedometer gear and shims (about 0.060 inch [1.5 mm]) on the output shaft.
    c. Position the rear-output-shaft-housing (rear) assembly on the rear-output-shaft housing (front). Be sure the O ring is in position and the vent is upward.
    d. Install the flange, washer, and retaining nut. Leave the nut slightly loose.
13. Install a shim pack on the shaft, in front of the rear

bearing, to control the end play to within 0.001 to 0.005 inch [0.025 to 0.12 mm].

14. Hold the rear flange and rotate the front-output shaft to check for binding of the shaft while checking the end play (Fig. 16-22).

15. Install the speedometer-driven gear in the housing.

16. Install the lockout-clutch shift-rail poppet ball, spring, and screw plug.

17. Install the shift levers on the range-box shifter shaft if they have been removed or left on the vehicle linkage.

18. Fill the transfer case to the proper level with the specified lubricant. Tighten the filler plug to specifications.

19. Adjust linkages (✱ 15-6) and road test the vehicle to make sure the transfer case is working properly.

Chapter 16 review questions

Select the *one* correct, best, or most probable answer to each question. Then check your answers against the correct answers given at the end of the book.

1. The purpose of installing a water-hose clamp on the input shaft during disassembly is to:
   a. keep the shaft from falling out
   b. keep from losing the bearings
   c. hold the gears in place
   d. prevent the two sides of the case from falling apart

2. The differential used in the transfer case is different from the other differentials because it has, including the rear output shaft, only
   a. four gears
   b. five gears
   c. six gears
   d. three gears

3. The purpose of using petroleum jelly when loading the rollers into the pinions is to:
   a. provide initial lubrication
   b. keep parts from rusting
   c. hold the pinions in place
   d. hold the bearing rollers in place

4. The transfer case described in this chapter is called a full-time unit because:
   a. it is on the vehicle full time
   b. it works full time
   c. it sends power to both the front and rear wheels all the time
   d. the vehicle is a heavy-duty unit

5. The transfer case discussed in this chapter has:
   a. one shift fork
   b. two shift forks
   c. three shift forks
   d. no shift forks

DRIVE-LINE CONSTRUCTION AND OPERATION

After studying this chapter, and with proper instruction and equipment, you should be able to:

1. Explain the purpose of drive lines.
2. Name the two basic types of joints in a drive line and explain the purpose, construction, and operation of each.
3. Describe the differences between the drive lines for rear-drive cars and those for the front-drive cars.
4. Identify and name the parts of disassembled drive lines.

 17-1 Basic actions in drive lines The drive lines used in both rear-drive and front-drive cars have certain basic actions. One is that the wheels move up and down, changing the angle between the transmission or transaxle and the wheel axles. This also changes the length of the drive line.

In addition, on front-drive cars, the front wheels swing to one side or the other when the car is steered. The driveshafts for front-drive cars must allow this movement while continuing to deliver power to the front wheels.

Two types of joints permit these movements, universal joints and slip joints. The universal joint, in effect, allows the shaft to bend. Because of the universal joint, power can flow through even though the two sections of the shaft are running at an angle. The slip joint allows the length of the shaft to change. Both of these joints are described in detail later in the chapter.

NOTE: In front-drive cars, the driveshafts connecting the transaxle to the front wheels are often called the *drive axles*. Also, the driveshaft used in front-engine, rear-wheel-drive cars is often called the *propeller shaft,* or "prop" shaft. In this book we will use the term *driveshaft* for this part.

✸ 17-2 Types of drive A variety of drive arrangements have been used on automobiles. The most common arrangement is front-engine rear-wheel drive. The engine is mounted at the front and drives the rear wheels through a driveshaft and differential.

Other arrangements include front-engine front-wheel drive, rear-engine rear-wheel drive, front-engine four-wheel drive, and rear-engine four-wheel drive.

1. Front-engine rear-wheel drive The front-engine rear-wheel drive uses a long driveshaft (Fig. 17-1). This has been the most common arrangement for many years. It is still widely used in larger cars. However, in recent years millions of smaller cars have been built with front-engine front-wheel drive.

A major problem with front-engine rear-wheel drive is the hump in the floor pan of the car that is required to make room for the driveshaft. Chapter 19 on differentials discusses how the gearing has been changed to lower the driveshaft and make the hump in the floor less noticeable.

2. Front-engine front-wheel drive Many new small cars have the engine mounted at the front and driving the front wheels. This arrangement often uses a transaxle (✸ 4-14 and 4-15 and Chaps. 11 to 14). The transaxle is connected by short driveshafts, or axles, to the front wheels. Figure 17-2 shows the arrangement, looking up from under the car. The transaxle is attached to the engine, and power flows through the transaxle to the two driveshafts (drive axles). Figure 17-3 shows the front suspension and driveshafts alone. The driveshafts have universal joints that permit the angle of drive and the length of the driveshafts to change.

3. Front-engine rear-wheel drive—rear wheels independently suspended This arrangement has been used on a few cars (Fig. 17-4). It is the same general arrangement as for the front-engine rear-wheel drive described above, except that each rear wheel is suspended independently. Therefore, each axle driveshaft must have universal and slip joints, just as in the driveshafts used in the front-engine front-wheel-drive car.

4. Rear-engine rear-wheel drive With this arrangement, the engine is mounted at the rear and the rear wheels are driven (Fig. 17-5). Each rear wheel is driven by a short axle shaft, or driveshaft. These driveshafts have universal and slip joints, just as in the

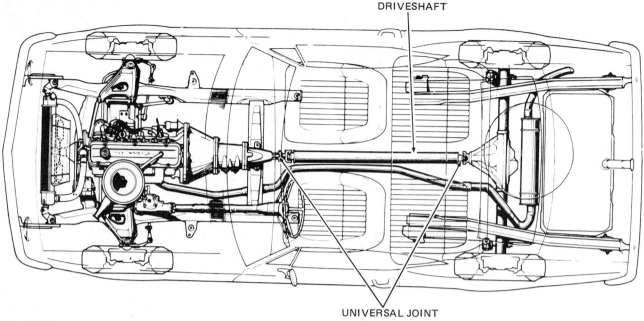

DRIVESHAFT

UNIVERSAL JOINT

Fig. 17-1 Location of the drive line in the power train.

driveshafts of front-drive cars and rear-drive cars with independently suspended wheels (Figs. 17-2 to 17-4).

5. **Front-engine four-wheel drive** The front-engine four-wheel drive is used in a small number of cars and in trucks and utility vehicles, such as the Jeep, Bronco, and Blazer. The engine is mounted at the front. Each of the four wheels is connected to a differential by means of driveshafts. Universal joints are used in the front-wheel driveshaft so that the front wheels can be steered.

Figure 17-6 is a bottom view of a four-wheel-drive vehicle showing the drive arrangement. Power passes from the transmission to a transfer case, which is connected to both front and rear driveshafts. These shafts, in turn, are connected to the front and rear differentials. The differentials are connected by drive axles or shafts to the front and rear wheels. Transfer cases are described in ✿ 4-17 and Chaps. 15 and 16.

6. **Rear-engine four-wheel drive** There is one other arrangement, the engine mounted at the rear and driving all four wheels. However, no manufacturer is producing this as a standard vehicle.

✿ **17-3 Function of the driveshaft** In automobiles in which the engine is at the front and the rear wheels drive the car, a driveshaft is required to connect the transmission main or output shaft to the differential at the rear-wheel axles (Fig. 17-1). Front-wheel-drive cars, which have the engine at the front, do not require this type of driveshaft. Instead, they use short driveshafts, or axles, that connect the transaxle to the front wheels (✿ 17-7 to 17-9). On front-engine, rear-wheel-drive cars, the rotary motion of the transmission main or output shaft carries through the driveshaft to the differential, causing the rear wheels to rotate.

Driveshaft design must take into consideration two

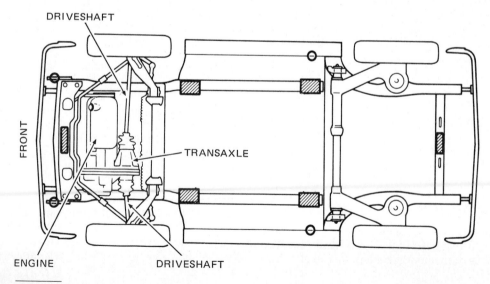

DRIVESHAFT

FRONT

TRANSAXLE

ENGINE   DRIVESHAFT

Fig. 17-2 View from underneath front end of car, showing location of the engine, transaxle, and driveshafts. *(Chrysler Corporation)*

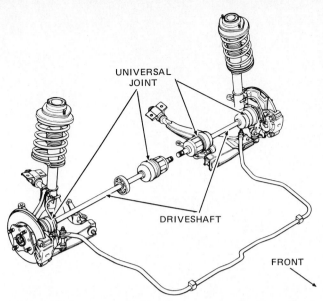

Fig. 17-3 Front suspension and driveshafts for a front-drive car. *(Toyo Kogyo, Ltd.)*

facts. First, the engine and transmission are more or less rigidly attached to the car frame. Second, the rear-axle housing (with wheels and differential) is attached to the frame by springs. As the rear wheels encounter irregularities in the road, the springs compress or expand. This changes the angle of drive between the driveshaft and transmission shaft. It also changes the distance between the transmission and the differential (Fig. 17-7).

In order for the driveshaft to take care of these two changes, it must have two separate devices. These are one or more universal joints to permit variations in the angle of drive and a slip joint that permits the effective length of the drive line to change.

The driveshaft is usually a hollow tube with universal joints at front and back and, on some models, at the center. Many driveshafts are one piece, as shown in Fig. 17-8. Some have two sections, as shown in Figs. 17-9 to 17-11. The two-section driveshaft has a center support bearing, as shown in Figs. 17-10 and 17-11.

Typical disassembled views of the two general types of driveshafts are shown in Figs. 17-12 and 17-13. Note, in Fig. 17-12, the several alternative constructions. The three alternative tube constructions to the lower right show methods of reducing noise and vibration. These include rubber elements in the tube, which act as vibration dampers. The two alternative universal-joint constructions to the upper left in Fig. 17-12 show variations in the way the bearings are installed in the universal joint (✿ 17-4).

Figure 17-13 shows a two-piece driveshaft in disassembled view. This is very similar to the driveshaft shown in Fig. 17-10.

✿ 17-4 Universal joints The universal joint is often called a *U joint*. A simple universal joint is shown in Fig. 17-14. It is essentially a double-hinged joint consisting of two Y-shaped yokes, one on the driving shaft and the other on the driven shaft, and a cross-shaped member called the *cross,* or *spider.*

The four arms of the cross, known as *trunnions,* are

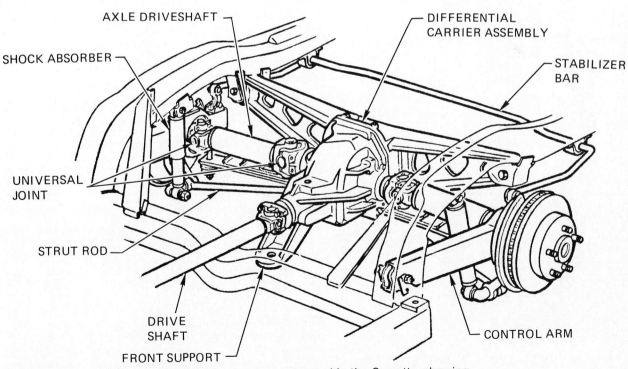

Fig. 17-4 Rear suspension and drive-line components used in the Corvette, showing the transverse leaf spring and the axle driveshafts, each of which has two universal joints. *(Chevrolet Motor Division of General Motors Corporation)*

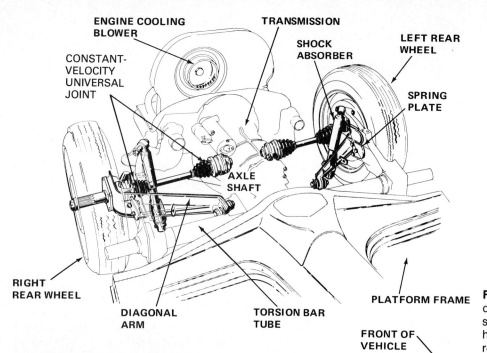

ENGINE COOLING BLOWER

TRANSMISSION

SHOCK ABSORBER

LEFT REAR WHEEL

CONSTANT-VELOCITY UNIVERSAL JOINT

SPRING PLATE

AXLE SHAFT

RIGHT REAR WHEEL

DIAGONAL ARM

TORSION BAR TUBE

PLATFORM FRAME

FRONT OF VEHICLE

Fig. 17-5 Rear suspension and drive-line components used in some Volkswagen models that have rear-mounted engines and rear-wheel drive. *(Volkswagen of America, Inc.)*

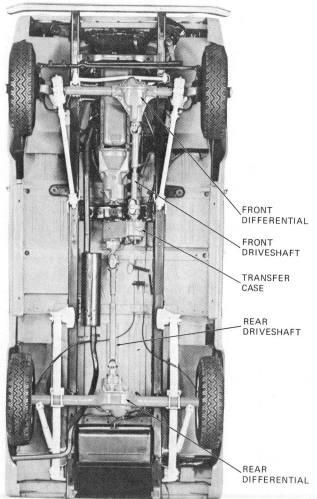

FRONT DIFFERENTIAL

FRONT DRIVESHAFT

TRANSFER CASE

REAR DRIVESHAFT

REAR DIFFERENTIAL

Fig. 17-6 View from the bottom of the Ford Bronco, showing the drive train between the transmission and the four wheels. This is a four-wheel-drive vehicle. *(Ford Motor Company)*

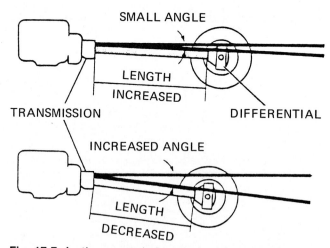

SMALL ANGLE

LENGTH INCREASED

TRANSMISSION

DIFFERENTIAL

INCREASED ANGLE

LENGTH DECREASED

Fig. 17-7 As the rear-axle housing, with differential and wheels, moves up and down, the angle between the axle housing and the transmission output shaft changes and the length of the shaft also changes. The reason the driveshaft shortens as the angle increases is that the rear axle and differential move in a shorter arc than the driveshaft. The center point of the axle-housing arc is the rear spring or control-arm attachment to the frame.

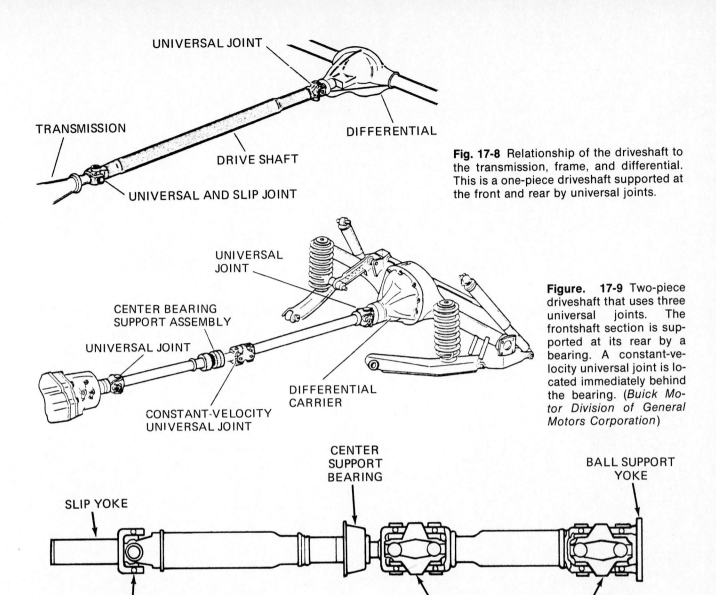

UNIVERSAL JOINT

TRANSMISSION

DRIVE SHAFT

DIFFERENTIAL

UNIVERSAL AND SLIP JOINT

Fig. 17-8 Relationship of the driveshaft to the transmission, frame, and differential. This is a one-piece driveshaft supported at the front and rear by universal joints.

UNIVERSAL JOINT

CENTER BEARING SUPPORT ASSEMBLY

UNIVERSAL JOINT

DIFFERENTIAL CARRIER

CONSTANT-VELOCITY UNIVERSAL JOINT

Figure. 17-9 Two-piece driveshaft that uses three universal joints. The frontshaft section is supported at its rear by a bearing. A constant-velocity universal joint is located immediately behind the bearing. (*Buick Motor Division of General Motors Corporation*)

CENTER SUPPORT BEARING

BALL SUPPORT YOKE

SLIP YOKE

FRONT UNIVERSAL JOINT

CONSTANT-VELOCITY UNIVERSAL JOINT

Fig. 17-10 Driveshaft with center-support bearing, two shafts, and three universal joints. The two rear joints are of the constant-velocity type. (*Cadillac Motor Car Division of General Motors Corporation*)

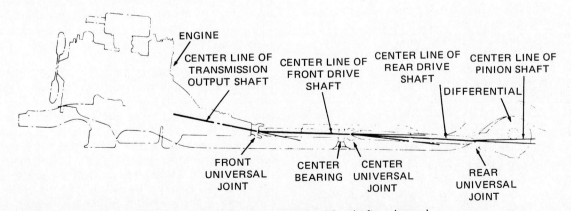

ENGINE

CENTER LINE OF TRANSMISSION OUTPUT SHAFT

CENTER LINE OF FRONT DRIVE SHAFT

CENTER LINE OF REAR DRIVE SHAFT

CENTER LINE OF PINION SHAFT

DIFFERENTIAL

FRONT UNIVERSAL JOINT

CENTER BEARING

CENTER UNIVERSAL JOINT

REAR UNIVERSAL JOINT

Fig. 17-11 Schematic drawing of the relationship of a two-piece driveshaft, universal joints, transmission, and differential. (*Chevrolet Motor Division of General Motors Corporation*)

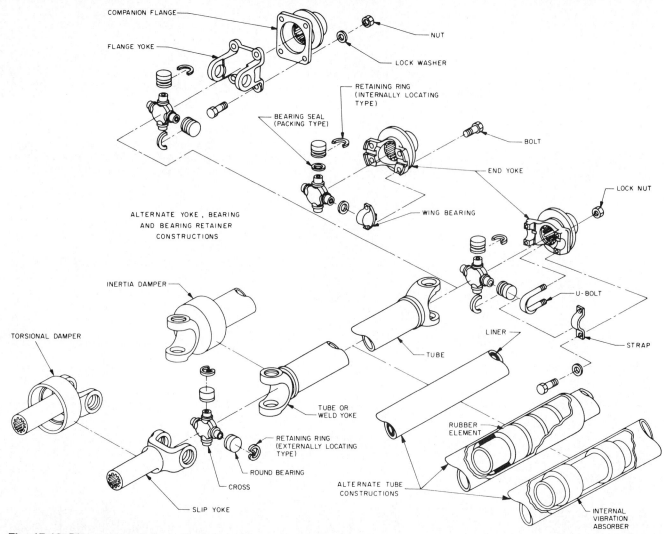

Fig. 17-12 Disassembled view of a one-piece driveshaft showing alternate tube and universal-joint construction. *(Society of Automotive Engineers)*

assembled into bearings in the ends of the two shaft yokes. The driving shaft causes the cross to rotate, and the other two trunnions of the cross cause the driven shaft to rotate. When the two shafts are at an angle to each other, the bearings in the yokes permit the yokes to swing round on the trunnions with each revolution.

A variety of universal joints have been used on automobiles. The types now in most common use for front-engine, rear-drive cars are the cross-and-two-yoke, ball-and-trunnion, and constant-velocity universal joints.

1. **Cross-and-two-yoke universal joint** The cross-and-two-yoke design is essentially the same as the simple universal joint discussed previously. However, the bearings are often of the needle type (Figs. 17-12, 17-13, and 17-15). There are four sets of needle bearings, one set for each trunnion of the cross. The bearings are held in place by snap rings that drop into undercuts in the yoke-bearing holes.

2. **Ball-and-trunnion universal joint** The ball-and-trunnion type of universal joint combines the universal and the slip joint in one assembly. A universal joint of this design is shown in Fig. 17-16 in exploded view. The shaft has a pin pressed through it, and around both ends of the pin are placed balls drilled out to accommodate needle bearings.

The other member of the universal joint consists of a steel casing or body that has two longitudinal channels into which the balls fit. The body is bolted to a flange on the mating shaft (not shown in Fig. 17-16). The rotary motion is carried through the pin and balls, according to the direction of drive. The balls can move back and forth in the channels of the body to compensate for varying angles of drive. At the same time, they act as a slip joint by slipping along the channels to change the effective length of the driveshaft.

3. **Constant-velocity universal joints** The simple universal joint shown in Fig. 17-14, consisting of a cross

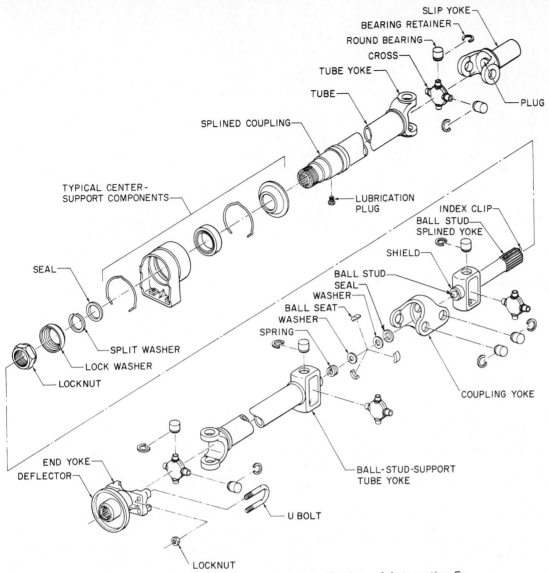

Fig. 17-13 Disassembled view of a two-piece driveshaft. *(Society of Automotive Engineers)*

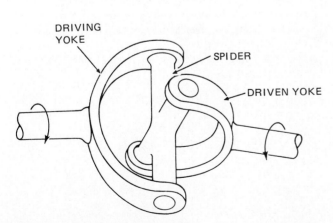

Fig. 17-14 Simple universal joint.

and two yokes, is not a constant-velocity joint. As the driveshaft rotates, the driven shaft will be given a variable velocity. The greater the angle between the two, the greater the variation. This causes drive-line problems because the loads on the bearings and gears will pulsate. Therefore, repeated reductions and increases in load occur with every revolution. This could cause increased wear of the affected parts. To eliminate this condition, constant-velocity universal joints are used on many cars.

Constant-velocity universal joints are shown on driveshafts in Figs. 17-9 and 17-10 and in sectional view in Fig. 17-17. Figure 17-18 is a simplified drawing of a constant-velocity universal joint. This joint, known as a *double-Cardan universal joint*, consists of two individual universal joints linked by a ball and socket.

The ball and socket splits the angle of the two drive-shafts between the two universal joints. Because the

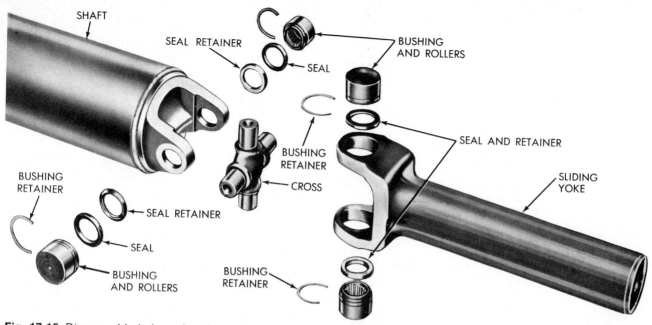

Fig. 17-15 Disassembled view of a universal joint. The cross is also called a *spider* by some manufacturers. *(Chrysler Corporation)*

two joints are operating at the same angle, the normal fluctuations that could result from the use of a single universal joint are canceled out. The acceleration or deceleration of one universal joint at any instant is nullified by the equal and opposite action of the other.

4. **Rzeppa universal joint** Another type of constant-velocity universal joint is the Rzeppa joint, shown in disassembled view in Fig. 17-19. The large balls roll in the curved grooves between the inner housing and the inner race. They are retained in position by the cage.

Regardless of the angle between the driving and driven shafts, the velocity imparted is constant because the balls roll between the two races. Note that the inner housing is positioned in the outer housing with a series of small balls in the straight grooves. This arrangement acts as a slip joint to permit the effective length of the driveshaft to change. Slip joints are described in ✿ 17-5.

5. **Three-ball-and-trunnion joint** The "tripot," or three-ball-and-trunnion universal joint, shown in Fig. 17-20, is also a constant-velocity universal joint. The balls on the ends of the three trunnions can move

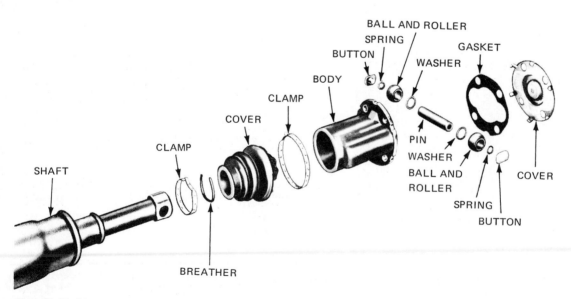

Fig. 17-16 Disassembled view of a ball-and-trunnion universal joint. *(Chrysler Corporation)*

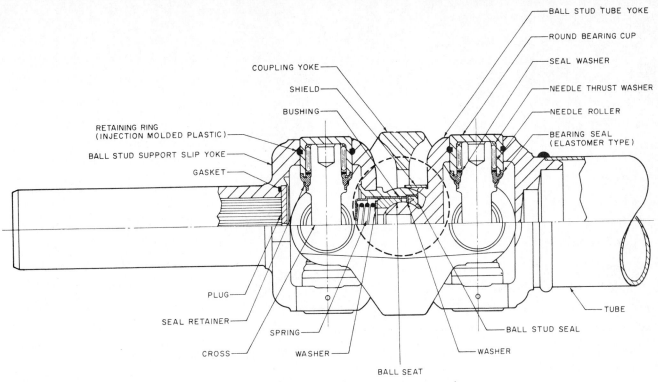

Fig. 17-17 Sectional view of a constant-velocity universal joint. *(Society of Automotive Engineers)*

back and forth in the grooves in the housing as required by the angle of drive.

⚙17-5 Slip joint Slip joints are shown in Figs. 1-20 and 17-21. The slip joint consists of external splines on one shaft and matching internal splines on the mating hollow shaft. The splines cause the two shafts to rotate together but permit the ends of the two shafts to slide along each other. Therefore, any effective change of length in the drive line as the drive axles move toward or away from the car frame is accommodated.

For certain high-performance or heavy-duty applications, special ball- or needle-bearing slip joints are used. Figure 17-22 shows a slip joint using spring-loaded balls that can roll between grooves cut in the shaft and hollow yoke. This design greatly reduces the frictional force necessary to allow the shaft and yoke to slide back and forth. Another design for special applications is shown in Fig. 17-23. In this design, a series of recirculating needle bearings is used. These bearings can roll back and forth as the yoke and shaft move with respect to each other.

⚙17-6 Rear-end torque Whenever the rear wheels are being driven by the power train, they rotate, as shown in Fig. 17-24. At the same time, the rear-axle housing tries to rotate in the opposite direction, as shown by the smaller curved arrow. The twisting motion applied to the axle housing is called *rear-end torque*. To understand why this happens, let us review briefly the operation of the differential. See ⚙1-11 and Fig. 1-25.

The ring gear on the differential is connected through other gears to the rear wheels. When torque is applied to the ring gear by the drive pinion, the ring gear and rear wheels rotate. It is the side thrust of the drive-pinion teeth against the rear-gear teeth that makes the ring gear rotate. This side thrust also causes the drive-

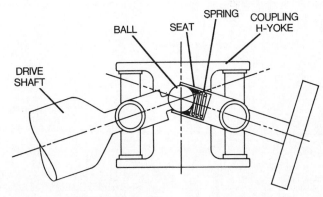

Fig. 17-18 Simplified drawing of a constant-velocity, or double-Cardan, universal joint. *(Service Parts Division of Dana Corporation)*

pinion shaft to push against the shaft bearing (Fig. 17-25). The drive-pinion-shaft bearing is part of the differential and axle housing. Therefore this push tries to make the axle housing rotate in a direction opposite the wheel rotation.

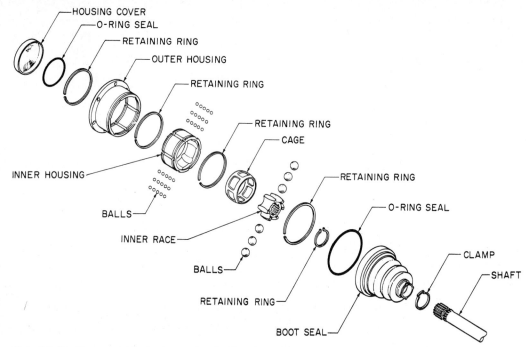

Fig. 17-19 Disassembled view of a ball-spline Rzeppa universal joint. *(Society of Automotive Engineers)*

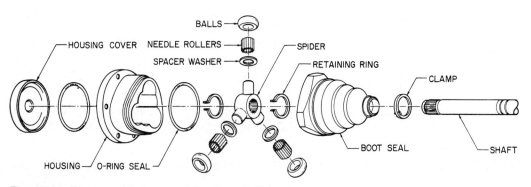

Fig. 17-20 Disassembled view of the three-ball-and-trunnion universal joint. *(Society of Automotive Engineers)*

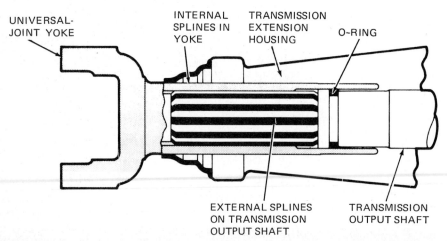

Fig. 17-21 Slip joint used to compensate for changes in length of the drive line.

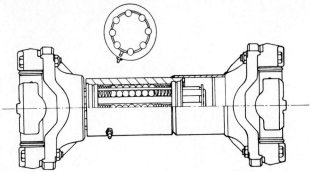

Fig. 17-22 Slip joint using balls that roll in grooves instead of splines.

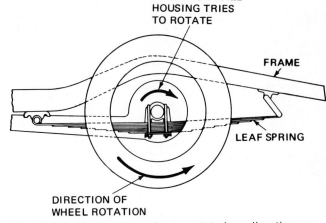

Fig. 17-24 Axle housing tries to rotate in a direction opposite to wheel rotation.

DIRECTION AXLE HOUSING TRIES TO ROTATE

FRAME

LEAF SPRING

DIRECTION OF WHEEL ROTATION

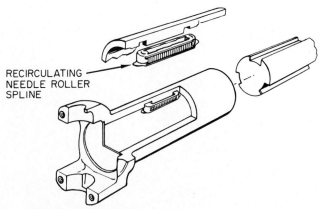

RECIRCULATING NEEDLE ROLLER SPLINE

Fig. 17-23 Slip joint using recirculating needles that roll in grooves cut in the shaft instead of splines.

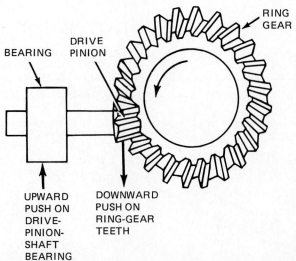

RING GEAR

BEARING

DRIVE PINION

UPWARD PUSH ON DRIVE-PINION-SHAFT BEARING

DOWNWARD PUSH ON RING-GEAR TEETH

Fig. 17-25 Side view of differential showing the ring gear and drive pinion.

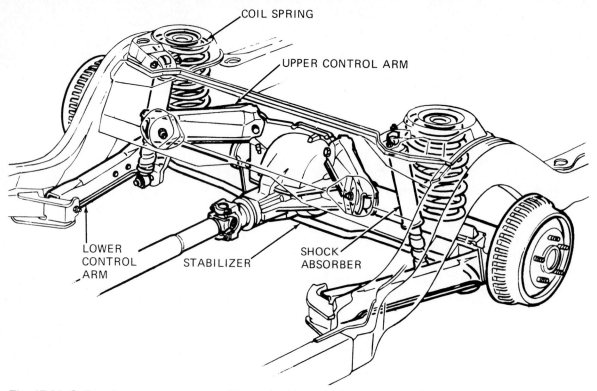

Fig. 17-26 Coil-spring rear suspension. *(Chevrolet Motor Division of General Motors Corporation)*

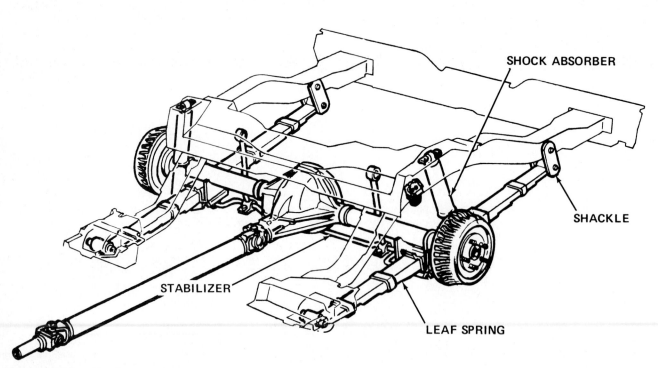

Fig. 17-27 Leaf-spring rear suspension. *(Chevrolet Motor Division of General Motors Corporation)*

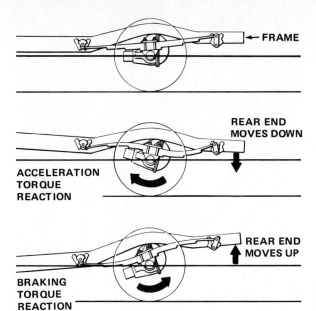

FRAME

REAR END
MOVES DOWN

ACCELERATION
TORQUE
REACTION

REAR END
MOVES UP

BRAKING
TORQUE
REACTION

Fig. 17-28 Actions of the spring and rear end when the car is accelerated or braked. *(Ford Motor Company)*

To prevent excessive axle-housing movement, the housing is braced by the rear-suspension system. There are two types of rear-suspension systems on front-engine, rear-wheel-drive cars. One type uses coil springs (Fig. 17-26). A pair of coil springs is located above the rear-axle housing, between the housing and the car frame. This system uses control arms to prevent excessive sideway or backward-and-forward movement of the axle housing.

The second type uses leaf springs (Fig. 17-27). This arrangement needs no extra control arms. The leaf springs control both sideway and backward-and-forward movement of the axle housing.

Rear-end torque is absorbed by the control arms on the coil-spring rear-suspension systems. Rear-end torque is absorbed by the leaf springs on the leaf-spring rear-suspension systems.

One effect of rear-end torque is rear-end "squat" when the car is accelerated (Fig. 17-28). When a car is accelerated from a standing start, the differential drive pinion tries to "climb" the teeth of the ring gear. Therefore, the drive pinion and the front of the differential carrier move upward. The result is that the rear springs are pulled downward, or compressed, so that the rear end of the car moves down, or squats. On braking, the rear end of the car moves up (Fig. 17-28). The weight of the car tries, in effect, to pivot about the front axle.

☼ 17-7 Driveshafts for front-wheel-drive vehicles
In front-engine, front-wheel-drive vehicles, the driveshafts that connect the transaxle to the front wheels are short and include two universal joints each. Figures 17-2 and 17-3 show the general arrangement. Figure 17-29 shows how the driveshafts are connected to the transaxle. Figure 17-30 shows the details of a driveshaft and how it is connected to a wheel.

Typically, there are two types of universal joints on each driveshaft. The inner, or inboard, universal joint is the double-offset type (☼ 17-9), and the outer, or outboard, universal joint is a constant-velocity joint (☼ 17-8).

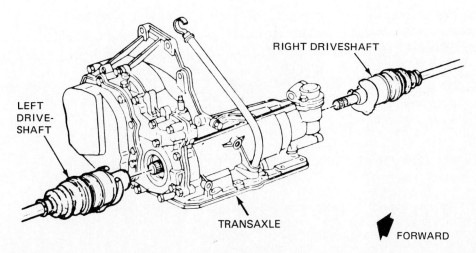

RIGHT DRIVESHAFT

LEFT
DRIVE-
SHAFT

TRANSAXLE

FORWARD

Fig. 17-29 Illustration showing how the driveshafts are connected to the transaxle. *(Chevrolet Motor Division of General Motors Corporation)*

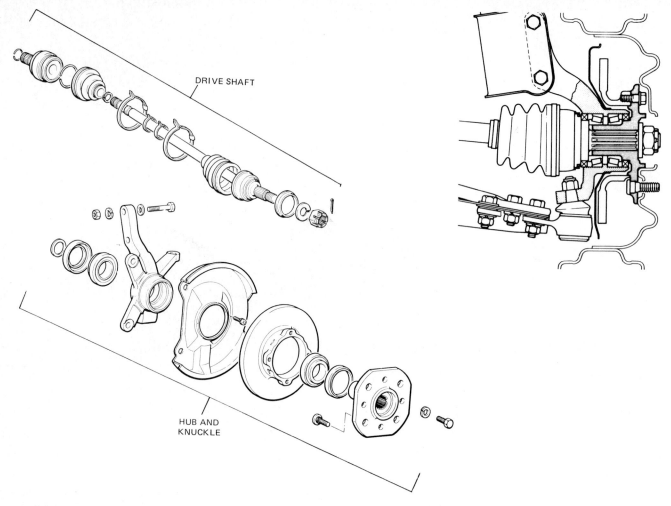

Fig. 17-30 Driveshaft assembly and disassembled front-wheel hub and associated parts. *(Chrysler Corporation)*

The two universal joints carry engine torque from the transaxle through the driveshaft to the wheel axles. They also permit the length of the driveshaft assembly to change because of the construction of the double-offset universal joint (✿ 17-9).

✿ 17-8 Constant-velocity universal joint On many front-wheel-drive cars, a constant-velocity, or Birfield, type of universal joint is used as the outboard joint. It is similar to the Rzeppa universal joint (Fig. 17-19). However, the Birfield universal joint does not have either the outer race with straight inside grooves and balls or the inner housing with straight grooves on its outside. These grooves form a slip joint, as explained in ✿ 17-4.

The outboard (Birfield) universal joint has just the outer race, cage, inner race, and balls (Fig. 17-31). The outer race has a splined shaft which enters the wheel hub (upper right in Fig. 17-30) when the driveshaft is installed in the vehicle. The outer race has curved grooves in which the balls can roll. The balls are held

in place by the cage. The inner race has grooves that match the grooves in the outer race. When the universal

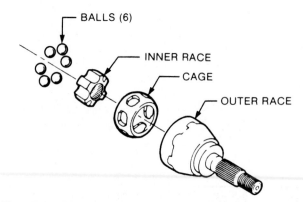

Fig. 17-31 Disassembled outboard joint. *(Chevrolet Motor Division of General Motors Corporation)*

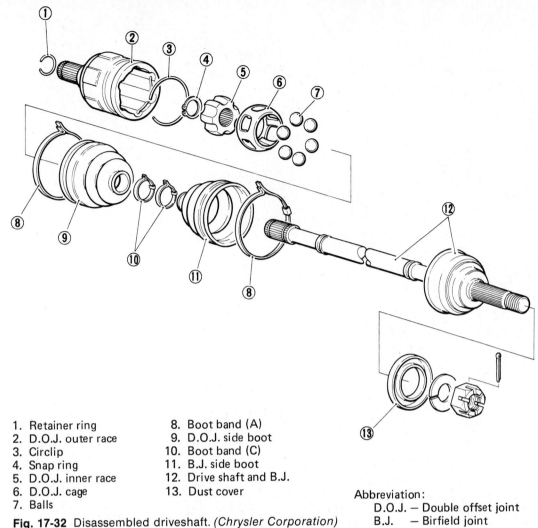

1. Retainer ring
2. D.O.J. outer race
3. Circlip
4. Snap ring
5. D.O.J. inner race
6. D.O.J. cage
7. Balls
8. Boot band (A)
9. D.O.J. side boot
10. Boot band (C)
11. B.J. side boot
12. Drive shaft and B.J.
13. Dust cover

Abbreviation:
   D.O.J. — Double offset joint
   B.J.   — Birfield joint

Fig. 17-32 Disassembled driveshaft. *(Chrysler Corporation)*

joint is transmitting torque at an angle, the balls can roll in their grooves to accommodate the angle.

✹ 17-9 Double-offset universal joint Figure 17-32 shows a disassembled inner, or inboard, universal joint on a driveshaft. This is also a modified Rzeppa universal joint (Fig. 17-19). The inner race is splined to the inner end of the driveshaft. The outer race has a splined shaft that enters the transaxle assembly. The balls, in their cage, are positioned so that they can roll in the outside grooves of the inner race and the inside grooves of the outer race. The grooves on the outer race are long. This allows the driveshaft with the inner race to move in or out of the outer race to change the effective length of the driveshaft. At the same time, the angle of drive can change. This universal joint is also a constant-velocity joint.

Figure 17-33 shows the two completely disassembled driveshaft assemblies for a front-drive car. On cars with automatic transaxles, the left-hand universal-joint

outer race is different to accommodate the bulkier transaxle assembly.

✹ 17-10 Front-end torque How applying torque to the rear wheels produces a reaction on the axle housing was covered in ✹ 17-6. When the wheels turn in one direction, the axle housing tries to turn in the opposite direction (Fig. 17-24). This produces rear-end squat (Fig. 17-28).

A similar action takes place at the front of the car in front-drive cars. When the car is accelerated, the front end squats. The wheels at which the torque is applied produce the reaction. However, the squat is less noticeable with the transaxle. The transaxle, with the attached engine, absorbs the torque more effectively.

Regardless of whether the car is front-drive or rear-drive, the same braking reaction occurs. The rear end of the car moves up and the front end moves down because of the inertia of the car.

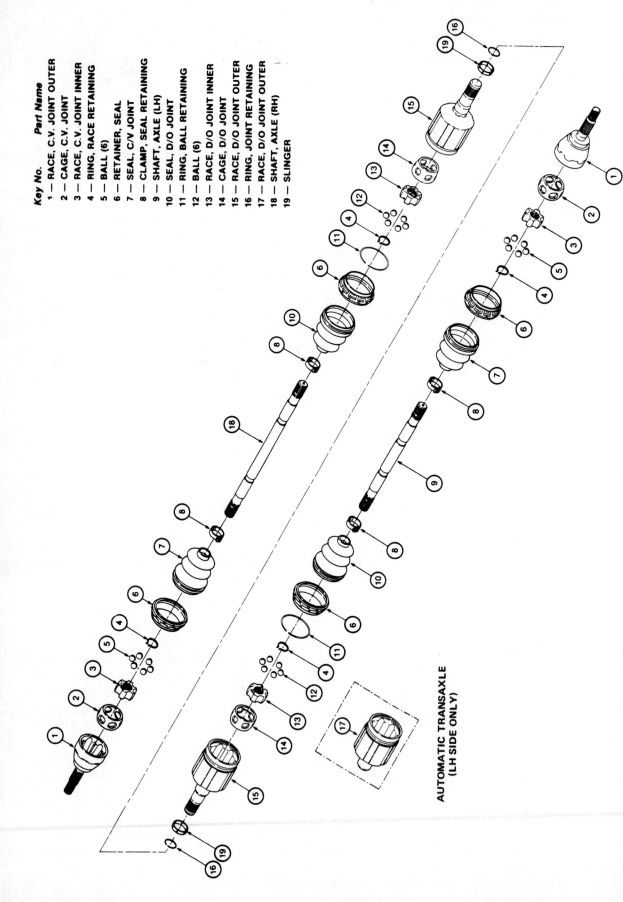

Key No.   Part Name
1 — RACE, C.V. JOINT OUTER
2 — CAGE, C.V. JOINT
3 — RACE, C.V. JOINT INNER
4 — RING, RACE RETAINING
5 — BALL (6)
6 — RETAINER, SEAL
7 — SEAL, C/V JOINT
8 — CLAMP, SEAL RETAINING
9 — SHAFT, AXLE (LH)
10 — SEAL, D/O JOINT
11 — RING, BALL RETAINING
12 — BALL (6)
13 — RACE, D/O JOINT INNER
14 — CAGE, D/O JOINT
15 — RACE, D/O JOINT OUTER
16 — RING, JOINT RETAINING
17 — RACE, D/O JOINT OUTER
18 — SHAFT, AXLE (RH)
19 — SLINGER

AUTOMATIC TRANSAXLE
(LH SIDE ONLY)

Fig. 17-33 Completely disassembled driveshafts for front-drive car with transaxle. *(Chevrolet Motor Division of General Motors Corporation)*

204

Chapter 17 review questions

Select the *one* correct, best, or most probable answer to each question. Then check your answers against the correct answers given at the end of the book.

1. The driveshaft in the front-engine, rear-wheel-drive car has at least:
   a. one universal joint
   b. two universal joints
   c. three universal joints
   d. two slip joints
2. The driveshaft must have two types of joints:
   a. U and universal
   b. transmission and differential
   c. slip and spline
   d. universal and slip
3. The basic purpose of the universal joint is to allow the:
   a. driveshaft length to change
   b. driveshaft to be supported at the middle
   c. drive angle between two parts of the driveshaft to change
   d. driveshaft to be removed and installed easily
4. The two most common types of drive arrangements are the front-engine, rear-wheel drive and the:
   a. front-engine, front-wheel drive
   b. rear-engine, rear-wheel drive
   c. front-engine, four-wheel drive
   d. rear-engine, four-wheel drive
5. In the slip joint, the slipping may occur between:
   a. splines
   b. balls and trunnion
   c. balls and grooves
   d. all of the above
6. Rear-end torque occurs in:
   a. front-engine, front-wheel-drive cars
   b. rear-engine, rear-wheel-drive cars
   c. front-engine, rear-wheel-drive cars
   d. all of the above
7. Front-end torque occurs in:
   a. front-engine, front-wheel-drive cars
   b. rear-engine, real-wheel-drive cars
   c. front-engine, rear-wheel-drive cars
   d. all of the above
8. Rear-end torque is absorbed by:
   a. control arms or coil springs
   b. control arms or leaf springs
   c. leaf springs or driveshaft
   d. none of the above
9. Rear-end squat occurs when the car is:
   a. accelerated
   b. braked
   c. driven around a curve
   d. all of the above
10. The typical front-engine, front-wheel-drive system has:
    a. one universal joint
    b. two universal joints
    c. three universal joints
    d. four universal joints

DRIVE-LINE AND UNIVERSAL-JOINT SERVICE

After studying this chapter, and with proper instruction and equipment, you should be able to:

1. List the eight complaints in the rear-wheel-drive trouble-diagnosis chart and explain what might cause each.
2. Check a driveshaft for balance and correct balance if necessary.
3. Check driveshaft runout.
4. Check and adjust universal-joint angles.
5. Service the drive lines for rear-drive cars, including the universal joint.
6. List the three complaints in the front-drive trouble-diagnosis chart and explain what might cause each.
7. Service a driveshaft for a front-wheel-drive car.

 Driveshaft and universal-joint service—front-engine, rear-wheel-drive cars

This part of the chapter deals with the servicing of the driveshafts and universal joints used in vehicles with the engine at the front and with the rear wheels being driven.

✷ 18-1 Driveshaft and universal-joint trouble diagnosis The chart on page 207 lists the various possible complaints that might be blamed on the driveshaft and universal joints in front-engine, rear-wheel-drive cars. The troubles and possible causes are not listed in frequency of occurrence. Item 1, or item a, does not necessarily occur more often than item 2, or item b.

NOTE: See ✷ 3-1 for a discussion of diagnosis theory.

✷ 18-2 Universal-joint and driveshaft service Most universal joints require no periodic maintenance. They are lubricated for life and cannot be lubricated on the car. If a universal joint becomes noisy or worn, it must be replaced. Manufacturers supply service kits that include all necessary parts to make the replacement.

The driveshaft is a balanced unit. On some cars, if the shaft becomes unbalanced and vibrates, it can be rebalanced by adding hose clamps, as will be discussed later. On other cars, the recommendation is to replace an out-of-balance driveshaft. The angle through which the universal joint is turning is important and should be measured. If the angle is excessive, it can cause wear of the joint. Also, if the angles of two universal joints in the driveshaft are different, vibration and noise can result. Typical servicing procedures follow.

NOTE: The driveshaft and universal joints are carefully balanced during original assembly. To ensure correct relationship after completing a service job, be sure to mark the parts before disassembly. Mark them so that you can see where the parts go and how they line up. Then, after reassembly and installation, they should still be in balance.

✷ 18-3 Driveshaft balancing If the driveshaft is out of balance and the vehicle has been undercoated, remove any undercoat from the driveshaft. Then road test the vehicle. If the problem still exists, disconnect the driveshaft and rotate it 180°. Then reconnect the shaft and road test the vehicle again. If these two steps do not eliminate the vibration, proceed as follows.

NOTE: There are two methods of checking for balance, a mechanical method and a method using a stroboscopic light. The mechanical method is described first.

1. **Mechanical method** The mechanical method uses a crayon or pencil to mark the driveshaft while it is rotating.

1. Raise the vehicle on a twin-post lift so that the rear of the vehicle is supported on the rear-axle housing with the wheels free to rotate. Remove the wheels to eliminate any wheel-and-tire-assembly unbalance.

Driveshaft and Universal-Joint Trouble-Diagnosis Chart

(For Front-Engine, Rear-Wheel-Drive Vehicles)

COMPLAINT	POSSIBLE CAUSE	CHECK OR CORRECTION
1. Leak at front slip yoke*	a. Rough outside surface on splined yoke	Replace seal if cut by burrs on yoke; minor burrs can be smoothed by careful use of crocus cloth or honing with a fine stone; replace yoke if outside surface is rough or burred badly
	b. Defective transmission rear oil seal	Replace transmission rear oil seal; bring transmission oil up to proper level after correction
2. Knock in drive line; clunking noise when car is operated under floating condition at 10 mph [16 km/h] in high gear or neutral	a. Worn or damaged universal joints	Disassemble universal joints, inspect and replace worn or damaged parts
	b. Side-gear-hub counterbore in differential worn oversize	Replace differential case and/or side gears as required
3. Ping, snap, or click in drive line†	a. Loose upper or lower control-arm bushing bolts	Tighten bolts to specified torque
	b. Loose companion flange	Remove companion flange, turn 180° from its original position, apply white lead to splines and reinstall; tighten pinion nut to specified torque
4. Roughness, vibration, or body boom at any speed‡	a. Bent or dented driveshaft	Replace
	b. Undercoating on driveshaft	Clean driveshaft
	c. Tire unbalance (30 to 80 mph [48 to 128 km/h], not throttle conscious)	Balance or replace as necessary
	d. Excessive U-bolt torque	Check and correct to specified torque
	e. Tight universal joints	Hit yokes with a hammer to free up; overhaul joint if unable to free up or if joint feels rough when rotated by hand
	f. Worn universal joints	Overhaul, replacing necessary parts
	g. Burrs or gouges on companion flange	Rework or replace companion flange; check snap ring, locating surfaces on flange yoke
	h. Driveshaft or companion flange unbalance	Check for missing balance weights on driveshaft; remove and reassemble driveshaft to companion flange, 180° from original position
	i. Excessive looseness at slip-yoke spline	Replace necessary parts
	j. Driveshaft runout (50 to 80 mph [80 to 128 km/h], throttle conscious)	Check driveshaft runout at front and rear; should be less than specified; if above, rotate shaft 180° and recheck; if still above specified, replace shaft
5. Roughness at low speeds, light load, 15 to 35 mph [24 to 56 km/h]	a. U-bolt clamp nuts excessively tight	Check and correct torque to that specified; if torque was excessive or if brinnelled pattern on trunnions, replace joints
6. Scraping noise	a. Oil slinger, companion flange, or end yoke rubbing on rear-axle carrier	Straighten slinger to remove interference
7. Roughness on heavy acceleration (short duration)	a. Double-Cardan joint ball seats worn; ball seat spring may be broken	Replace joint and shaft assembly
8. Roughness above 35 mph [56 km/h] felt and/or heard	a. Tires unbalanced or worn	Balance or replace as required

*An occasional drop of lubricant leaking from the splined yoke is normal and requires no attention.
†Usually occurs on initial load application after transmission has been put into gear, either forward or reverse.
‡With tachometer installed in car, determine whether driveshaft is cause of complaint by driving through speed range and note the engine speed (rpm) at which vibration (roughness) is most pronounced. Then, shift transmission to a different gear range and drive car at same engine speed (rpm) at which vibration occurred before. Note the effect on the vibration. If vibration occurs at the same engine rpm, regardless of transmission gear range, the driveshaft assembly is not at fault, since the shaft speed (rpm) varies. If vibration is decreased or is eliminated in a different gear but at the same engine rpm, check the possible causes.

2. With the transmission in high gear, open the throttle until the speedometer is showing about 50 mph [80 km/h].

3. Carefully bring a crayon or colored pencil up close to the rear end of the driveshaft (Fig. 18-1). Use a stand to support and steady your hand, or use a suitable runout indicator for making the mark.

CAUTION: Avoid working near any balance weights on the driveshaft. A balance weight on the rear of the driveshaft can be seen in Fig. 18-1. The spinning weight can easily cut your hand.

Careful: Do not run the engine faster than 55 mph [89 km/h] and do not run it for long periods. This will overheat the engine and transmission.

4. When the crayon barely touches the rotating drive-shaft, it will make a mark which indicates the heavy spot or runout of the driveshaft.

5. Stop the engine and install two screw-type hose clamps on the driveshaft so that their heads are just opposite the crayon mark (Figs. 18-2 and 18-3). The hose-clamp heads should now be the heavy spot on the driveshaft.

6. Tighten the clamps. With the transmission in high gear, increase the speed to 60 mph [97 km/h] for a short time. If no vibration is felt, install the wheels, lower the vehicle, and road test it.

7. If unbalance still exists, rotate the two clamp heads away from each other in equal amounts (step 3 in Fig. 18-3) and try again. Repeat this until balance is achieved. Install the wheels and road test the vehicle to make sure the problem has been eliminated.

Careful: You cannot install hose clamps on the drive-shaft of some cars. There is too little clearance between the driveshaft and the tunnel in the floorpan when the wheels are in the full, or up, position. For example,

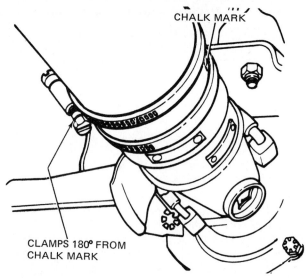

Fig. 18-2 Installing hose clamps on the driveshaft. *(Ford Motor Company)*

Ford warns against using hose clamps on the drive-shafts of their Fairmount and Zephyr models. On these, to correct for out of balance, disconnect the driveshaft at the rear and turn the shaft 45°. If this does not cure the problem, turn it another 45°. If this procedure does not solve the problem, a new driveshaft may be required.

2. Stroboscopic method This method uses a wheel balancer that has a stroboscopic light.

1. First, clean the driveshaft and road test the vehicle. If this does not eliminate the problem, disconnect the driveshaft and reconnect it at 180° from its original position. If this does not solve the problem, raise the car on a hoist. Then position stands under the rear of the frame as shown in Fig. 18-4.

   Lower the lift just enough to allow the lift supports to clear the axle housing. This takes all weight off the axle housing. If the hoist supports the car weight through the axle housing, the sensitivity of the test could be ruined.

2. Put the vibration detector (the transducer) of the wheel balancer in contact with the nose of the differential case, just in back of the universal joint. This is done by making up an extension as shown in Fig. 18-4, or by raising the transducer on a stand as shown in Fig. 18-5.

3. Put four equally spaced marks on the driveshaft, 90° apart. Number each mark (Fig. 18-6).

4. Run the engine with the transmission in high gear at the speed where maximum vibration is felt. Point the stroboscopic light at the spinning shaft and note the position of one of the numbered marks. The stroboscopic light flashes only when the transducer "feels" the heavy spot, which is the point of maximum unbalance. This pinpoints the heavy spot. Stop the engine and turn the driveshaft to the same position you saw. You now know where the heavy spot is.

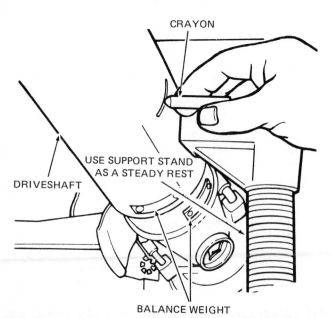

Fig. 18-1 Marking a rotating driveshaft. *(Ford Motor Company)*

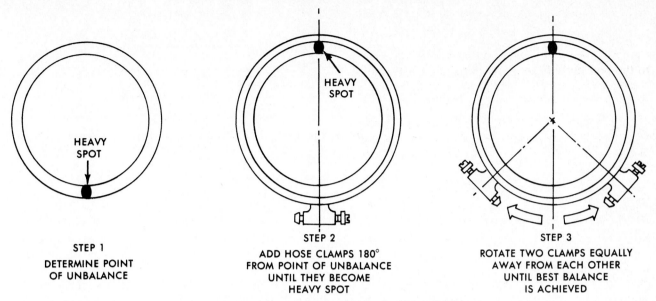

STEP 1	STEP 2	STEP 3
DETERMINE POINT OF UNBALANCE	ADD HOSE CLAMPS 180° FROM POINT OF UNBALANCE UNTIL THEY BECOME HEAVY SPOT	ROTATE TWO CLAMPS EQUALLY AWAY FROM EACH OTHER UNTIL BEST BALANCE IS ACHIEVED

Fig. 18-3 Positioning hose clamps to balance the driveshaft. *(Chevrolet Motor Division of General Motors Corporation)*

5. Install hose clamps as in the mechanical method described above. If necessary, move the clamp heads apart little by little, retesting each time, until the driveshaft is balanced.

✿ 18-4 Checking driveshaft runout If the car has been in a collision, and the driveshaft may have been damaged, you should check the shaft for straightness. A driveshaft that has been bent may not cause vibration except at high speeds.

1. Raise the vehicle on a twin-post lift so that the rear of the vehicle is supported on the rear-axle housing. The wheels are free to rotate.
2. Mount a dial indicator on a movable support so that the dial-indicator button will touch the driveshaft (Fig. 18-7). The dial indicator can also be attached to a magnetic base and the base attached to a suitable smooth place on the underbody of the vehicle.
3. Readings should be taken at the front, center, and rear of the driveshaft as indicated in Fig. 18-8.

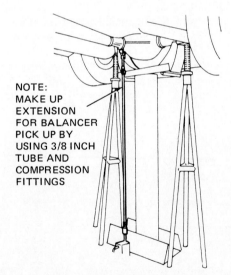

NOTE: MAKE UP EXTENSION FOR BALANCER PICK UP BY USING 3/8 INCH TUBE AND COMPRESSION FITTINGS

DROP TWIN POST HOIST JUST ENOUGH TO ALLOW THE VEE OF THE HOIST TO CLEAR THE AXLE. THIS PLACES THE WEIGHT OF THE CAR ON THE STANDS. THE SYSTEM WILL THEN BE RELEASED AND FREE TO RESPOND TO DRIVESHAFT.

Fig. 18-4 Supporting rear end of vehicle on stands and using an extension to raise balancer pickup to nose of differential. *(Chevrolet Motor Division of General Motors Corporation)*

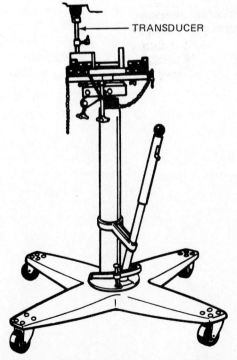

TRANSDUCER

Fig. 18-5 Using a transmission jack to support and position the transducer, or balancer pickup. *(Ford Motor Company)*

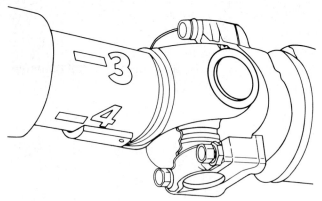

Fig. 18-6 Making numbered marks on the driveshaft, 90° apart. *(Chevrolet Motor Division of General Motors Corporation)*

4. With the transmission in neutral, rotate one rear wheel so that the driveshaft rotates. Note the amount the dial-indicator needle moves. This indicates how much the driveshaft is out of round. Specifications for some Chevrolet models are that the runout should not exceed 0.040 inch [1 mm] at the front, center, or rear.
5. If the runout is excessive, disconnect the driveshaft, rotate it 180°, and reconnect it. If this does not solve the problem, and the noise and vibration are objectionable, install a new driveshaft.
6. If the new driveshaft does not solve the problem, check for a bent universal-joint-coupling flange or slip yoke.

✿ 18-5 Checking universal-joint angles In normal operation, the universal joints allow changes in the angle of drive between the driving member and the driven member. Any variation in the drive angle from

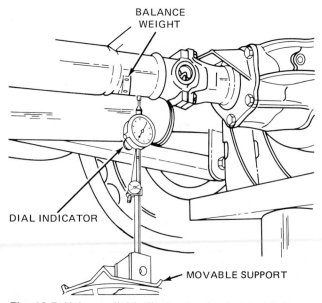

Fig. 18-7 Using a dial indicator to check driveshaft runout. *(ATW)*

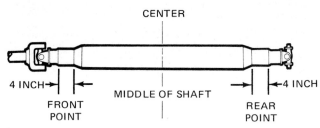

Fig. 18-8 Places to check for driveshaft runout. *(Chevrolet Motor Division of General Motors Corporation)*

straight causes the driven member to speed up and slow down twice during each shaft revolution (✿ 17-4). However, if the two universal joints in the drive line are set so that the driving yokes are 90° apart, the fluctuation effect can be eliminated—provided that the two joints are transmitting torque through the same angle. If the angles are different, there will still be speed fluctuations at the differential.

Because of this, the universal-joint angles should be checked and adjusted, if necessary, if the vehicle has been in a collision or if there is a vibration on acceleration at low car speeds.

1. **Leaf-spring rear-suspension vehicles** When testing leaf-spring rear-suspension vehicles, the vehicle should be empty except for a full tank of gasoline.

1. Raise the vehicle on a lift that does *not* lift on the frame.
2. Rotate the driveshaft so that the cross and roller bushings on the axle and the transmission yokes are facing straight down (Figs. 18-9 and 18-10).
3. Clean the bushing surface so that the magnet of the tool to be used (an inclinometer) will have a smooth surface to adhere to.
4. Install the universal-joint inclinometer as shown in Figs. 18-9 and 18-10 to check the universal-joint angles at front and back. Note that the magnet is placed on the universal-joint bushing of the differential yoke and the transmission yoke.
5. With both the front and rear tabs of the driveshaft-

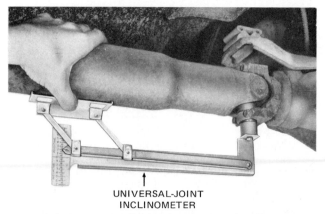

Fig. 18-9 Using a special measuring instrument called a *universal-joint inclinometer* to check the universal-joint angle at the rear of the driveshaft. *(Chrysler Corporation)*

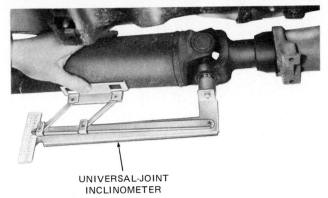

Fig. 18-10 Using a universal-joint inclinometer to check the universal-joint angle at the transmission. (*Chrysler Corporation*)

contact shoe touching the driveshaft, read the angle shown on the inclinometer.

6. To correct the front universal-joint angle, install shims between the rear mount and the transmission-extension housing (Fig. 18-11).

7. To correct the angle at the rear, install tapered shims between the springs and spring seats (Fig. 18-12).

2. **Coil-spring rear-suspension vehicles** On coil-spring rear-suspension systems, the height D shown in Fig. 18-13 must be measured and corrected, if necessary. Height D is the distance between the top of the axle-housing tube and the frame. Weight may have to be added to make the test. Note that the dimensions given are in millimeters [mm] followed by the dimensions in inches. The table is for various Chevrolet models.

1. With the vehicle supported on a lift, use the inclinometer as shown in Fig. 18-14 to measure the inclination of the rear universal-joint bearing cup on the driveshaft yoke. The bearing cup must be clean and straight up and down.

2. Rotate the driveshaft 90° and measure the inclination of the bearing cup on the differential yoke. Subtract the smaller angle from the larger angle to get the real universal-joint angle.

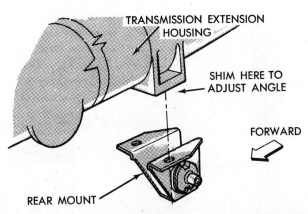

Fig. 18-11 Where shims are installed at the front of the driveshaft to correct universal-joint angle. (*Chrysler Corporation*)

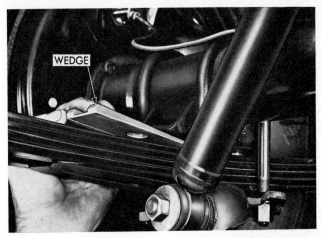

Fig. 18-12 Installing a tapered wedge above the spring to correct the rear universal-joint angle. (*Chrysler Corporation*)

3. Repeat the procedure at the front to measure the front universal-joint angle.

4. To change the angle at the front, add shims between the transmission extension and the transmission rear mount.

5. To change the angle at the rear, you must change the control arm. The longer control arm, as one example, will change the rear universal-joint angle about 2°. This action will also change the angle of the front universal joint.

⚙ **18-6 Chevrolet driveshaft service** To remove the driveshaft assembly, mark the relationship of the

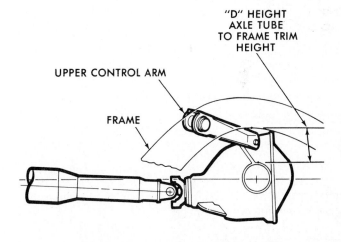

MODEL	±6.35 mm	[0.250 INCH]
MALIBU	107 mm	[4.210 INCH]
IMPALA	151.9 mm	[5.980 INCH]
IMPALA WAGON	111.6 mm	[4.390 INCH]

Fig. 18-13 Trim-height specifications. (*Chevrolet Motor Division of General Motors Corporation*)

Fig. 18-14 Measuring the angle at the rear driveshaft bearing cup. *(Chevrolet Motor Division of General Motors Corporation)*

shaft to the companion flange at the differential. Then disconnect the rear universal joint by removing the trunnion bearing straps or flange attaching bolts (Figs. 18-15 and 18-16). On the strap-attachment type, tape the bearing cups to the trunnion to keep the bearing rollers from falling out.

Then withdraw the driveshaft front yoke from the transmission by moving the shaft to the rear. Pass it under the axle housing. Watch for leakage from the transmission extension housing.

1. **Universal-joint service** Three universal-joint designs have been used on Chevrolet: the Cleveland, Saginaw, and double-Cardan constant-velocity types. Special service kits are supplied for all these designs (Figs. 18-17 and 18-18). Service procedures for the Cleveland and constant-velocity universal joints are discussed below.

a. Cleveland universal-joint disassembly Remove the bearing lock rings from the trunnion yoke. Support the trunnion yoke on a piece of 1¼-inch [31.75-mm] inside-diameter pipe in a shop press (Fig. 18-19).

NOTE: A bench vise can also be used to exert the force necessary to press the bearing cup loose.

Apply force to the trunnion until the bearing cup is almost out. It cannot be pressed all the way out. Grasp the cup in a vise or pliers and work it out of the yoke. Reverse the position of the trunnion and remove the other bearing cup.

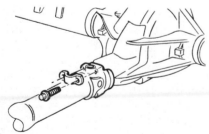

Fig. 18-15 Strap attachment at the rear end of the drive-shaft. *(Chevrolet Motor Division of General Motors Corporation)*

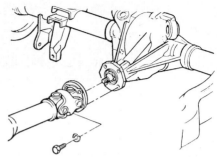

Fig. 18-16 Flange attachment at the rear end of the drive-shaft. *(Chevrolet Motor Division of General Motors Corporation)*

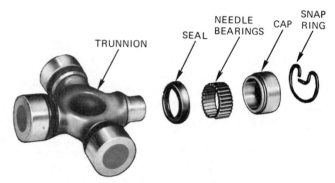

Fig. 18-17 Cleveland-type universal-joint repair kit. *(Chevrolet Motor Division of General Motors Corporation)*

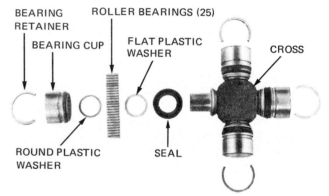

Fig. 18-18 Saginaw-type universal-joint repair kit. *(Chevrolet Motor Division of General Motors Corporation)*

Fig. 18-19 Remove the bearing cup. *(Chevrolet Motor Division of General Motors Corporation)*

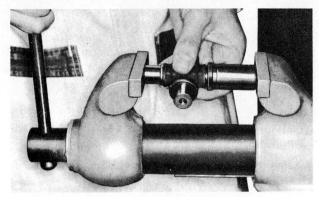

Fig. 18-20 Installing the trunnion seal. *(Chevrolet Motor Division of General Motors Corporation)*

Clean and inspect dust seals, bearing rollers, and trunnion. If all parts can be reused, repack the bearings and lubricate the reservoirs at the ends of trunnions with high-melting-point wheel-bearing lubricant. Use new seals. Then reassemble as explained in *b*. If any part is defective, install all the new parts included in the repair kit (Fig. 18-17).

b. Cleveland universal-joint reassembly When packing lubricant into the lubricant reservoirs at the ends of the trunnions, make sure that they are completely filled from the bottom. The use of a squeeze bottle is recommended to prevent air pockets at the bottom.

Use a trunnion-seal installer, as shown in Fig. 18-20, to install the seals. Then position the trunnion in the yoke. Partly install one bearing cup, as shown in Fig. 18-21. Then partly install the other bearing cup. Align the trunnion in the cups and press the cups into the yoke.

c. Constant-velocity universal-joint service In Fig. 18-22, the constant-velocity universal joint is shown with the bearing cups numbered in the order in which they should be removed. Figure 18-23 shows the universal joint with alignment punch marks, which serve as a guide for reassembly. Figure 18-24 shows a sectional view and Fig. 18-28 shows a disassembled view of this type of joint.

Remove the bearing cups in the same way they were removed on the Cleveland joint. Then disengage the flange yoke and trunnion from the centering ball. Figure 18-25 shows how the ball socket is part of the flange-

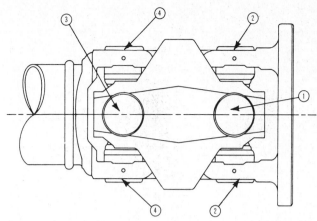

Fig. 18-22 Sequence in which bearing cups should be removed from a constant-velocity universal joint. *(Chevrolet Motor Division of General Motors Corporation)*

yoke assembly. The centering ball is pressed onto a stud and is part of the ball-stud yoke. Pry the seal from the ball socket and remove the washers, spring, and ball seats. Replace everything with a service kit.

Remove all plastic from the groove of the coupling yoke. To replace the centering ball, use a special puller (Fig. 18-26). The fingers of the puller are placed under the ball. Then a collar is put on the tool and a nut tightened on the puller screw threads. This pulls the ball off the ball stud. Then a new ball can be driven onto the stud.

Careful: When installing the new ball, it must seat firmly against the shoulder at the base of the stud.

Use the grease furnished with the service kit to lubricate all parts. Install parts in the clean ball-seat cavity in this order: spring, small washer, three ball seats with largest opening outward to receive the ball, large washer, and seal. Press the seal flush with a special tool. Fill the cavity with the grease provided in the kit. Install the flange yoke on the centering ball, making sure that the alignment marks line up. Install the

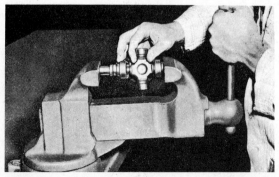

Fig. 18-21 Installing the bearing cup and trunnion. *(Chevrolet Motor Division of General Motors Corporation)*

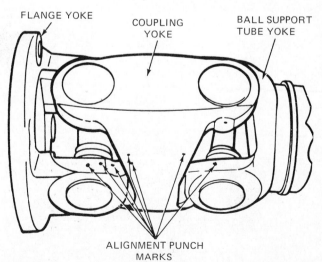

FLANGE YOKE   COUPLING YOKE   BALL SUPPORT TUBE YOKE

ALIGNMENT PUNCH MARKS

Fig. 18-23 Alignment punch marks in a constant-velocity universal joint, which aid in proper reassembly of the joint. *(Ford Motor Company)*

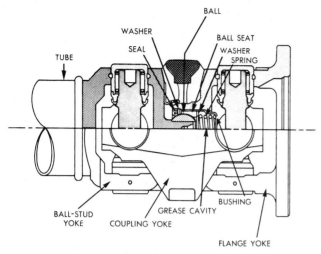

Fig. 18-24 Sectional view of a constant-velocity universal joint. *(Chevrolet Motor Division of General Motors Corporation)*

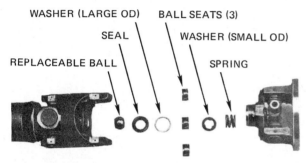

Fig. 18-25 Disassembled view of the centering-ball mechanism in a constant-velocity universal joint. *(Chevrolet Motor Division of General Motors Corporation)*

Fig. 18-26 Special tool for removing the ball from the stud. When a collar is placed over the tool and a nut is tightened on the screw threads, the ball is pulled off the stud. *(Chevrolet Motor Division of General Motors Corporation)*

trunnion and bearing caps in the same way they were installed on the Cleveland unit.

d. Driveshaft installation Inspect the yoke seal at the transmission extension. Replace it if necessary. Apply a light coating of transmission oil to the transmission-shaft splines. Insert the front shaft yoke into the transmission extension, making sure that the output-shaft splines mate with the shaft splines. Align the driveshaft with the companion flange using the reference marks made during removal. Remove the tapes used to retain the bearing caps. Connect the exposed bearing caps to the companion flange by installing the retainer strap and screws or bolts (Figs. 18-15 and 18-16).

✿ **18-7 Ford driveshaft service** Ford Motor Company vehicles use two types of universal joints: the single- and double-Cardan constant-velocity universal joints (Figs. 18-27 and 18-28). The driveshaft is re-

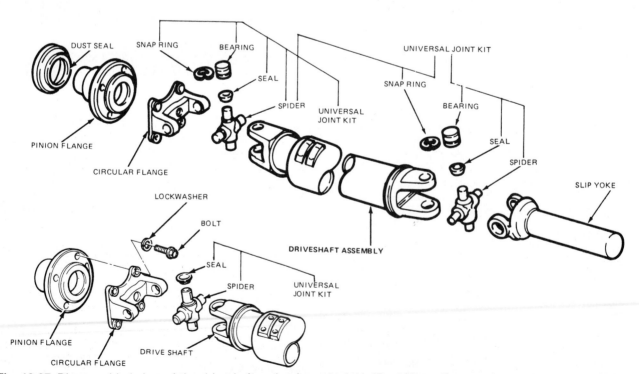

Fig. 18-27 Disassembled view of the driveshaft and universal joints. *(Ford Motor Company)*

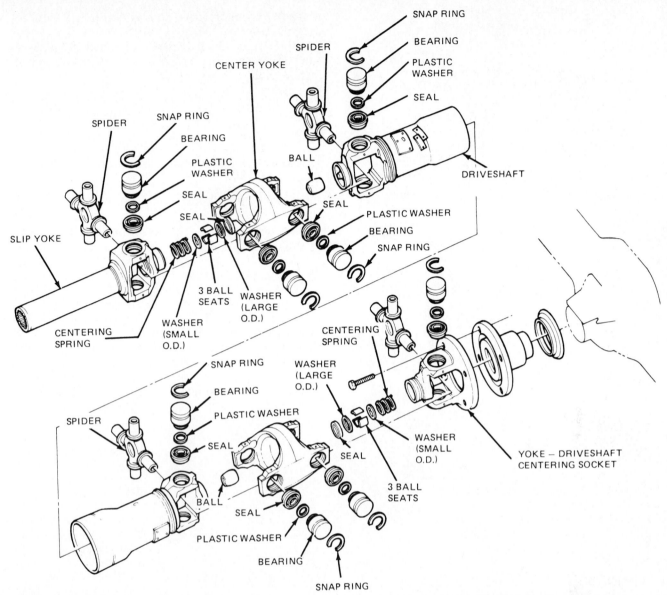

Fig. 18-28 Disassembled view of the driveshaft and constant-velocity universal joint. *(Ford Motor Company)*

moved and installed in the same manner as for the Chevrolet units, described in ✿ 18-6. Disassembly and reassembly are also similar. However, Ford recommends the use of a U-joint press, which is a special type of C clamp (Figs. 18-29 and 18-30).

✿ 18-8 Plymouth driveshaft service The driveshaft service recommended for Chrysler Corporation cars is very similar to that already described for Chevrolet. However, disassembly of the universal joint is different.

Mark parts to show the original relationship. Apply penetrating oil to the universal-joint bushings and remove the snap rings. Then, with one yoke supported in a vise, place on the yoke a socket that is larger in diameter than the bushing (Fig. 18-31). Tap on the socket with a soft hammer. This will drive the yoke down and the bushing up inside the socket.

After removing one bushing, turn the parts in the vise one half turn. Then remove the other bushing in the same way.

To reassemble the universal joint, align the marks made before disassembly. Hold the cross in position with one hand and start one bushing assembly into the yoke with the other hand (Fig. 18-32). Hammer the bushing into the yoke and install the snap ring. Install the opposite bushing and snap ring in the same way. Then repeat the procedure to install the other yoke.

Driveshaft and universal-joint service—front-engine, front-wheel-drive cars

This part of the chapter deals with the servicing of the driveshafts and universal joints used in vehicles with the engine in front and with the front wheels being driven.

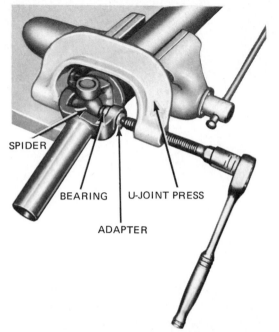

Fig. 18-29 Removing a universal-joint bearing with a special U-joint press. *(Ford Motor Company)*

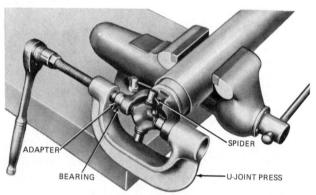

Fig. 18-30 Installing a universal-joint bearing with a U-joint press. *(Ford Motor Company)*

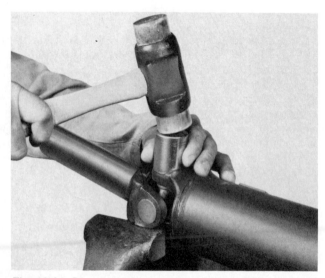

Fig. 18-31 Disassembling a universal joint. *(Chrysler Corporation)*

✿ **18-9 Driveshaft and universal-joint trouble-diagnosis chart** The chart that follows lists various possible complaints that might be blamed on the driveshafts or universal joints on front-wheel-drive cars with front engines.

NOTE: See ✿ 3-1 for a discussion of diagnostic theory.

Driveshaft and Universal Joint Trouble-Diagnosis Chart

(For Front-Engine, Front-Wheel-Drive Vehicles)

COMPLAINT	POSSIBLE CAUSE
1. Vehicle pulls to one side	a. Bent driveshaft b. Defective universal joint c. Dragging wheel system d. Trouble in front suspension or steering
2. Steering-wheel shimmy; front-end instability	a. Bent driveshaft b. Worn or loose universal joint c. Worn or loose wheel bearing d. Worn driveshaft or wheel hub splines e. Worn driveshaft or differential-side-gear splines f. Tight or locked universal joint g. Tight wheel bearing h. Trouble in front suspension or steering
3. Noises	a. Clicking noise in turns could be due to a damaged or worn outboard universal joint; check for damaged seals b. A clunk when accelerating from coast to drive could be due to a worn inboard universal joint c. Shudder or vibration when accelerating could be due to excessive joint angles, worn or damaged universal joints, or a sticking inboard joint d. Vibration at highway speed is probably due to out-of-balance front wheels or tires, or to out-of-round front tires e. Other noises can come from worn or loose splines in the drive line, or from troubles in the suspension or steering

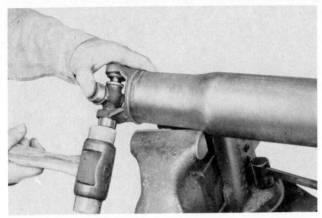

Fig. 18-32 Assembling a universal joint. *(Chrysler Corporation)*

❀ 18-10 Servicing the driveshaft on Chrysler Corporation front-drive cars Figures 17-30 and 17-32 show the details of a driveshaft and how it is attached to the wheel hub. The inboard double-offset universal joint can be disassembled. However, Chrysler specifies that you should not disassemble the outboard (Birfield) joint. (General Motors does supply disassembling information on this joint.) The servicing procedure for the Chrysler driveshaft follows:

1. Removal   To remove the driveshaft, proceed as follows:

1. Remove the hub cap and loosen the driveshaft nut.
2. Raise the vehicle on a lift that supports the front end on the suspension system, not on the tires.
3. Remove the wheels.
4. Remove the undercover.
5. Remove the lower-arm ball joint and strut bar from the lower control arm.

Careful: Put the lower-arm ball joint on the lower arm to prevent damage to the ball-joint dust boot.

6. Drain the transaxle lubricant.
7. Insert the wheel-nut wrench carried in the car or a similar tool between the transaxle case and the uni-

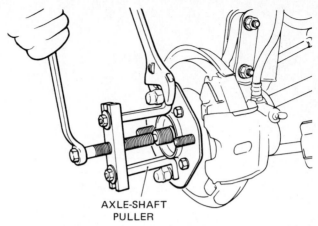

Fig. 18-34 Separating the driveshaft from the wheel hub. *(Chrysler Corporation)*

versal-joint outer case (Fig. 18-33). Pry the universal-joint case away from the transaxle.

Careful: Pry against the rib of the transaxle. Do not insert the tool too deep or you will damage the oil seal. Do not let the driveshaft assembly drop as you pull it out. Put a clean cloth in the hole in the transaxle to keep out dust.

8. Use the special axle-shaft puller as shown in Fig. 18-34 to force the driveshaft out of the wheel hub. Keep the spacer from falling out of place. Place the assembly on the bench for service.

2. Inspection before disassembly Inspect the boots for damage and check the two universal joints for looseness and smoothness of operation. Examine the splines for wear.

3. Disassembly   Do not disassemble the outboard (Birfield) universal joint. Use only the specified grease on reassembly. To continue with the disassembly:

1. Remove the inboard boot band and discard it (Fig. 18-35).
2. Remove the Circlip (snap ring) (Fig. 18-36).
3. Separate the driveshaft from the joint. Remove the

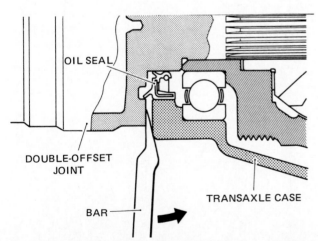

Fig. 18-33 Separating the driveshaft from the transaxle. *(Chrysler Corporation)*

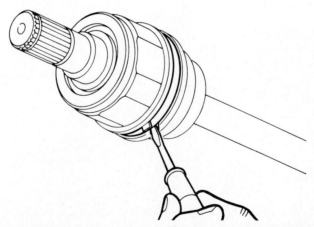

Fig. 18-35 Removing the boot band. *(Chrysler Corporation)*

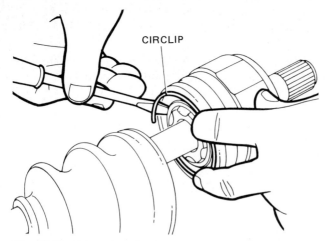

Fig. 18-36 Removing the Circlip. *(Chrysler Corporation)*

snap ring and the inner race, cage, and balls (Fig. 18-37).

4. Wash the inner race, cage, and balls in clean solvent without separating them. Lubricate immediately.
5. If you are going to reuse a boot, wrap some tape around the shaft splines to protect the boot before removing it (Fig. 18-38).
6. Inspect parts for damage and discard defective or worn parts.
7. When parts need replacement, get the proper repair kit (Fig. 18-39).

To reassemble the driveshaft:

1. Wrap tape around the shaft splines to protect the boots (Fig. 18-38).

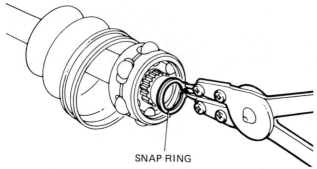

Fig. 18-37 Removing the inner race, cage, and balls. *(Chrysler Corporation)*

Repair kit name	Illustration
Driveshaft and Birfield joint	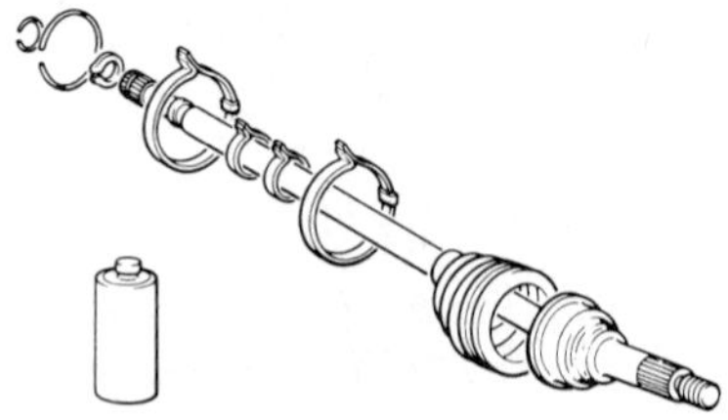
Double offset joint	
Birfield joint	
Double offset joint boot	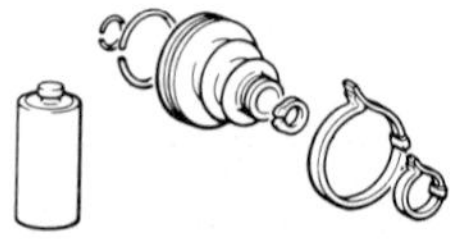

Fig. 18-39 Driveshaft repair kits. *(Chrysler Corporation)*

2. Lubricate the shaft and install the boots. Figure 18-40 shows the difference between the two boots.
3. Grease the inner cage with the grease included with the repair kit. If the balls have come out, position the inner race in the cage (Fig. 18-41) and install the balls.
4. Install the inner race on the driveshaft, chamfered side facing out as shown in Fig. 18-42. Secure with the snap ring.
5. Install the outer race on the shaft after applying some of the specified grease to the ball grooves.

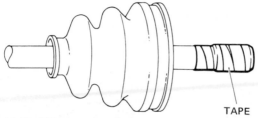

Fig. 18-38 Wrap the splined end of the driveshaft with tape before removing the boot if the boot is to be reinstalled. *(Chrysler Corporation)*

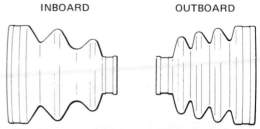

Fig. 18-40 Identification of the inboard and outboard boots. *(Chrysler Corporation)*

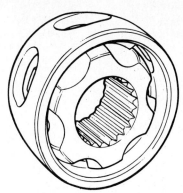

Fig. 18-41 Inner race and cage, put together. *(Chrysler Corporation)*

6. Install the outboard universal joint.
7. Attach the boots and tighten the boot bands.
8. The two bands on the inboard universal joint should be separated by the distance shown in Fig. 18-43.
9. When installing the assembly on the vehicle, use a new retainer ring on the inboard side. The washer on the driveshaft outer end should be installed in the position shown in Fig. 18-44. Secure the nut with a cotter pin.

✿ 18-11 Servicing the driveshaft on General Motors front-drive cars Figure 17-33 shows both front driveshafts completely disassembled. Note that both the inboard and outboard universal joints can be disassembled.

To remove the shafts from the vehicle:

1. Remove the hub nut and raise the car on a lift.
2. Remove the front wheels.
3. Install a shaft-boot-seal protector (Fig. 12-5).
4. Disconnect the brake-line clip at the strut.
5. Remove the brake caliper and support it away from the work area with a wire.
6. Mark the cam bolt to ensure proper camber alignment on reinstallation. Remove the cam bolt and upper attaching bolt.
7. Pull the steering knuckle out of the strut bracket.
8. Use the special slide hammer as shown in Fig. 12-5 to remove the driveshaft from the transaxle.

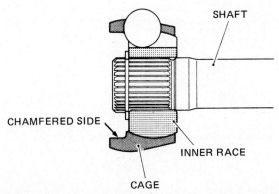

Fig. 18-42 Proper position of the inner race on the driveshaft. *(Chrysler Corporation)*

9. Use the special spindle remover as shown in Fig. 18-45 to separate the shaft from the hub-and-bearing assembly.

Figures 18-46 to 18-49 show, step by step, how to service the inner and outer joints and joint seals.

The installation procedure follows. If the driveshaft is being replaced, replace the knuckle seal.

1. Loosely install the driveshaft on the steering knuckle and transaxle.

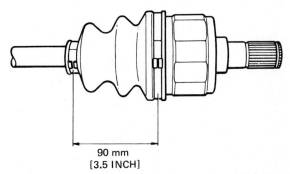

Fig. 18-43 Installing the inboard boot. *(Chrysler Corporation)*

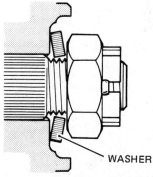

Fig. 18-44 Proper position of the washer. *(Chrysler Corporation)*

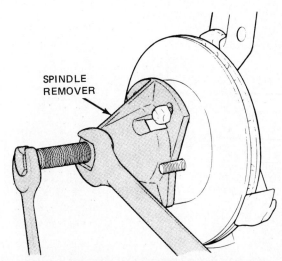

Fig. 18-45 Removing the driveshaft from the hub. *(Chevrolet Motor Division of General Motors Corporation)*

219

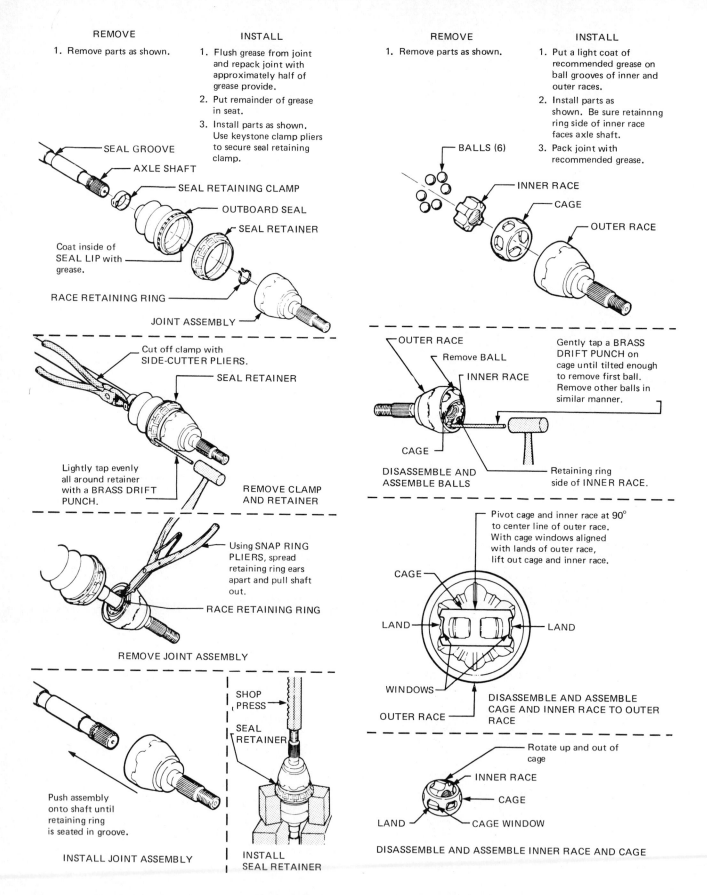

REMOVE

1. Remove parts as shown.

INSTALL

1. Flush grease from joint and repack joint with approximately half of grease provide.
2. Put remainder of grease in seat.
3. Install parts as shown. Use keystone clamp pliers to secure seal retaining clamp.

SEAL GROOVE

AXLE SHAFT

SEAL RETAINING CLAMP

OUTBOARD SEAL

SEAL RETAINER

Coat inside of SEAL LIP with grease.

RACE RETAINING RING

JOINT ASSEMBLY

Cut off clamp with SIDE-CUTTER PLIERS.

SEAL RETAINER

Lightly tap evenly all around retainer with a BRASS DRIFT PUNCH.

REMOVE CLAMP AND RETAINER

Using SNAP RING PLIERS, spread retaining ring ears apart and pull shaft out.

RACE RETAINING RING

REMOVE JOINT ASSEMBLY

Push assembly onto shaft until retaining ring is seated in groove.

INSTALL JOINT ASSEMBLY

SHOP PRESS

SEAL RETAINER

INSTALL SEAL RETAINER

Fig. 18-46 Removing and installing the outboard universal-joint seal. *(Chevrolet Motor Division of General Motors Corporation)*

REMOVE

1. Remove parts as shown.

INSTALL

1. Put a light coat of recommended grease on ball grooves of inner and outer races.
2. Install parts as shown. Be sure retainnng ring side of inner race faces axle shaft.
3. Pack joint with recommended grease.

BALLS (6)

INNER RACE

CAGE

OUTER RACE

OUTER RACE

Remove BALL

INNER RACE

CAGE

Gently tap a BRASS DRIFT PUNCH on cage until tilted enough to remove first ball. Remove other balls in similar manner.

Retaining ring side of INNER RACE.

DISASSEMBLE AND ASSEMBLE BALLS

Pivot cage and inner race at 90° to center line of outer race. With cage windows aligned with lands of outer race, lift out cage and inner race.

CAGE

LAND

LAND

WINDOWS

OUTER RACE

DISASSEMBLE AND ASSEMBLE CAGE AND INNER RACE TO OUTER RACE

Rotate up and out of cage

INNER RACE

CAGE

LAND

CAGE WINDOW

DISASSEMBLE AND ASSEMBLE INNER RACE AND CAGE

Fig. 18-47 Disassembling and reassembling the outboard universal joint. *(Chevrolet Motor Division of General Motors Corporation)*

220

1. Remove parts as shown.

1. Flush grease from joint. Repack joint with approximately half of grease provided.

2. Put remainder of grease in seal.

3. Install parts as shown. Use keystone clamp pliers to secure seal retaining clamp.

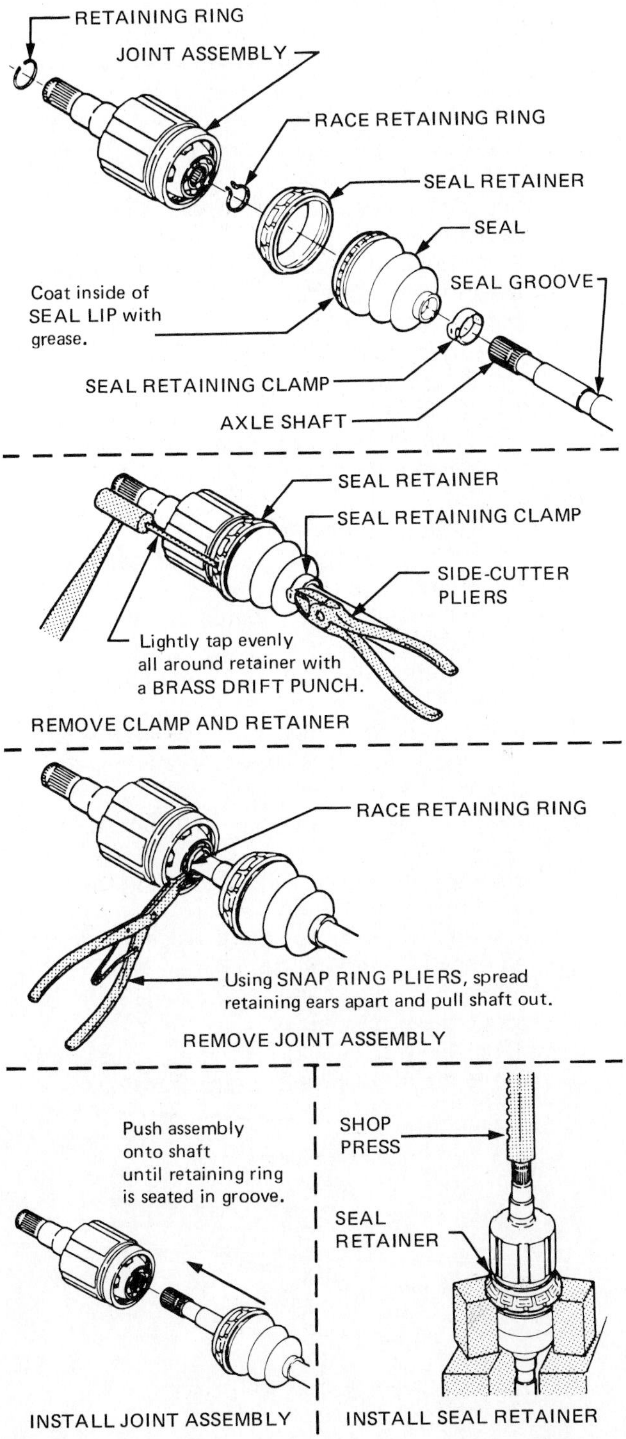

RETAINING RING

JOINT ASSEMBLY

RACE RETAINING RING

SEAL RETAINER

SEAL

Coat inside of SEAL LIP with grease.

SEAL GROOVE

SEAL RETAINING CLAMP

AXLE SHAFT

SEAL RETAINER

SEAL RETAINING CLAMP

SIDE-CUTTER PLIERS

Lightly tap evenly all around retainer with a BRASS DRIFT PUNCH.

REMOVE CLAMP AND RETAINER

RACE RETAINING RING

Using SNAP RING PLIERS, spread retaining ears apart and pull shaft out.

REMOVE JOINT ASSEMBLY

Push assembly onto shaft until retaining ring is seated in groove.

SHOP PRESS

SEAL RETAINER

INSTALL JOINT ASSEMBLY

INSTALL SEAL RETAINER

Fig. 18-48 Removing and installing the inboard universal-joint seal. *(Chevrolet Motor Division of General Motors Corporation)*

REMOVE

INSTALL

1. Remove parts as shown.

1. Install parts as shown. Retaining ring side of inner race and small end of cage face axle shaft.

2. Pack joint with recommended grease

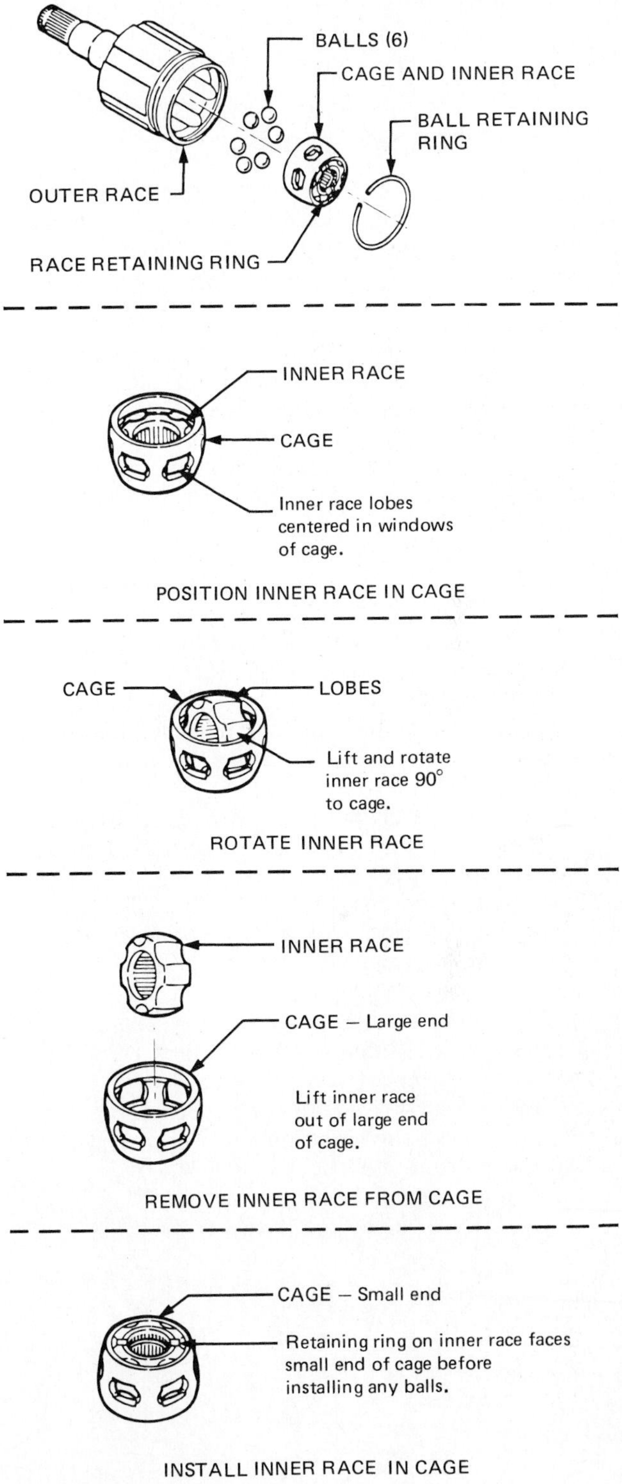

BALLS (6)

CAGE AND INNER RACE

BALL RETAINING RING

OUTER RACE

RACE RETAINING RING

INNER RACE

CAGE

Inner race lobes centered in windows of cage.

POSITION INNER RACE IN CAGE

CAGE

LOBES

Lift and rotate inner race 90° to cage.

ROTATE INNER RACE

INNER RACE

CAGE — Large end

Lift inner race out of large end of cage.

REMOVE INNER RACE FROM CAGE

CAGE — Small end

Retaining ring on inner race faces small end of cage before installing any balls.

INSTALL INNER RACE IN CAGE

Fig. 18-49 Disassembling and reassembling the inboard universal joint. *(Chevrolet Motor Division of General Motors Corporation)*

2. Install the brake caliper.
3. Install the driveshaft on the steering knuckle. The drive axle is an interference fit. Install the hub nut. When the shaft begins to turn, stick a drift punch into a slot in the brake rotor to hold it while the nut is tightened.
4. Apply a load to the hub assembly by lowering the vehicle onto a safety stand. Align the cam bolt with the alignment marks and torque the nuts to specifications.
5. Connect the brake-line clip to the strut bracket.
6. Install the wheels.
7. Lower the vehicle and tighten the hub nut to specifications.

Chapter 18 review questions

Select the *one* correct, best, or most probable answer to each question. Then check your answers against the correct answers given at the end of the book.

1. Most universal joints:
   a. require periodic maintenance
   b. require periodic lubrication
   c. require no periodic maintenance
   d. should be replaced at periodic intervals
2. Hose clamps can be installed on a driveshaft to:
   a. prevent the loss of bearing rollers
   b. hold in the universal-joint lubricant
   c. prevent excessive slippage
   d. balance the driveshaft
3. The first service step to correct driveshaft balance is to:
   a. disconnect the driveshaft and turn it 180°
   b. check for and remove all traces of undercoat from the driveshaft
   c. rotate the driveshaft and check for high spots with a crayon
   d. none of the above.
4. If the two hose clamps overbalance the driveshaft:
   a. move the clamp heads away from each other
   b. remove one clamp
   c. install lighter clamps
   d. install a clamp 180° from the original two
5. Hose clamps cannot be installed on the driveshafts of some cars because:
   a. the clamps could damage the driveshaft
   b. there is too little clearance between the driveshaft and the tunnel in the floorpan
   c. it would further unbalance the driveshaft
   d. some cars have no tunnel
6. Driveshaft runout is checked:
   a. with the engine running at low speed
   b. with the transmission in low gear
   c. with a dial indicator
   d. after disconnecting the driveshaft from the transmission
7. If the car pulls to one side in a front-wheel-drive car, the trouble could be:
   a. a bent driveshaft
   b. a defective universal joint
   c. a dragging wheel bearing
   d. all of the above
8. One difference between General Motors and Chrysler Corporation service instructions is that:
   a. Chrysler specifies you should not disassemble the Birfield universal joint
   b. Chrysler specifies you should disassemble the Birfield universal joint
   c. Chrysler specifies you should not disassemble either universal joint
   d. General Motors specifies you should not disassemble the Birfield joint
9. The first step in servicing the driveshaft of transaxle-equipped Chrysler Corporation cars is to:
   a. remove the wheels
   b. remove the hub cap and loosen the driveshaft nut
   c. disconnect the lower-arm ball joint and strut bar
   d. drain the transaxle fluid
10. The first step in servicing the driveshaft of transaxle-equipped General Motors cars is to:
   a. remove the hub nut and raise the car on a lift
   b. remove the front wheels
   c. pull the steering knuckle out of the strut bracket
   d. remove the driveshaft from the transaxle

CHAPTER 19

DIFFERENTIALS AND DRIVE AXLES

After studying this chapter, and with proper instruction and equipment, you should be able to:

1. Explain the construction and operation of a differential.
2. Identify the parts of a disassembled differential.
3. Describe the operation of a limited-slip differential.
4. Discuss the construction of and action in the front axle of a four-wheel-drive vehicle.
5. Name three types of hubs used on the front wheels of four-wheel-drive vehicles, and describe the operation of each.

19-1 Differential applications This chapter describes the construction and operation of different types of differentials and drive axles used on automobiles. These are differentials that are used in front-engine, rear-wheel-drive cars.

Differentials are also used in transaxles (⚙ 4-14 to 4-16) and in transfer cases (⚙ 4-17). The basic differential discussed in this chapter is located at the back of the vehicle, between the two rear wheels (Fig. 19-1). It is connected to the transmission through the driveshaft.

NOTE: Manufacturers' service manuals usually refer to the differential as the *rear axle* or *rear-axle assembly*. By this, they mean the rear-axle housing, the wheel axles, and the differential assembly. In this book, the term *differential* is used to designate the differential assembly itself.

⚙ **19-2 Function of the differential** A differential is required to compensate for the difference in distance that the drive wheels travel when the car rounds a curve. If a right-angle turn were made with the inner rear wheel turning on a 20-foot (6.1-m) radius, the inner rear wheel would travel about 31 feet (9.5 m) while the outer rear wheel would travel about 39 feet (12 m) (Fig. 19-2). The differential permits application of power to both drive wheels while allowing the wheels to turn different amounts when the car is rounding a curve.

DIFFERENTIAL

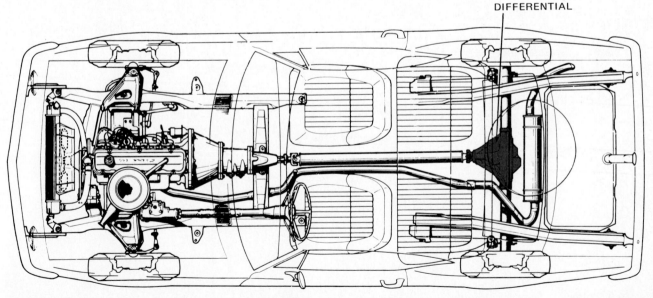

Fig. 19-1 Location of the differential in the power train.

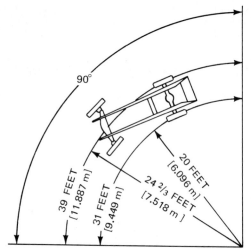

Fig. 19-2 Difference in wheel travel as car makes a 90° turn with the inner wheel turning on a 20-foot [6.1-m] radius.

⚙ **19-3 Construction of the differential** Figure 19-3 is a cutaway view of a rear-axle-and-differential assembly. Figure 19-4 shows a similar axle and differential in exploded view.

To study differential construction and operation, we build up, gear by gear, a simple differential. The two rear wheels are mounted on axles. On the inner end of each axle is a small bevel gear called a differential side gear (Fig. 19-5). All the teeth of a bevel gear are at an angle. When any two bevel gears are put together so that their teeth mesh, the driving and driven shafts can be at a 90° angle (Fig. 19-6).

Figure 19-7 shows all the essential parts of a differential. The parts are separated so that they can be seen clearly. Keep referring to Fig. 19-7 as we put the parts of the differential together.

First, we add the differential case to the two wheel axles and differential side gears (Fig. 19-8). The differential case has bearings that permit it to rotate independently of the two axles. Next, we add the two pinion gears and their supporting shaft, called the *pinion shaft* (Fig. 19-9). The shaft fits into the differential case. The two pinion gears are held in place by the pinion shaft. They mesh with the two differential side gears, which are attached to the inner ends of the axle shafts.

Next, we add the ring gear, as shown in Fig. 19-10. The ring gear is bolted to a flange on the differential case. The differential case is rotated by the ring gear which is attached to it. Finally, we add the drive pinion (Fig. 19-11). The drive pinion is at the end of the drive-shaft and meshes with the ring gear. When the drive-shaft turns, the drive pinion rotates, which rotates the ring gear.

⚙ **19-4 Operation of the differential** When the car is on a straight road, the ring gear, differential case, differential pinion gears, and two differential side gears all turn as a unit without any relative motion.

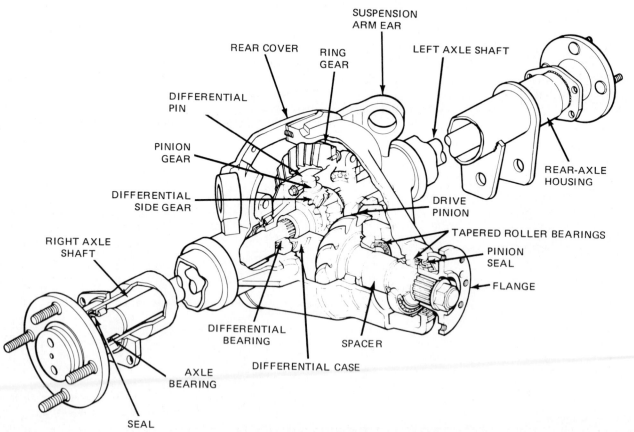

Fig. 19-3 Cutaway view of a differential and rear axle. *(Ford Motor Company)*

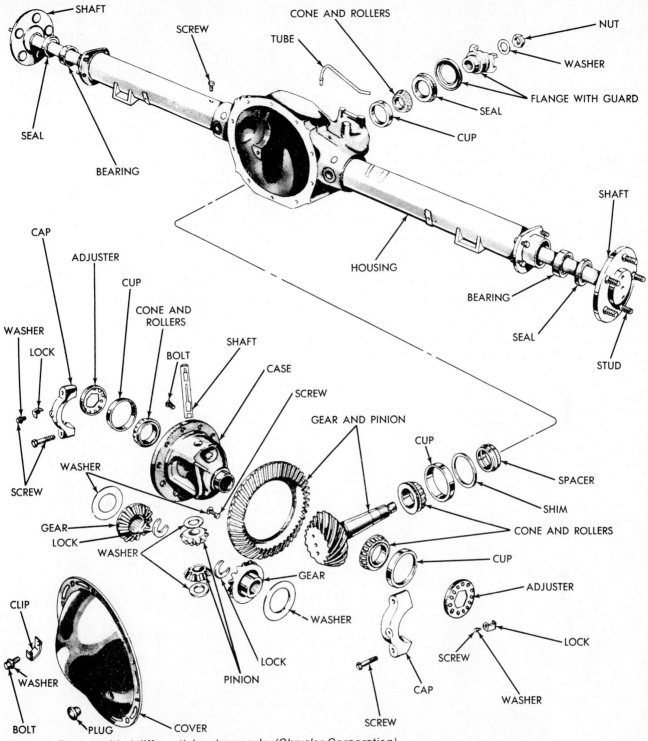

Fig. 19-4 Disassembled differential and rear axle. *(Chrysler Corporation)*

When the car is on a straight road, the two differential pinion gears do not rotate on the pinion shaft. This is because they exert equal force on the two side gears. As a result, the side gears turn at the same speed as the ring gear, which causes both drive wheels to turn at the same speed also.

However, when the car begins to round a curve, the differential pinion gears rotate on the pinion shaft. This permits the outer wheel to turn faster than the inner wheel.

Suppose one wheel turns slower than the other as the car rounds a curve. As the differential case rotates, the pinion gears must rotate on their shaft. The reason for this is that the pinion gears must "walk around" the slower-turning differential side gear. Therefore, the pinion gears carry additional rotary motion to the

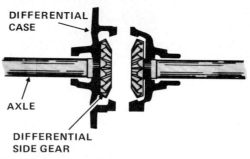

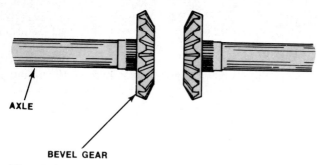

Fig. 19-5 Inner ends of the rear axles with bevel gears (differential side gears) installed on them.

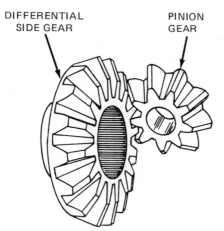

Fig. 19-6 Two meshing bevel gears.

Fig. 19-8 Adding the differential case.

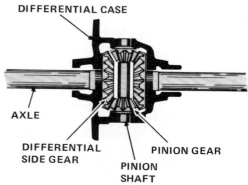

Fig. 19-9 Adding the two pinion gears and supporting shaft.

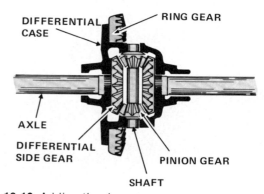

Fig. 19-10 Adding the ring gear.

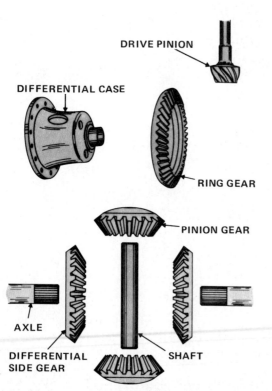

Fig. 19-7 Basic parts of a differential.

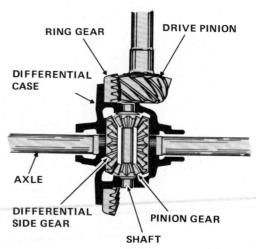

Fig. 19-11 To complete the basic differential, the drive pinion is added. The drive pinion is meshed with the ring gear.

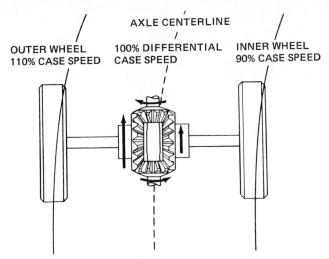

OUTER WHEEL
110% CASE SPEED

AXLE CENTERLINE

100% DIFFERENTIAL
CASE SPEED

INNER WHEEL
90% CASE SPEED

Fig. 19-12 Differential action on turns. *(Chevrolet Motor Division of General Motors Corporation)*

faster-turning outer wheel on the turn. The action in a typical situation is shown in Fig. 19-12. The differential-case speed is considered to be 100 percent. The rotating action of the pinion gears carries 90 percent of this speed to the slower-rotating inner wheel. It sends 110 percent of the speed to the faster-rotating outer wheel.

This is how the differential allows one drive wheel to turn faster than the other. Whenever the car goes around a turn, the outer drive wheel travels a greater distance than the inner drive wheel. The two pinion gears rotate on their shaft and send more rotary motion to the outer wheel.

When the car moves down a straight road, the pinion gears do not rotate on their shaft. They apply equal torque to the differential side gears. Therefore, both drive wheels rotate at the same speed.

☸ 19-5 Differential gearing Since the ring gear has many more teeth than the drive pinion, a considerable gear reduction is produced in the differential. The gear ratios vary on different cars, depending on car and engine design. Ratios from 2:1 upward to about 4:1 are used on passenger cars. This means that the ring gear has from two to four times as many teeth as the drive pinion. Therefore the drive pinion must rotate from two to four times (according to gear ratio) to cause the ring gear to rotate once. For heavy-duty trucks, ratios of about 9:1 may be used. Such high ratios are secured by use of *double-reduction* gearing (Fig. 19-13), as described in ☸ 19-7.

The gear ratio in the differential is usually referred to as the *axle ratio*. However, it would be more accurate to call it the *differential ratio*.

Early cars used simple spur-gear-type drive pinions and ring gears (Fig. 19-14). In this type of gearing, the lines of the gear teeth are straight and all point toward the center of the gear. The center line of the drive-

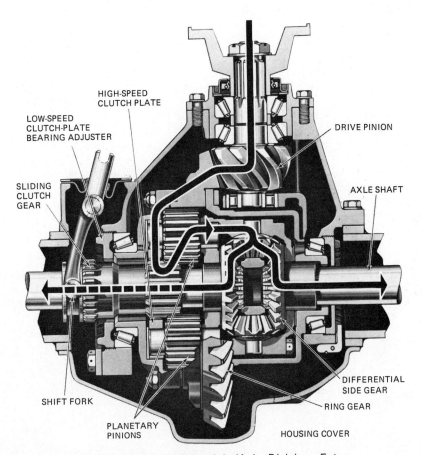

LOW-SPEED
CLUTCH-PLATE
BEARING ADJUSTER

HIGH-SPEED
CLUTCH PLATE

DRIVE PINION

SLIDING
CLUTCH
GEAR

AXLE SHAFT

DIFFERENTIAL
SIDE GEAR

RING GEAR

SHIFT FORK

PLANETARY
PINIONS

HOUSING COVER

Fig. 19-13 Sectional view of a double-reduction differential. *(Axle Division, Eaton Corporation)*

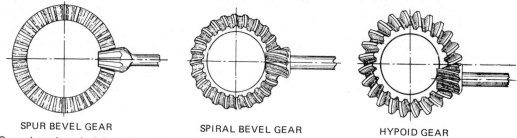

SPUR BEVEL GEAR   SPIRAL BEVEL GEAR   HYPOID GEAR

Fig. 19-14 Spur bevel, spiral bevel, and hypoid differential drive pinions and ring gears.

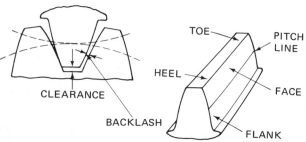

Fig. 19-15 Gear-tooth nomenclature. *(Chrysler Corporation)*

pinion shaft, if extended, would intersect the center line of the axles.

A later design made use of spiral bevel gears (Fig. 19-14), in which the teeth have a curved, or spiral, shape. This shape permits contact between more than one pair of teeth at a time. More even wear and quieter operation result. Extension of the center line of the drive-pinion shaft would intersect the axle center line.

Modern front-engine, rear-drive cars with low bodies created the problem of interference between the driveshaft and the floor of the car body. To permit further lowering of the car body without interference with the driveshaft, hypoid differential gears are used (Figs. 19-3 and 19-4). These gears are similar to spiral bevel gears except that the tooth formation allows the drive-pinion shaft to be lowered. In this type of gear, a wiping action takes place between the teeth as the teeth mesh and unmesh. This wiping action, which is characteristic of hypoid gears, makes the use of special hypoid-gear lubricants necessary.

Figure 19-15 shows gear-tooth nomenclature. The mating teeth to the left illustrate clearance and backlash, and the tooth to the right has the various tooth parts named. *Clearance* is the distance between the top of the tooth of one gear and the valley between adjacent teeth of the mating gear. *Backlash* is the distance between adjacent meshing teeth in the driving and driven gears. It is the distance one gear can rotate backward, or backlash, before it will cause the other gear to move. The *toe* is the smaller section of the gear tooth which is nearest the center of the ring gear. The *heel* is the larger section of the gear tooth, which is farthest from the center of the ring gear. The toe end of the tooth is smaller than the heel end.

⚙ 19-6 Hunting and nonhunting gear sets The gear set (consisting of the drive pinion and the ring gear) may be of either the *hunting* or the *nonhunting* type. Hunting refers to the number of ring-gear teeth that each drive-pinion tooth makes contact with. In a hunting-type gear set, any one pinion-gear tooth will come into contact with every ring-gear tooth. In a non-hunting gear set, any one pinion-gear tooth will come into contact with only a few gear teeth.

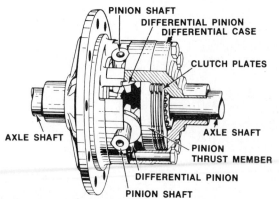

Fig. 19-16 Cutaway view of a limited-slip differential. *(Chrysler Corporation)*

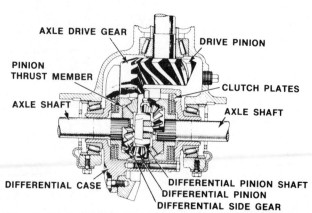

Fig. 19-17 Sectional view of a limited-slip differential. *(Chrysler Corporation)*

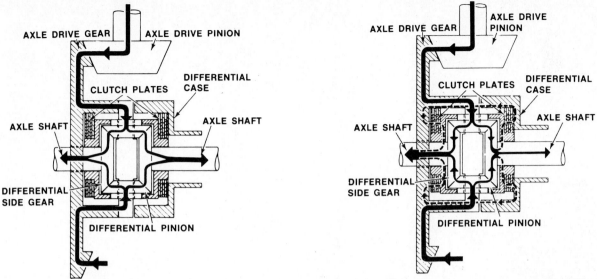

Fig. 19-18 Power flow through a limited-slip differential on a straightaway. *(Chrysler Corporation)*

Fig. 19-19 Power flow through a limited-slip differential when rounding a turn. Heavy arrows show greater torque to the left axle shaft. *(Chrysler Corporation)*

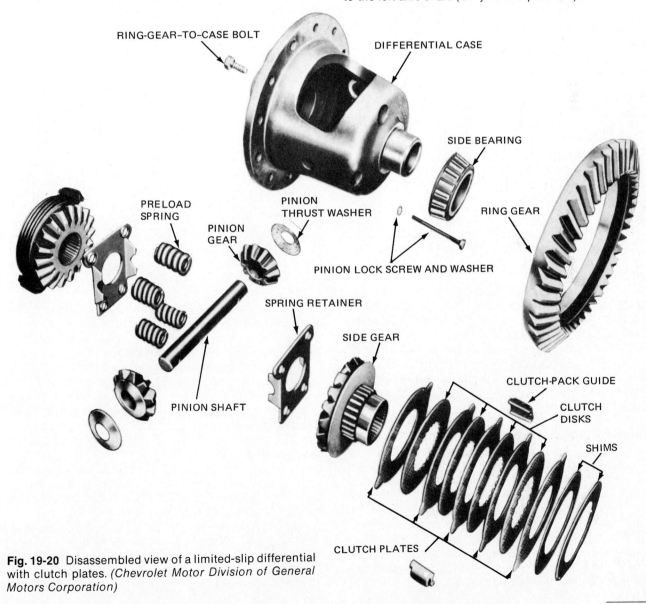

Fig. 19-20 Disassembled view of a limited-slip differential with clutch plates. *(Chevrolet Motor Division of General Motors Corporation)*

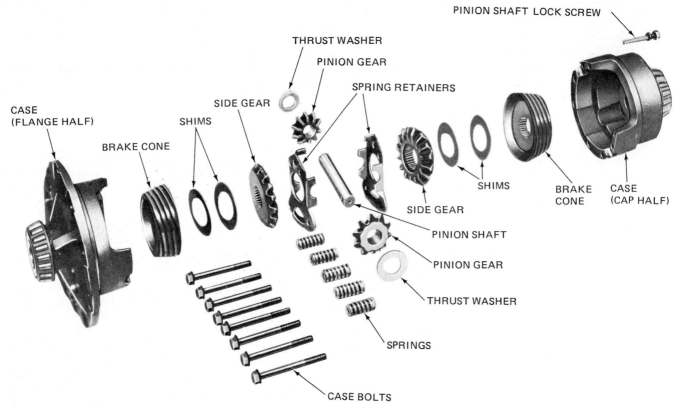

Fig. 19-21 Disassembled view of a limited-slip differential with clutch cones. *(Chevrolet Motor Division of General Motors Corporation)*

For example, suppose a differential drive pinion has 13 teeth and the ring gear has 39 teeth. In this gear set, each drive-pinion tooth would make contact with only three ring-gear teeth. Every three drive-pinion revolutions, each drive-pinion tooth would meet with the same three ring-gear teeth. This is a nonhunting gear set because the drive-pinion teeth do not "hunt out" different ring-gear teeth. They mesh with the same teeth.

However, suppose that the drive pinion has 9 teeth and the ring-gear has 37 teeth. Now, each drive-pinion tooth will mesh with all ring-gear teeth as the gear set revolves. This is a hunting gear set.

When checking the gear-tooth-contact pattern, it is important to know whether the gear set is hunting or nonhunting. The checking procedures and acceptable patterns are different for the two gear sets, as we will explain later.

✿ 19-7 Double-reduction differentials To secure additional gear reduction through the differential and provide a higher gear ratio between the engine and the rear wheels, some large trucks use *double-reduction differentials* (Fig. 19-13). In this type of differential the drive pinion meshes with a ring gear assembled to a straight shaft on which there is a reduction-drive gear set. The reduction-drive gear set drives a driven gear set that has a greater number of gear teeth. Gear reduction is obtained between the drive pinion and the ring gear and also between the reduction gear sets.

The driven gear set is attached to the differential case, the case being supported by bearings in the differential housing in a manner similar to the differential discussed earlier. Notice that the differential illustrated in Fig. 19-13 has a four-pinion differential instead of the two-pinion differential shown in Figs. 19-3 and 19-4. Otherwise, the construction and principle of operation are the same as on the differential described earlier.

✿ 19-8 Limited-slip differentials The standard differential delivers the same amount of torque to each rear wheel when both wheels have equal traction. When one wheel has less traction than the other—for example, when one wheel slips on ice—the other wheel cannot deliver torque. All the turning effort goes to the slipping wheel. To provide good traction even though one wheel is slipping, a *limited-slip differential* is used in many cars. It is very similar to the standard unit but has some means of preventing wheel spin and loss of traction.

To sum up, the standard differential delivers maximum torque to the wheel with minimum traction. However, the limited-slip differential delivers maximum torque to the wheel with maximum traction.

One of the older types of limited-slip differentials is shown in Fig. 19-16. It has two sets of clutch plates. Also, the ends of the pinion-gear shafts lie loosely in notches in the two halves of the differential case. Figure 19-17 is a sectional view of the limited-slip differential.

During normal straight-road driving, the power flow is as shown in Fig. 19-18.

NOTE: In Figs. 19-16 to 19-18, the ring gear is called the *axle drive gear*. The pinion gears are called the *differential pinions*. Different manufacturers use different names for the same parts.

Note that the rotating differential case carries the pinion-gear shafts around with it. Since there is considerable side thrust, the pinion shafts tend to slide up the sides of the notches in the two halves of the differential case. As the pinion shafts slide up, they are forced outward. This force is carried to the two sets of clutch plates. The clutch plates lock the axle shafts to the differential case. Therefore both wheels turn.

Suppose one wheel encounters a patch of ice or snow and loses traction, or tends to slip. Then the force is released on the clutch plates feeding power to that wheel. The torque goes to the other wheel, and the wheel on the ice does not slip.

During normal driving, if the car rounds a curve, force is released on the clutch feeding the inner wheel. Just enough force is released to permit some slipping. Figure 19-19 shows the action. This release of force permits the outer wheel to turn faster than the inner wheel.

Late-model limited-slip differentials are of two types: those with spring-loaded clutch plates and those with spring-loaded clutch cones. These differentials are shown in disassembled views in Figs. 19-20 and 19-21. The action is the same as in the limited-slip differential described above. The essential difference is that the plates or cones are preloaded by springs to give a more positive action.

⚙ 19-9 Four-wheel-drive drive axles The most widely used application of four-wheel drive is in multipurpose utility vehicles, such as the American Motors Jeep, Chevrolet Blazer, and Ford Bronco. These vehicles, along with most other four-wheel-drive cars and trucks, normally drive the rear axle. But when four-wheel drive is engaged through the transfer case, power is also delivered to the front axle.

Chapter 4 covers various types of transfer cases used in these vehicles. Differential action is covered earlier in this chapter. The design and operation of various types of universal joints are discussed in Chap. 17.

The front axle on most four-wheel-drive vehicles is very similar to the rear axle on vehicles with rear-wheel drive. As in the rear axle, the differential and front-axle shafts are carried inside a rigid tube, or axle housing (Fig. 19-22). However, the front axle is the steering axle. Therefore universal joints are located in the right and left axle shafts so that the outer ends can swing with the steering knuckle. Figure 19-23 shows a front-drive axle in exploded view.

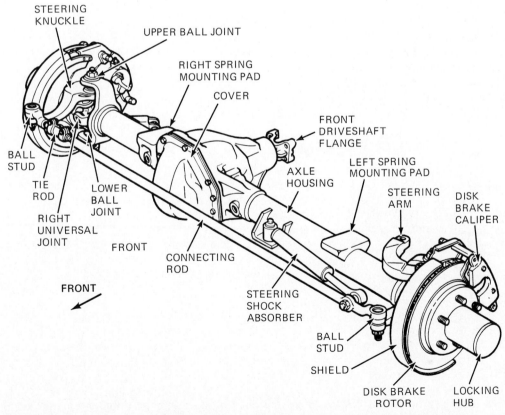

Fig. 19-22 A typical front-drive axle used on a four-wheel-drive vehicle. *(Chevrolet Motor Division of General Motors Corporation)*

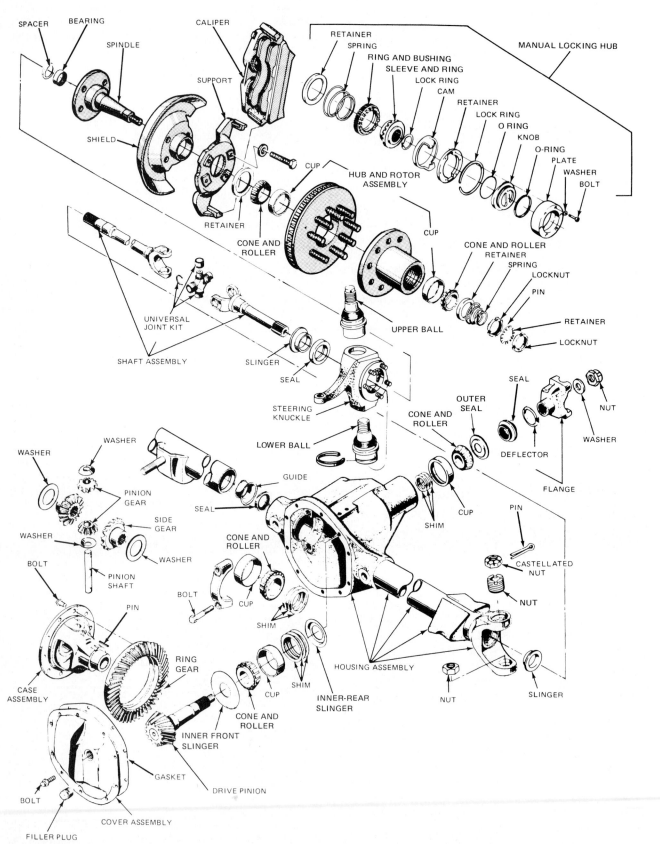

Fig. 19-23 A complete front-drive axle assembly in exploded view. *(Ford Motor Company)*

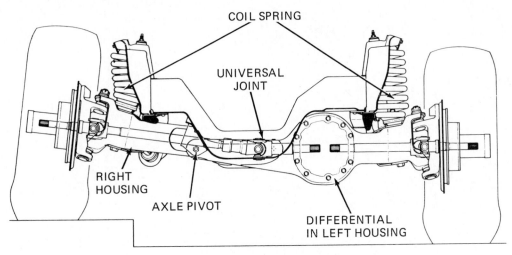

COIL SPRING

UNIVERSAL JOINT

RIGHT HOUSING

AXLE PIVOT

DIFFERENTIAL IN LEFT HOUSING

Fig. 19-24 A front axle for a four-wheel-drive vehicle with independent front suspension. *(Ford Motor Company)*

In 1980, Ford introduced full-size 4×4 trucks with independent front suspension (Figs. 19-24 and 19-25). This system utilizes two steel axle carriers and an additional universal joint in the right axle shaft next to the differential. These permit each front wheel to move up and down independently of the other. Instead of a one-piece tube to serve as the axle housing, the tube is split into two shorter sections and joined together at a pivot point. The axle shaft can flex near the pivot because of the centrally located third universal joint. Off road, this provides greater wheel control and traction than with a solid axle, according to Ford.

With independent front suspension, each wheel is able to react to the road conditions individually. Bumps are absorbed by each front wheel instead of some of the shock being transmitted through the solid front axle to the other wheel. Front coil springs are used on Ford F-150 models (Fig. 19-24). For heavier loads and handling requirements of F-250 and F-350 models, tapered leaf springs are used. Figure 19-22 shows the spring mounting pads for leaf springs on the axle housing.

NOTE: A designation such as *4×4* is used with vehicles to indicate the number of wheels on the ground and the number of wheels that can be driven. For example, the typical car has four wheels on the ground and two-wheel drive. Therefore, it is designated *4×2*. A four-wheel-drive vehicle is a *4×4*. There are four wheels on the ground and all four can be driven.

✿ 19-10 Locking hubs Three different types of hubs are used on front-drive axles. These are the manual locking hub (Fig. 19-23), the automatic locking hub with manual override, and the hub used with full-time four-wheel drive. Locking hubs are also called *free-wheeling hubs*. This is because their purpose is to disengage the front axle and allow it to free-wheel while the vehicle is operating in two-wheel drive. As a result, fuel economy and tire life are increased, and engine and transfer case wear is decreased.

A locking hub is a type of clutch that disengages the outer ends of the axle shafts from the wheel hub. Then the axle shafts do not turn, or *back-drive*, the differential. Therefore the differential does not turn the driveshaft from the transfer case to the front differential. The engine power saved by not having to turn these parts has been reported to increase fuel economy by as much as two miles per gallon.

Many vehicles with part-time four-wheel drive are equipped with manual locking hubs (Figs. 19-23 and 19-26). These hubs are locked or unlocked by the driver, who must turn a knob or lever at each wheel (Fig. 19-27). Another type of hub is the automatic locking hub (Fig. 19-28). It engages when the transfer-case shift lever is moved to four-wheel drive. When the lever is in two-wheel drive, the front axle and driveshaft are not engaged and do not rotate. Automatic locking hubs can be manually locked by turning the control knob at each wheel. (Fig. 19-27).

Locking hubs are not needed with full-time front-wheel drive. The hubs are always engaged with the axle shafts. The interaxle differential in the transfer case prevents damage and undue wear of power-train parts.

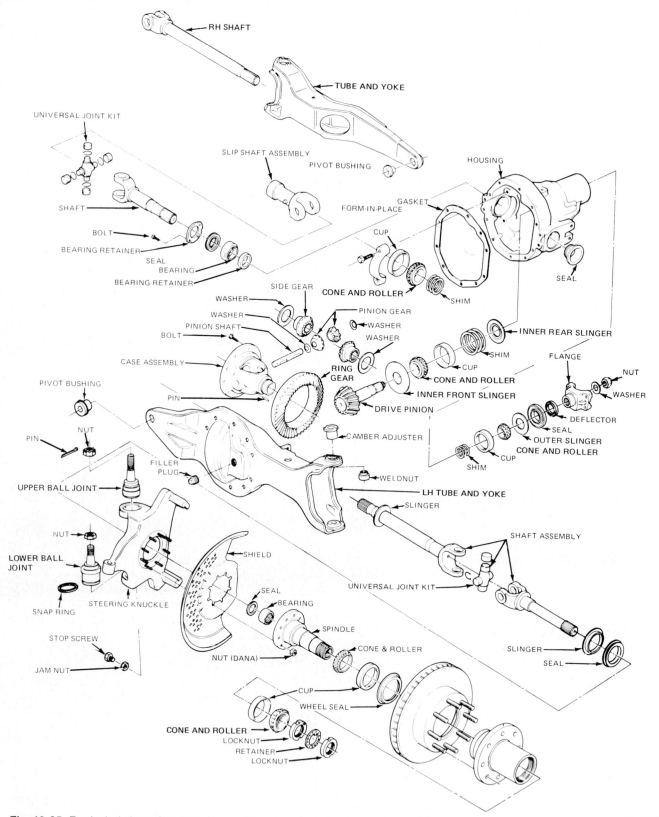

Fig. 19-25 Exploded view of an independent-front-suspension type of front-drive axle. (*Ford Motor Company*)

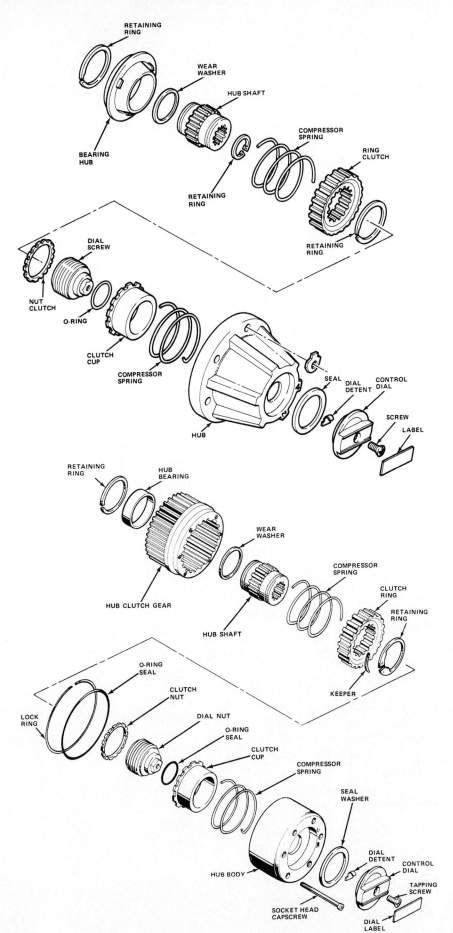

Fig. 19-26 Two types of manual locking hubs. (*American Motors Corporation*)

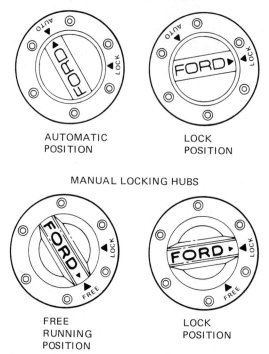

AUTOMATIC LOCKING HUBS

AUTOMATIC
POSITION

LOCK
POSITION

MANUAL LOCKING HUBS

FREE
RUNNING
POSITION

LOCK
POSITION

Fig. 19-27 Positions of control knob on manual and automatic locking hubs. *(Ford Motor Company)*

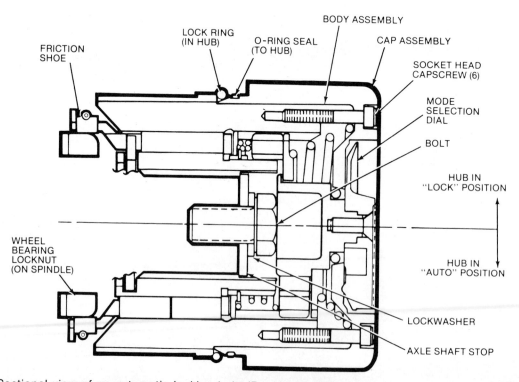

FRICTION
SHOE

LOCK RING
(IN HUB)

O-RING SEAL
(TO HUB)

BODY ASSEMBLY

CAP ASSEMBLY

SOCKET HEAD
CAPSCREW (6)

MODE
SELECTION
DIAL

BOLT

HUB IN
"LOCK" POSITION

WHEEL
BEARING
LOCKNUT
(ON SPINDLE)

HUB IN
"AUTO" POSITION

LOCKWASHER

AXLE SHAFT STOP

Fig. 19-28 Sectional view of an automatic locking hub. *(Ford Motor Company)*

Chapter 19 review questions

Select the *one* correct, best, or most probable answer to each question. Then check your answers against the correct answers given at the end of the book.

1. In the differential, the ring gear is bolted to the:
   a. differential housing
   b. differential case
   c. drive pinion
   d. axle shaft
2. The drive pinion is assembled into the:
   a. carrier or axle housing
   b. differential case
   c. axle case
   d. driveshaft
3. In the differential on modern cars with front engine and rear-wheel drive, the type of gearing used for the drive pinion and ring gear is:
   a. spur
   b. spiral bevel
   c. hypoid
   d. parabolic
4. The distance between adjacent meshing teeth of mating gears is called:
   a. clearance
   b. pitch line
   c. backlash
   d. flank
5. In a differential with a gear ratio of 4:1, the drive pinion would revolve four times to cause the ring gear to rotate:
   a. one time
   b. two times
   c. four times
   d. 16 times
6. If the drive pinion has 10 teeth and the ring gear has 30 teeth, the gear set is of the:
   a. hunting type
   b. nonhunting type
   c. hypoid type
   d. spiral-bevel type
7. If the drive pinion has 10 teeth and the ring gear has 33 teeth, the gear set is of the:
   a. hunting type
   b. nonhunting type
   c. hypoid
   d. spur-bevel type
8. In the basic differential illustrated in Fig. 19-11, there are:
   a. nine gears
   b. eight gears
   c. seven gears
   d. six gears
9. Universal joints are used at the outer ends of the front-axle shafts of a four-wheel-drive vehicle to allow:
   a. normal spring action
   b. normal steering action
   c. normal braking action
   d. none of the above
10. The use of locking hubs on a four-wheel-drive vehicle improves:
    a. ride quality
    b. traction
    c. fuel economy
    d. all of the above

DIFFERENTIAL AND DRIVE-AXLE SERVICE

After studying this chapter, and with proper instruction and equipment, you should be able to:

1. Describe the various troubles that a differential might have and how to check out noise to find its cause.
2. Remove and install rear axle shafts, wheel bearings, and oil seals.
3. Remove and install a differential.
4. Describe the various contact patterns between ring-gear teeth and drive-pinion-gear teeth and explain what they mean.
5. Explain how to make adjustments if the gear-teeth-contact patterns are not correct.
6. Disassemble, reassemble, and adjust a standard differential.
7. Disassemble, reassemble, and adjust a limited-slip differential.

20-1 Differential and drive-axle service This chapter describes the servicing of differentials and drive axles used in front-engine, rear-wheel-drive cars. This includes removal, disassembly, inspection, cleaning, reassembly, adjustment, and reinstallation. The operation and servicing of other types of differentials, such as in transaxles and transfer cases, are covered in earlier chapters. Transaxles are covered in Chaps. 11 to 14. Transfer cases are covered in Chaps. 15 and 16.

NOTE: For rear-wheel-drive cars, manufacturers' service manuals usually refer to the differential as the *rear axle* or the *rear-axle assembly*. By this, they may mean the rear-axle housing, the wheel axles, or the differential assembly. *Rear axle* is also used in shop manuals to mean the differential itself. To avoid confusion, the term *differential* is used in this book to designate the differential assembly itself, as described in the previous chapter.

20-2 Rear-axle and differential trouble diagnosis See 3-1 for a discussion of diagnosis theory. Most often, it is noise that draws attention to troubles in the rear axles or differential. It is sometimes difficult to pin down the cause of the trouble. Road testing the car and determining the type of noise being heard under different driving conditions will help in the diagnosis.

To check the vehicle for noise and where it is coming from, drive the car on a level asphalt road. First make sure the differential has sufficient lubricant, and drive fast enough to warm up the lubricant.

Drive and coast the car at various speeds and note the speeds at which the noise occurs. Stop the car, and with the clutch disengaged (manual transmission) or the transmission in neutral (automatic transmission), run the engine slowly up and down through the engine speeds at which the noise was loudest. If you hear the same noise, it is not due to the differential or rear axle but to the exhaust system or other engine condition.

Repeat this test (car stationary) while engaging and disengaging the clutch with the transmission in neutral (manual transmission). If you get the noise with the clutch disengaged, then the problem is not in the power train at all. However, the clutch throwout bearing or the pilot bushing in the end of the crankshaft might cause the noise if they are worn or lack lubricant.

Some clue as to the cause of trouble may be gained by noting whether the noise is a hum, growl, or knock; whether it is obtained when the car is operating on a straight road or on turns only; and whether the noise is most noticeable when the engine is driving the car or when the car is coasting.

A humming noise in the differential is often caused by improper drive-pinion or ring-gear adjustment, which prevents normal tooth contact between the gears. This condition produces rapid gear-tooth wear, and so the noise will gradually take on a growling characteristic.

If the noise is loudest when the car accelerates, there probably is heavy heel contact on the gear teeth. The ring gear must be moved nearer the drive pinion. If the noise is loudest when the car coasts in gear with the throttle closed, there probably is heavy toe contact on the gear teeth. The ring gear must be moved away from the drive pinion.

If the noise occurs only when the car is rounding a curve, the trouble is due to some condition in the differential. Possibilities are differential pinion gears tight on the pinion shaft, differential side gears tight in the differential case, damaged gears or thrust washers, or excessive backlash between gears. If the noise is a knock, the bearings or gears probably are badly worn or damaged.

To make sure a noise is not tire noise, drive on different types of pavement. If the noise changes, it is probably tire noise. If the noise stays essentially the same, it is not being caused by the tires.

NOTE: Other conditions to consider which are not related to the differential or axles are:

1. Pitch of the exhaust can sound like gear whine or even a wheel-bearing rumble at times.
2. Roof racks on station wagons can sometimes cause a roaring or rumbling sound. Even trim and moldings can sometimes cause a whistling or whining noise. You can test for rack noise by strapping some packages on the roof and road testing the car again.

✿ 20-3 Diagnosing the noise Various types of noise and their possible causes include the following:

1. Howling or whining is probably a gear noise. It is most audible at speeds of 20 to 55 mph [32 to 89 km/h] under acceleration or a heavy pull, at constant speed with the engine just pulling the car, or while coasting with the throttle closed and the car in gear. This type of noise is usually due to improper ring-gear or drive-pinion adjustment, gear damage, or incorrect bearing preload.
2. Chuckle is a rattling noise that sounds like a stick being held against the spokes of a spinning bicycle wheel. You can hear it when slowing down and stopping from about 40 mph [64 km/h]. The frequency varies with car speed. It is caused by excessive gear-tooth clearance in the differential gears or by damaged teeth on the coast side of the drive pinion or ring gear (Figs. 20-1 to 20-3). This is discussed further in ✿ 20-6.
3. Knock is similar to chuckle, but it may be louder and occur on both acceleration and deceleration. If you hear it every revolution of the rear wheels, it is due to a damaged rear-wheel bearing. If it is a continuous knock, the probable cause is drive-side gear-tooth damage (Fig. 20-4). Also, loose ring-gear bolts can knock against the carrier case. Excessive end play of the axle shafts can cause knock. End play can be measured with a dial indicator. If the end play is excessive, Ford recommends that a shim be cut in

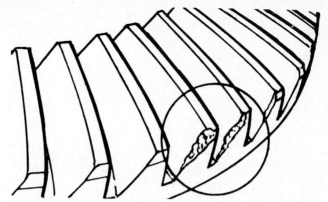

Fig. 20-1 Damaged ring-gear tooth. *(Ford Motor Company)*

Fig. 20-2 Scored gear teeth. *(Ford Motor Company)*

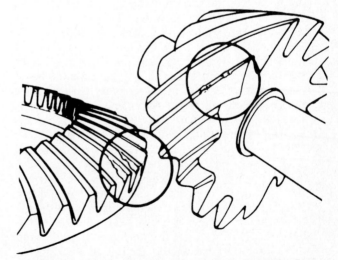

Fig. 20-3 Damaged teeth on gear set. *(Ford Motor Company)*

239

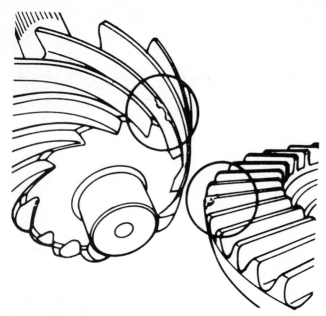

Fig. 20-4 Damaged gear teeth that will produce a drive-side knock. *(Ford Motor Company)*

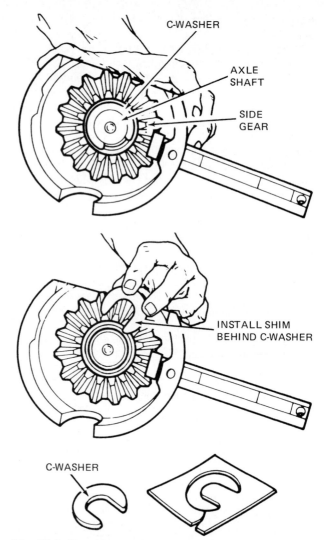

Fig. 20-5 Checking and correcting axle-shaft end play. *(Ford Motor Company)*

the pattern of the C washer that holds the shaft in the pocket of the differential side gear (Fig. 20-5). Install the shim in back of the C washer.

4. Low-speed knock can be caused by worn universal joints or by a worn side-gear-hub bore in the differential case.

5. Clunk is a metallic noise heard when the car is accelerated or decelerated. It is usually caused by excessive backlash in the drive train resulting from trouble in the differential or elsewhere.

To check backlash in the differential, raise the vehicle on a frame or twin-post lift so that the drive wheels are free. Clamp a bar between the companion flange and the housing as shown in Fig. 20-6 so that the flange cannot move. Lock the left rear wheel so that it cannot turn. Turn the right wheel slowly until you take up all the backlash. Hold a chalk marker on the side of the tire, 12 inches [305 mm] from the wheel center. Slowly turn the wheel the other way until you again take up all backlash. Measure the length of the chalk mark. It should be 1 inch [25.4 mm] or less. If the backlash is more than this, the trouble is in the differential. If the differential backlash is not excessive, the clunk is due to excessive backlash in the drive line or transmission, or to loose axle-shaft or side-gear splines.

6. Bearing whine sounds much like a whistle. It could come from the differential-drive-pinion bearings or from the wheel bearings. This noise occurs at all car speeds, which distinguishes it from gear whine, which usually comes and goes as car speed changes.

7. Bearing rumble sounds like marbles being tumbled. It is usually caused by a bad wheel bearing. It is lower in pitch because the wheel turns at only about one third of the driveshaft speed.

8. Chatter on corners is a condition where the whole rear end vibrates when the car is moving. It can be

felt as well as heard. In standard differentials, it may be caused by too many differential thrust washers. This could cause a partial lockup to produce the condition.

On limited-slip differentials, the condition may be caused by extended highway driving during very hot weather. The clutch plates tend to stick during tight turns. If the condition persists, it may often be cured by flushing the differential and filling it with new lubricant and additive.

⚙ **20-4 Rear-axle and differential service** A variety of differentials are used in late-model cars. The major automotive manufacturers describe from 4 to 12 differentials in their shop manuals. Although the construction and operation are quite similar, the basic service procedures for each differential model can be special.

As examples of typical service procedures, the procedures for the two basic types of standard differentials—integral carrier and removable carrier—are given here. Also included are the servicing procedures for

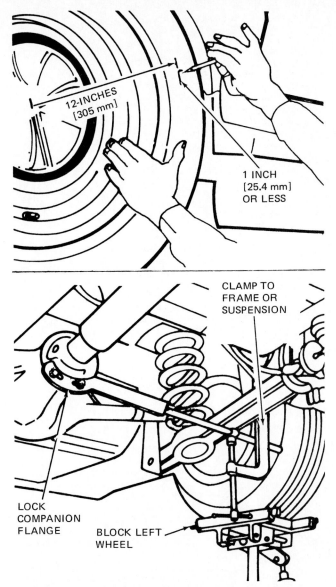

Fig. 20-6 Checking backlash in the differential. *(Ford Motor Company)*

three types of limited-slip differentials—the two-clutch-pack type, the spring-loaded single-clutch-pack type, and the spring-loaded cone type. In addition, rear-axle and wheel-bearing service is covered in later sections.

Careful: Always refer to the service manual that covers the model of differential you are working on before servicing and adjusting it. Procedures and specifications vary from model to model.

✿ 20-5 Differential pre-repair diagnosis Before any work is started on the differential-and-rear-axle assembly, certain checks should be made to determine what needs to be done. The results of these checks, plus the condition that brought the vehicle in for service, will often reveal just what is required. Frequent causes of differential noise are incorrect backlash, incorrect pinion-bearing or side-bearing preload, and incorrect tooth contact (ring-gear-to-drive-pinion

teeth). Therefore, a few simple adjustments may be all that are required to eliminate the trouble. The pre-repair checks are described first, followed by the servicing procedures on the differentials and axles.

There are two basic types of differentials, integral-carrier and removable-carrier. The servicing procedures are somewhat different for the two types. In the integral type (Fig. 20-7), the differential housing, or *carrier,* is part of the axle-housing assembly. The outer races of the roller bearings that support the differential case fit into recesses in the differential housing, which is part of the axle housing.

The removable-carrier type (Fig. 20-8) is mounted in a large recess in the axle housing. The carrier housing, which supports the differential-case assembly and the drive pinion, is bolted onto the axle housing. In this design, a pinion retainer bolts to the carrier housing to hold the drive pinion and bearings in place. There are other differences, which are described with the servicing procedures for the two types.

There are also two basic types of rear-suspension systems, coil spring and leaf spring (Figs. 20-9 and 20-10). Some of the rear-end servicing procedures are different for the two types of suspension systems, as explained later in the chapter.

✿ 20-6 Gear-tooth nomenclature The side of the ring-gear tooth which is curved outward, or is convex, is the drive side. The concave side is the coast side (Fig. 20-11). The names of other parts of the gear tooth are given in ✿ 19-5.

✿ 20-7 Gear-tooth-contact patterns It is important to check the contact between the drive-pinion teeth and the ring-gear teeth before proceeding with any other differential service. If the drive pinion is too far away from, or too near, the ring gear, the gears probably will be noisy. Also, wear will increase.

The contact between the ring-gear teeth and the drive-pinion teeth should be centered on the teeth (Fig. 20-12). On the integral differential (Fig. 20-7) the drive-pinion location is adjusted by installing or removing shims (Fig. 20-13). These shims are located between the pinion and the inner race of the rear drive-pinion bearing. Adjusting nuts or shims are then used to correct the backlash between the ring gear and the drive pinion.

On the removable-carrier differential (Fig. 20-8), the shims to adjust the drive pinion are located between the pinion retainer and the carrier housing (Figs. 20-14 and 20-15). Backlash is adjusted on the removable-carrier differential in the same way as on the integral-carrier differential. This is done by turning the side adjusting nuts or changing the shim thickness.

The gear-tooth-contact pattern is checked by coating the gear teeth with a gear-marking compound and then rotating the drive pinion. This shows where the teeth are making contact. Here is a typical procedure for an integral-carrier differential (Fig. 20-7):

1. Raise the car on a lift so that the rear wheels are free. Disconnect the rear universal joint from the

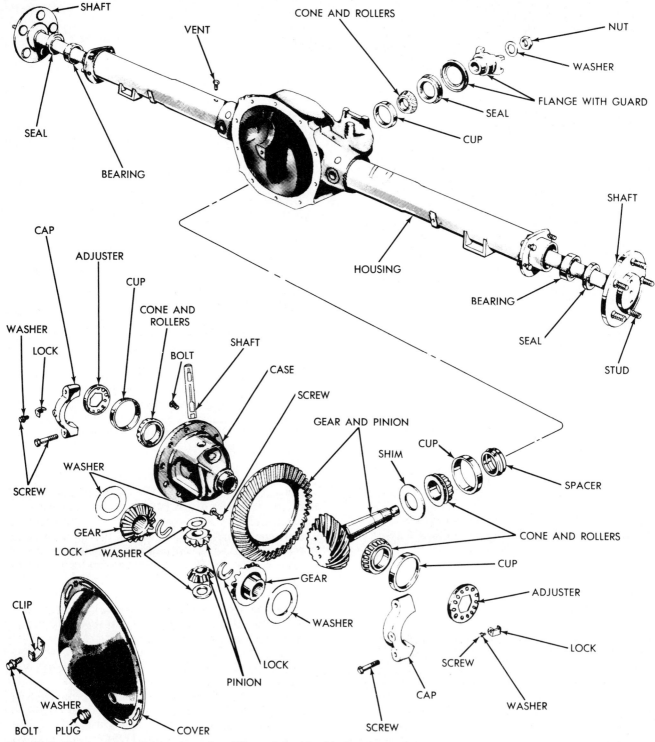

Fig. 20-7 Disassembled integral-carrier differential with side-bearing adjuster nuts. *(Chrysler Corporation)*

drive-pinion flange after marking the parts so that they can be reattached in the same relative positions (Fig. 20-16).

2. Loosen the cover bolts and pry the bottom of the cover loose enough to allow the lubricant to drain. Catch it in a container.

3. Remove the cover. Wipe out the oil and carefully clean each tooth of the ring gear.

4. Brush a gear-marking compound onto all ring-gear teeth with a medium-stiff brush. When this is properly done, you will see the gear-tooth-contact pattern develop as you rotate the ring gear.

5. Tighten the bearing-cup bolts to the specified torque.

6. Expand the brake shoes at both rear wheels to load the differential.

7. Turn the drive-pinion flange with a wrench to mark

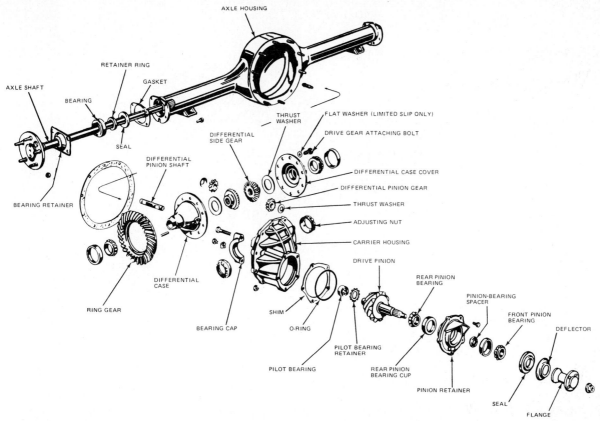

Fig. 20-8 Disassembled rear-axle assembly with removable-carrier differential. *(Ford Motor Company)*

the gear pattern under load. Observe the contact pattern. It will be different for nonhunting and hunting gear sets (☀ 19-6). Figure 20-27 shows how the hunting gear set is marked by Ford.

8. To check the pattern on a removable-carrier differential, the differential must be off the car and installed in a holding fixture (Fig. 20-17). The drive pinion is then rotated as shown in Fig. 20-18 to produce the contact pattern.

☀ 20-8 Reading gear-tooth-contact patterns In the nonhunting gear set, only a few rotations are required to get the full pattern of gear-tooth contact. In the hunting gear set, more rotations are required. The ideal pattern is shown in Fig. 20-12. However, the pattern can vary somewhat from this ideal and still be satisfactory. Generally, the drive pattern should be centered on the tooth. The coast pattern should also be centered, but it can be slightly toward the toe.

When the load on the ring gear and drive pinion is increased, as during strong acceleration, the tooth contact will tend to spread out. Under very heavy load, it will extend from near the toe to near the heel on the drive side. The entire contact also tends to shift toward the heel under increasingly heavier loads. It will become somewhat broader from top to bottom of the teeth.

In general, the nonhunting gear set can have a more eccentric pattern than the hunting gear set and still be

satisfactory. For example, Figs. 20-19 and 20-20 show two acceptable patterns for nonhunting gear sets.

If the gear-tooth-contact pattern is not correct, it could be because of ring-gear runout. Check with a dial indicator as shown in Fig. 20-21 for integral-carrier differentials and in Fig. 20-22 for detached removable-carrier differentials. Turn the drive pinion and note the amount of ring-gear runout. If it is excessive, the differential must be disassembled so that the defective parts can be replaced. Ring-gear runout is probably due to worn case bearings or loose ring-gear bolts.

Improper gear-tooth contact could also be due to excessive backlash between the ring gear and the drive pinion. This can be checked with a dial indicator set up as shown in Fig. 20-23 for integral-carrier differentials and Fig. 20-24 for detached removable-carrier differentials. See how much the ring gear can be turned backward before the coast sides of the teeth meet the drive-pinion teeth.

☀ 20-9 Improving gear-tooth-contact patterns If the gear-tooth-contact pattern or backlash is incorrect, making the necessary adjustments may eliminate the noise problem. However, before this step, make sure that the gears and bearings are in good condition. If the gears all look good and the bearings show no roughness when the drive pinion is rotated, then adjustments should be tried. Figure 20-13 shows the shim location and adjusting nuts involved in making adjustments to

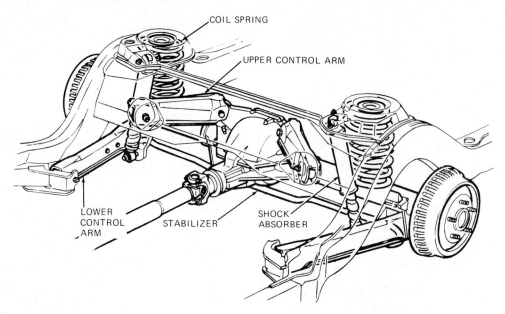

Fig. 20-9 Coil-spring rear suspension. *(Chevrolet Motor Division of General Motors Corporation)*

the integral-carrier differential. Figures 20-14 and 20-15 show the shim location and adjusting nuts involved in making adjustments to the removable-carrier differential.

Figures 20-25 and 20-26 show gear-tooth patterns requiring corrections and what to do to correct them. These are the patterns marked on the ring gear. Here are procedures for correcting drive-pinion location and backlash. Drive-pinion location is adjusted first, followed by the backlash adjustment.

1. **Adjusting the drive-pinion location** This adjustment is made by installing or removing shims. Figures 20-13 to 20-15 show the locations of the shims in the two types of differential.

a. Removable-carrier differential The positions of the

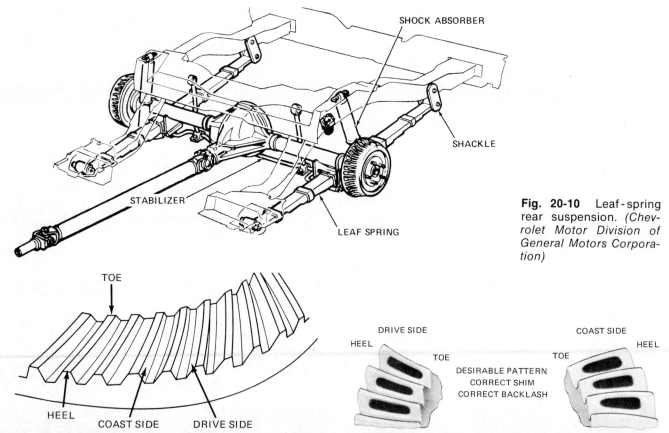

Fig. 20-10 Leaf-spring rear suspension. *(Chevrolet Motor Division of General Motors Corporation)*

Fig. 20-11 Ring-gear teeth, showing concave (coast) and convex (drive) sides and location of toe and heel.

DESIRABLE PATTERN
CORRECT SHIM
CORRECT BACKLASH

Fig. 20-12 Ideal gear-tooth-contact pattern. *(Ford Motor Company)*

244

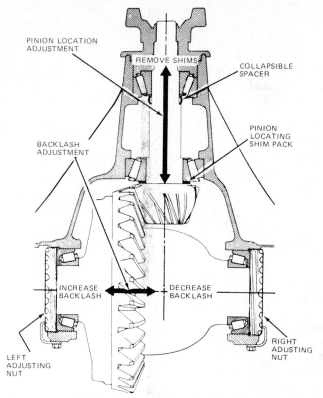

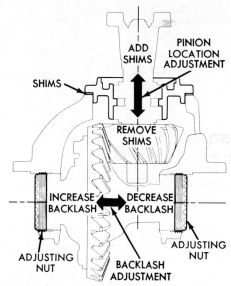

Fig. 20-14 Adjusting procedures to correct gear-tooth contact and backlash on a removable-carrier differential. *(Ford Motor Company)*

Fig. 20-13 Adjusting procedures to correct gear-tooth contact and backlash on an integral-carrier differential. This differential has adjusting nuts to adjust backlash and bearing preload. Another type uses shims to make the adjustment. *(Ford Motor Company)*

shims used to adjust the drive-pinion location in the removable-carrier differential are shown in Figs. 20-14 and 20-15. The shims are located between the carrier housing and the pinion retainer. Adding shim thickness moves the drive pinion away from the ring gear. Reducing shim thickness moves the drive pinion toward the ring gear.

To change shim thickness in the removable-carrier differential, unbolt the pinion retainer from the carrier housing and remove it to get at the shim. When reinstalling the pinion retainer, be careful not to pinch or

twist the O ring. Coat it with lubricant and snap it into the groove. Do not roll it in.

When removing the pinion retainer from the carrier, determine if the gear set is of the hunting or the nonhunting type. If it is of the nonhunting type, it will be identified by painted timing marks on the gear teeth (Fig. 20-27). If the gear set is of the hunting type, mark the drive pinion and ring gear when removing the drive pinion. Then align these marks when reinstalling the drive pinion.

b. Integral-carrier differential In the integral-carrier differential (Fig. 20-13), the drive-pinion-locating shims are between the pinion and the inner race of the rear bearing. Reducing the shim thickness will move the drive pinion away from the ring gear. Increasing the shim thickness will move the drive pinion toward the ring gear.

It is difficult to make this adjustment because the differential case must be removed so that the drive pinion can be taken out. Then shim thickness can be

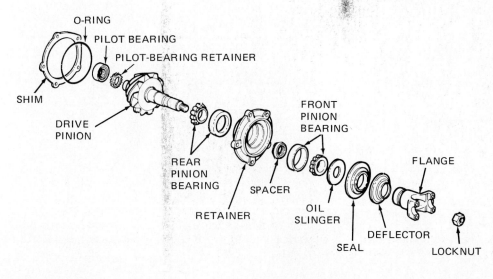

Fig. 20-15 Drive pinion, bearing retainer, shim, and related parts for a removable-carrier differential. *(Ford Motor Company)*

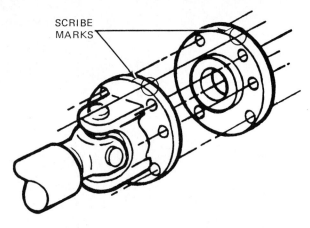

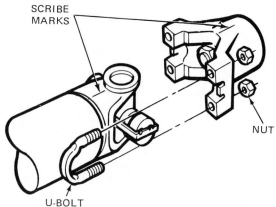

Fig. 20-16 Two types of driveshaft-to-differential-drive-pinion connections and how to mark them so they can be reconnected in their correct positions. (Ford Motor Company)

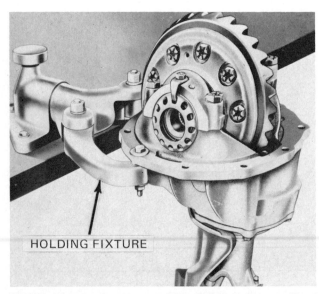

Fig. 20-17 Differential mounted in a bench fixture. (Ford Motor Company)

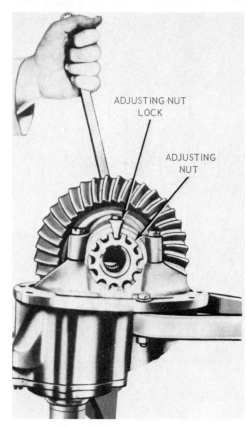

Fig. 20-18 Rotating the ring gear to check gear-tooth contact. (Ford Motor Company)

changed. If you were trying to set the drive-pinion location by checking the tooth pattern and then guessing how the shim should be changed, you might have to do this disassembly-reassembly procedure several times.

To avoid this, there is an alternate procedure which can be used to preset the drive-pinion depth before the differential case is mounted in the housing. This procedure is covered in ✪ 20-17 as part of the complete service procedure on integral-carrier differentials.

2. Adjusting backlash Backlash is adjusted in

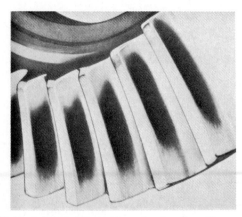

Fig. 20-19 Acceptable pattern for nonhunting gear set showing center-to-toe-to-center eccentricity. (Ford Motor Company)

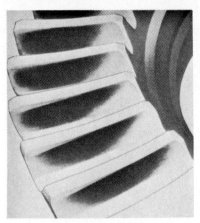

Fig. 20-20 Acceptable pattern for nonhunting gear set showing center-to-heel-to-center eccentricity. *(Ford Motor Company)*

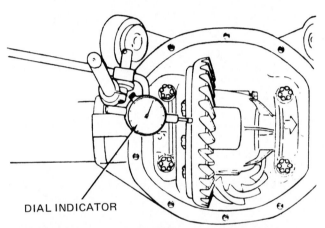

Fig. 20-21 Checking ring-gear runout with a dial indicator on an integral-carrier differential. *(Ford Motor Company)*

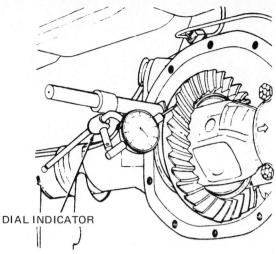

Fig. 20-23 Checking ring-gear backlash on an integral-carrier differential. *(Ford Motor Company)*

either of two ways, according to the design of the differential. One design has adjustment nuts on the outside of the case side bearings (Figs. 20-7, 20-13, and 20-14). On these, turning one adjustment nut in and the other adjustment nut out moves the case so that the backlash and gear-tooth-contact pattern are changed. Also, turning both adjustment nuts in will increase the side-bearing preload.

The other adjusting procedure is used on those differentials which have shims outside the two side bearings (Fig. 20-57). On this design, increasing shim thickness on one side and decreasing it on the other will move the differential case and change the backlash and gear-tooth-contact pattern. Increasing shim thickness on both sides will increase the bearing preload.

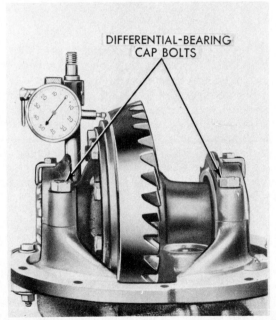

Fig. 20-22 Checking ring-gear runout with a dial indicator on a removable-carrier differential. *(Ford Motor Company)*

Fig. 20-24 Checking ring-gear backlash on a removable-carrier differential. *(Ford Motor Company)*

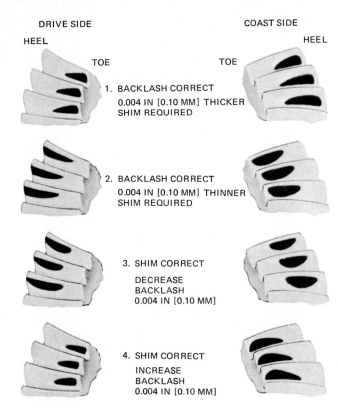

DRIVE SIDE   COAST SIDE

HEEL   HEEL

TOE   TOE

1. BACKLASH CORRECT
0.004 IN [0.10 MM] THICKER SHIM REQUIRED

2. BACKLASH CORRECT
0.004 IN [0.10 MM] THINNER SHIM REQUIRED

3. SHIM CORRECT
DECREASE BACKLASH
0.004 IN [0.10 MM]

4. SHIM CORRECT
INCREASE BACKLASH
0.004 IN [0.10 MM]

Fig. 20-25 Typical gear-tooth-contact patterns requiring correction. *(Ford Motor Company)*

a. Adjusting backlash on differential with adjustment nuts To adjust the backlash between the drive pinion and the ring gear, one adjustment nut is tightened while the other is loosened (Figs. 20-13 to 20-15). To make the adjustment, remove the adjustment-nut locks, loosen the differential bearing-cup bolts, and then torque the bolts to specifications.

The left adjustment nut is on the ring-gear side of the carrier, and the right nut is on the pinion side (as shown in Fig. 20-13). As a first step, loosen the right nut until it is away from the bearing cup. Tighten the left nut until the ring gear is just forced into the pinion with no backlash. Recheck the right nut to make sure it is still loose. Now, tighten the right nut two notches beyond the position where it first contacts the bearing cup. Rotate the ring gear several revolutions in each direction while the bearings are being loaded so that the bearings will seat.

Again loosen the right nut to release the preload. If there is any backlash between the gears, tighten the left nut just enough to remove the backlash. Carefully tighten the right nut until it just makes contact with the bearing cup. Then tighten it further two and one half to three notches to apply correct preload. This should force the ring gear away from the drive pinion, giving the proper backlash.

Tighten the differential cap bolts to the correct specifications and check the backlash (Figs. 20-23 and 20-24). If the backlash is uneven, the ring gear has runout (see Figs. 20-21 and 20-22 for the setup to check runout). If backlash is even but incorrect, readjust the adjustment nuts.

Careful: Always make the final adjustment in a tightening direction to make sure that the nut is in contact with the bearing race.

b. Adjusting backlash on differential with shims On this design, the shim thicknesses on the two side bearings must be changed to move the case and ring gear closer to, or farther away from, the drive pinion. At the same time, the final adjustment must leave the side bearings correctly preloaded. The procedure of adjusting backlash and bearing preload with shims is described in ✿ 20-19.

✿ 20-10 Removing the rear-axle housing There are two basic rear-suspension systems—coil spring and leaf spring (Figs. 20-9 and 20-10). The removal procedure requires disconnecting the driveshaft, control arms or leaf springs, brakes, and shock absorbers. A basic procedure follows:

1. Raise the rear of the car high enough so that you can work under it. Put a floor jack under the center of the differential housing so that the jack just starts to lift the axle housing. Put safety stands firmly under the frame members on both sides.
2. Mark the rear universal joint and pinion flange so that they can be reconnected in the same position (Fig. 20-16). Then disconnect the universal joint from the pinion flange.
3. Disconnect the parking-brake cables by removing the adjusting nuts at the equalizer. Slide the center cable to the rear and disconnect the two rear cables at the connectors to free them from the body.
4. Disconnect the rear-brake hose at the floor pan. Cover the brake hose and pipe openings to keep out dirt.
5. On coil-spring rear suspensions, disconnect the shock absorbers at the lower ends and push the shock absorbers up out of the way. Remove the upper control-arm bolts and loosen the lower control-arm bolts. Lower the jack under the differential housing until the rear springs can be removed. Then remove the lower control-arm bolts.
6. On the leaf-spring suspension, support the car by placing the safety stands under the frame in front of the leaf springs and at the rear of the leaf springs at the bumper. Then remove the lower spring-plate-attaching nuts and the front and rear attaching bolts and remove the springs.
7. With the axle assembly disconnected, roll it out from under the car.

✿ 20-11 Installing the rear-axle assembly On coil-spring rear suspensions (Fig. 20-9), after rolling the rear-axle assembly into position, connect the lower and upper control arms to the assembly. Final tightening of the bolts is made later with the suspension supporting the car. Put the coil springs in place and jack the axle housing up until the shock absorbers can be connected.

On leaf-spring rear suspensions (Fig. 20-10), connect the leaf springs at the front. Attach the lower spring plate. You may need to lower the axle housing to make

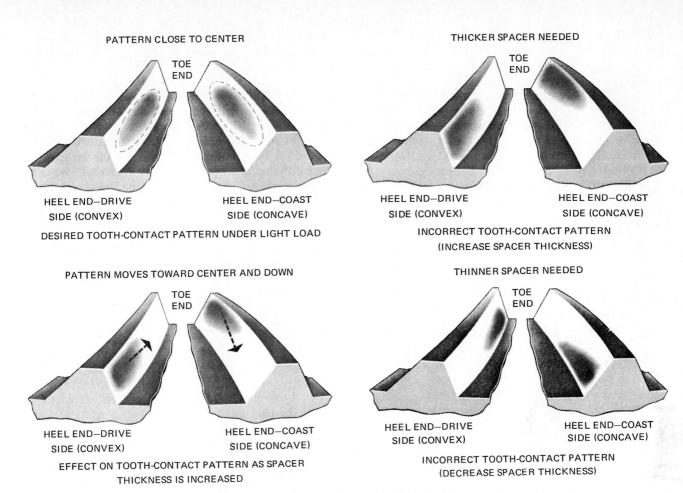

PATTERN CLOSE TO CENTER

TOE END

HEEL END—DRIVE
SIDE (CONVEX)

HEEL END—COAST
SIDE (CONCAVE)

DESIRED TOOTH-CONTACT PATTERN UNDER LIGHT LOAD

THICKER SPACER NEEDED

TOE END

HEEL END—DRIVE
SIDE (CONVEX)

HEEL END—COAST
SIDE (CONCAVE)

INCORRECT TOOTH-CONTACT PATTERN
(INCREASE SPACER THICKNESS)

PATTERN MOVES TOWARD CENTER AND DOWN

TOE END

HEEL END—DRIVE
SIDE (CONVEX)

HEEL END—COAST
SIDE (CONCAVE)

EFFECT ON TOOTH-CONTACT PATTERN AS SPACER
THICKNESS IS INCREASED

THINNER SPACER NEEDED

TOE END

HEEL END—DRIVE
SIDE (CONVEX)

HEEL END—COAST
SIDE (CONCAVE)

INCORRECT TOOTH-CONTACT PATTERN
(DECREASE SPACER THICKNESS)

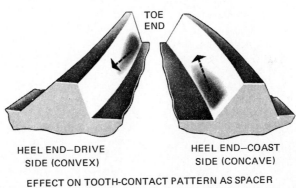

PATTERN MOVES INWARD AND UP

TOE END

HEEL END—DRIVE
SIDE (CONVEX)

HEEL END—COAST
SIDE (CONCAVE)

EFFECT ON TOOTH-CONTACT PATTERN AS SPACER
THICKNESS IS DECREASED

Fig. 20-26 Correct and incorrect gear-tooth-contact patterns and adjustments to correct them. *(Chrysler Corporation)*

this attachment. Be sure the spring cushions are in place. Then attach the leaf spring at the rear shackle. Tighten the nuts at the front of the springs. Connect the shock absorbers.

Proceed as follows for both types of suspensions:

1. Connect the driveshaft (Fig. 20-16).
2. Connect and adjust the parking-brake cable.
3. Connect the rear-brake hose. Bleed both rear brakes and refill the master cylinder.
4. Lower the car to the floor and tighten the control-arm bolts or nuts.
5. Fill the differential with the specified lubricant.

PAINT MARKING INDICATES POSITION
IN WHICH GEARS WERE LAPPED

Fig. 20-27 Gear-timing marks on a nonhunting gear set. *(Ford Motor Company)*

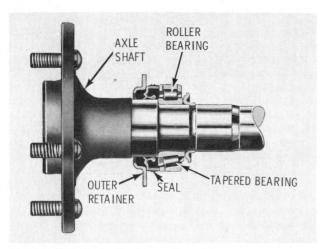

Fig. 20-28 Axle shaft with retainer, oil seal, and bearing. *(Chevrolet Motor Division of General Motors Corporation)*

✿ 20-12 Removing and installing rear-wheel axles Rear-wheel axles are held in position in various ways so that they will not move axially (sideways). One way of holding the axles is with a C washer that fits into a groove in the inner end of the axle (inside the differential case). This is the design used in the integral-carrier differential (Fig. 20-7). Another design, used in the removable-carrier differential (Fig. 20-8), has a re-

tainer at the outer end of the axle which is bolted to a backing plate (Fig. 20-28). Both types are explained below.

1. Removal of C-lockwasher-type axle The outer bearing and oil seal are shown in sectional view in Fig. 20-29. Use care in removing the axle to avoid damaging the oil seal. To remove the axle, raise the car and support it on a safety stand. Remove the wheel and brake drum.

1. Clean all dirt from the area of the differential cover.
2. Remove the cover to drain the lubricant.
3. Remove the differential-pinion-shaft lock bolt and shaft (Fig. 20-30).
4. Push the axle in far enough so that you can remove the C washer (Fig. 20-31). This releases the axle so that it can be pulled out.

Careful: Do not allow the splines in the end of the axle to damage the oil seal. If the seal is damaged in any way, it will leak lubricant. It must be replaced (✿ 20-13).

2. Installation of C lockwasher-type axle Slide the axle into place, using care to avoid damaging the oil seal. The splines in the end of the axle must engage easily with the splines of the differential side gear.

1. Install the C lockwasher (Fig. 20-31). Push the axle out so that the washer seats in the counterbore of the side gear.
2. Put the differential-pinion shaft into position. Secure it with the lock bolt (Fig. 20-30), torqued to specifications.
3. Use a new gasket and install the differential cover, torquing the bolts to specifications. Be sure the mating surfaces of the cover and case are clean before installing the cover and gasket.

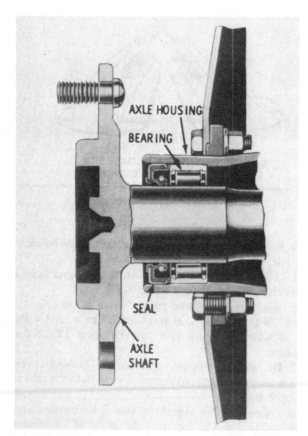

Fig. 20-29 Axle shaft with bearing and oil seal held in place by C lockwashers at the inner ends of the axles. *(Chevrolet Motor Division of General Motors Corporation)*

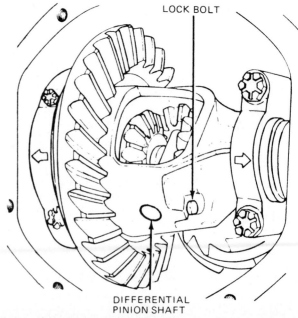

Fig. 20-30 Differential-pinion shaft and lock bolt. *(Ford Motor Company)*

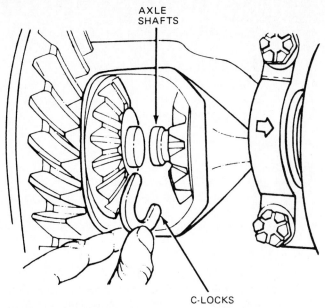

AXLE
SHAFTS

C-LOCKS

Fig. 20-31 Removing or installing a C lockwasher in the groove at the inner end of the axle shaft. *(Ford Motor Company)*

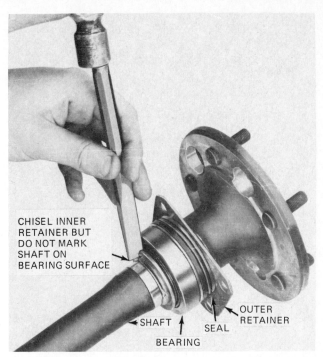

CHISEL INNER
RETAINER BUT
DO NOT MARK
SHAFT ON
BEARING SURFACE

SHAFT SEAL

BEARING

OUTER
RETAINER

Fig. 20-32 Cutting the bearing inner retainer on retainer-type axle. *(Chevrolet Motor Division of General Motors Corporation)*

4. Fill the differential with the proper amount and type of lubricant.
5. Install the brake drum and wheel.
6. Lower the car and test the operation of the differential.

3. Removal of retainer-type axle The rear axle of the retainer-type design (Fig. 20-28) is held in position by an outer retainer that is bolted to a backing plate. When the axle is removed, the retainer, seal, and bearing come off with it (Fig. 20-32). To remove the axle:

1. Remove the wheel and brake drum.
2. Remove the nuts holding the retainer to the backing plate. Pull the retainer clear of the bolts and reinstall the two lower nuts finger-tight to hold the backing plate in position.
3. Pull out the axle as shown in Fig. 20-33. A slide hammer may be required to loosen the axle splines from the differential-side-gear splines.
4. The inner part of the bearing, the seal, and the retainer will come out with the axle. The outer race and inside retainer will remain in the axle housing.

4. Installation of retainer-type axle Apply a thin coat of wheel-bearing grease in the bearing recess of the housing.

1. Insert the axle assembly carefully, making sure the axle splines engage with the splines in the side gear.
2. Remove the nuts holding the backing plate. Push the axle into position, placing the outer retainer over the studs.
3. Install the nuts and torque to specifications.
4. Install the drum and wheel.

✿ 20-13 Servicing rear-axle bearings and seals
On the C-lockwasher type of axle, the bearing and seal

remain in the axle housing when the axle is removed (Fig. 20-29). On the retainer-type axle the bearing, seal, and outer retainer come off with the axle (Fig. 20-28).

1. To remove the axle bearing and seal from a C-lockwasher axle, a slide hammer is required (Fig. 20-34). The tool hooks in back of the bearing outer race so that when the slide is operated, it pulls the bearing off the axle housing.
2. To remove the axle bearing and seal from a retainer-type axle, the inner retainer must be spread so that it can be slipped off the axle, or else broken off. Two procedures are covered below, one for ball bearings and one for tapered-roller bearings.

a. Ball bearings. The Ford and Chevrolet recommendation for removing ball bearings is to nick the retainer in two places with a chisel and hammer (Fig. 20-32). This spreads the retainer enough that it can be slipped off the axle.

After the inner bearing retainer is off, the axle can be pressed out of the bearing either in a shop press or with a special bearing press as shown in Fig. 20-35. Turning the bolts puts force on the outer retainer and bearing race. When a shop press is used, the remover/installer tool catches in back of the inner race of the bearing (Fig. 20-36) so that the force is exerted on the bearing only. With the bearing removed, the seal and outer retainer will then slip off the axle.

NOTE: The bearing should be removed only if it or the seal is defective. When a new bearing is installed, new inner and outer retainers and a new seal are required. If only the seal is replaced, then only new inner and outer retainers are needed.

251

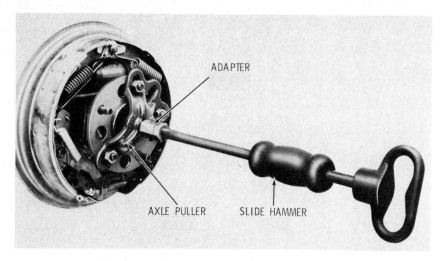

Fig. 20-33 Using a slide hammer to remove an axle shaft. *(Chevrolet Motor Division of General Motors Corporation)*

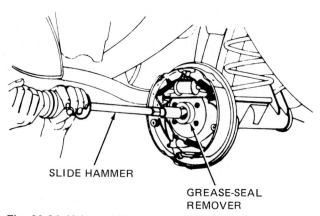

Fig. 20-34 Using a slide hammer to remove an axle seal or bearing. *(Ford Motor Company)*

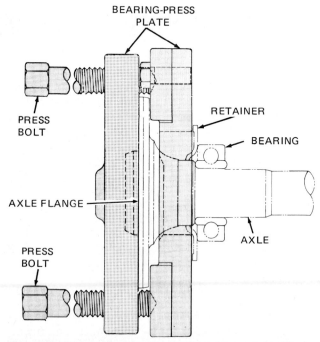

Fig. 20-35 Removing the axle-shaft bearing with a special bearing press. *(Chrysler Corporation)*

b. Tapered-roller bearings. To remove a tapered-roller bearing, Ford recommends drilling a ¼-inch [6.4-mm] hole in the outside diameter of the inner retainer ring (Fig. 20-37). Drill about three fourths of the way through the ring, but not into the axle, as this would damage the axle. Then use a chisel as shown in Fig. 20-38 to break the retainer ring.

The tapered-bearing outer race can be removed with a reversible-jaw puller attached to a slide hammer (Fig. 20-39). The inner race of the tapered bearing is removed with the special tools shown in Fig. 20-40. For disk brakes, use the tools and procedure shown in Fig. 20-41.

Careful: Do not damage the axle shaft. Never use heat on the axle, retainer, or bearings. This can weaken the axle so that it could fail in normal highway operation.

3. To install the axle bearing and seal on the C-lock-washer-type axle, first install the outer retainer over the axle. Lubricate the new bearing with gear lubricant and use the special bearing-installer tool as shown in Fig. 20-42 to install it. The bearing should

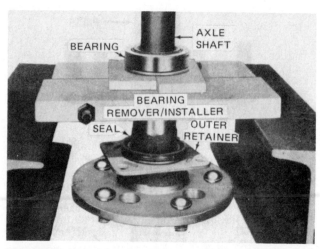

Fig. 20-36 Removing the axle-shaft bearing in a shop press. *(Chevrolet Motor Division of General Motors Corporation)*

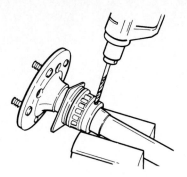

Fig. 20-37 Drilling the retainer ring. *(Ford Motor Company)*

be driven in until the tool bottoms against the shoulder in the axle housing.

Lubricate the seal lips with gear lubricant. Position the seal on the special oil-seal installer as shown in Fig. 20-43. Tap the seal into place until it is flush with the axle housing.

4. To install the retainer-type axle bearing and seal, first install the outer retainer over the axle. Lubricate the inner lips and outer diameter of the seal with lithium-soap grease. Then position the seal on the seal surface of the axle (Fig. 20-44).

Careful: Do not allow the seal to rub against the axle or it could be damaged.

a. Use the special tools as shown in Fig. 20-45 or the bearing press as shown in Fig. 20-46 to press the bearing up against the shoulder of the axle.

b. If a tapered-roller bearing is being installed, it should be positioned so that the manufacturer's coding is visible (Fig. 20-44).

c. Position a new inner retainer over the shaft and press the retainer in up against the inner face of the bearing (Fig. 20-45).

✿ **20-14 Drive-pinion oil-seal replacement** The drive-pinion oil seal is located at the front end of the differential housing and drive pinion. The removal procedure is different for the integral-carrier differential and the removable-carrier differential. The integral carrier (Fig. 20-7) is the more common. In this type, the

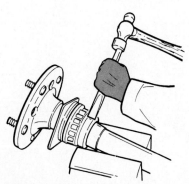

Fig. 20-38 Removing the bearing retainer ring. *(Ford Motor Company)*

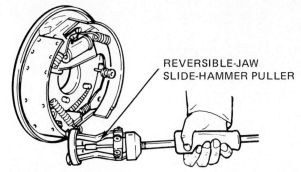

Fig. 20-39 Removing a tapered-bearing race with a reversible-jaw puller attached to a slide hammer. *(Ford Motor Company)*

differential housing is part of the axle-housing assembly.

With a removable carrier (Fig. 20-8), the axle housing has a large recess into which the differential assembly can be installed. The carrier housing bolts onto the axle housing. In this design, a pinion retainer bolts to the carrier housing to hold the drive pinion and bearings in place (Fig. 20-15).

To replace the drive-pinion oil seal in the integral-carrier differential, proceed as follows:

1. Mark the driveshaft and pinion flange so that they can be reconnected in the same position (Fig. 20-16). Disconnect the driveshaft and support it by wiring it to the exhaust pipe.

2. Mark the position of the pinion flange, pinion shaft, and nut so that the same bearing preload can be restored on reassembly.

3. Remove the pinion-flange nut and washer (Fig. 20-47). Use a flange holder to hold the flange while the nut is loosened.

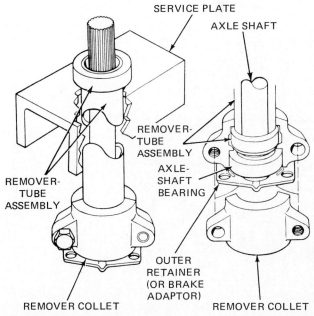

Fig. 20-40 Tool setup to press bearing off the axle shaft. *(Ford Motor Company)*

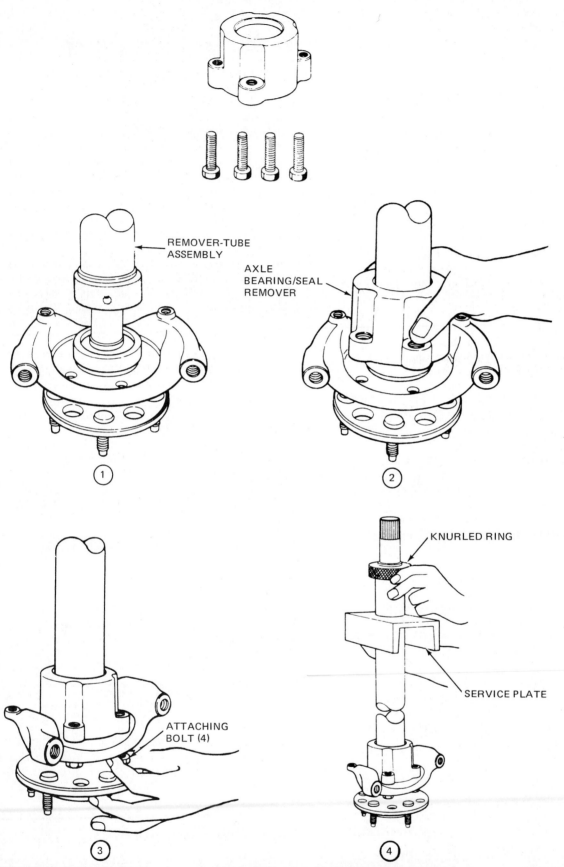

AXLE BEARING/SEAL REMOVER

REMOVER-TUBE
ASSEMBLY

AXLE
BEARING/SEAL
REMOVER

① ②

ATTACHING
BOLT (4)

KNURLED RING

SERVICE PLATE

③ ④

Fig. 20-41 Removing bearing from an axle shaft of a car with disk brakes. *(Ford Motor Company)*

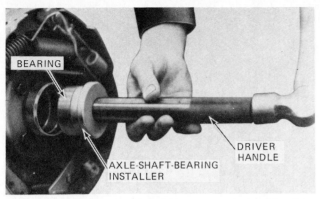

Fig. 20-42 Installing axle bearing in an axle housing having a C lockwasher. *(Chevrolet Motor Division of General Motors Corporation)*

Fig. 20-43 Installing axle seal in an axle housing having a C lockwasher. *(Chevrolet Motor Division of General Motors Corporation)*

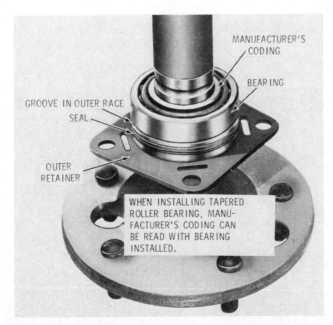

Fig. 20-44 Bearing installation on a retainer-type axle. *(Chevrolet Motor Division of General Motors Corporation)*

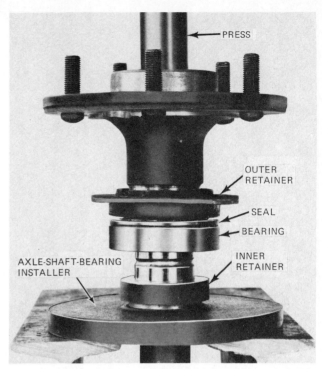

Fig. 20-45 Installing bearing retainer on a retainer-type axle. *(Chevrolet Motor Division of General Motors Corporation)*

4. Remove the pinion flange using the special flange-remover tools as in Fig. 20-48. Have a container ready to catch any lubricant that drains out.
5. Drive the oil seal out of the housing with a chisel (Fig. 20-49). Do not damage the housing.
6. Install a new seal after applying special seal lubricant to the outside diameter of the seal and to the seal lip (Fig. 20-50).

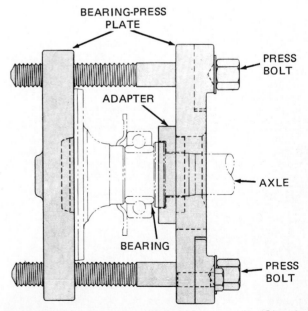

Fig. 20-46 Installing rear-axle bearing on axle. *(Chrysler Corporation)*

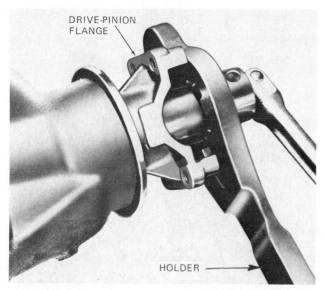

Fig. 20-47 Removing drive-pinion flange nut. *(Chevrolet Motor Division of General Motors Corporation)*

7. Install the pinion flange and tighten the nut to the position marked before disassembly. Then tighten the nut 1/16 inch [1.5 mm] beyond the alignment marks.

8. If the flange is damaged and requires replacement, another procedure is required. First check the bearing preload with a torque wrench as shown in Fig. 20-51. Then remove the pinion flange nut and washer, and the flange (Figs. 20-47 and 20-48).

When reinstalling the flange nut, tighten it a little at a time and turn the drive pinion several revolutions to set the bearing rollers. Check the preload of the bearing each time until the proper preload is obtained. This should be 3 to 5 pound-inches (3.45 to 5.75 kg-cm) more than the value recorded before disassembly (Fig. 20-5).

9. Reconnect the driveshaft, install the drum and wheel, and add lubricant as necessary.

The drive-pinion oil seal on a removable-carrier differential may be replaced with the differential either on

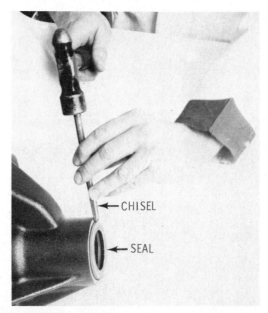

Fig. 20-49 Removing drive-pinion oil seal. *(Chevrolet Motor Division of General Motors Corporation)*

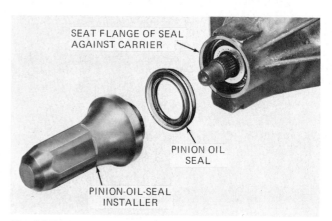

Fig. 20-50 Installing drive-pinion oil seal. *(Chevrolet Motor Division of General Motors Corporation)*

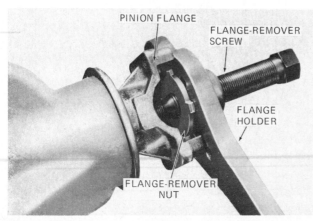

Fig. 20-48 Removing drive-pinion flange. *(Chevrolet Motor Division of General Motors Corporation)*

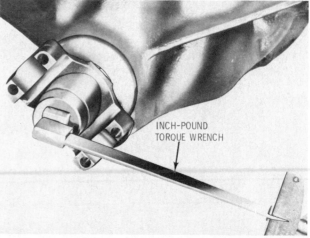

Fig. 20-51 Checking differential-bearing preload. *(Chevrolet Motor Division of General Motors Corporation)*

the car or removed for bench service. The procedure recommended by Ford follows:

1. Raise the vehicle and support it on safety stands.
2. Mark the driveshaft universal joint and drive-pinion flange so that they can be reconnected in the same relative position (Fig. 20-16). Be careful not to drop the loose universal-joint bearing cups. Mark them so that you can put them back in their original positions.
3. Remove the driveshaft from the transmission extension housing. Install an oil-seal-replacer tool in the extension housing to prevent loss of transmission fluid.
4. Use a torque wrench on the drive-pinion nut to measure the bearing preload (Fig. 20-51). Rotate the drive pinion several times and take several readings to get an accurate figure.
5. Mark the pinion shaft and universal-joint flange so they can be reattached in the same relative position. Then hold the flange with the flange holder and remove the drive-pinion-shaft nut and washer (Fig. 20-52).
6. Clean the area around the pinion-bearing retainer. Put a drain pan underneath to catch the lubricant. Then use the flange remover as shown in Fig. 20-53 to remove the flange.
7. Use the seal remover to remove the seal (Fig. 20-54).
8. Clean the oil-seal bore or seat. Apply a light coat of sealer to the bore. Then install the new seal with the seal installer (Fig. 20-55).
9. Check the splines on the pinion shaft. If they have burrs, use fine crocus cloth to remove them. Wipe the pinion shaft clean.
10. Apply a small amount of lubricant to the flange splines. Align the marks on the flange and pinion shaft and install the flange (Fig. 20-56).
11. Use a new nut and washer and hold the flange while tightening the nut (Fig. 20-52). Rotate the pinion occasionally to ensure proper bearing seating. Take frequent preload readings.

Careful: Do not run past the specified preload. If you overtighten, you must install a new pinion-bearing spacer.

12. Remove the oil-seal-replacer tool from the transmission-extension housing. Then install the front end of the driveshaft on the transmission output shaft.
13. Connect the rear end of the driveshaft to the universal-joint flange, aligning the marks.

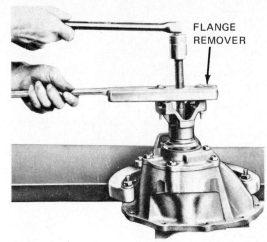

Fig. 20-53 Universal-joint-flange removal. *(Ford Motor Company)*

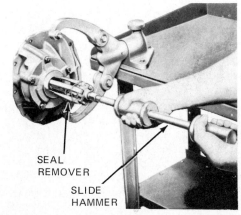

Fig. 20-54 Pinion-oil-seal removal. *(Ford Motor Company)*

Fig. 20-52 Pinion-shaft-nut removal. *(Ford Motor Company)*

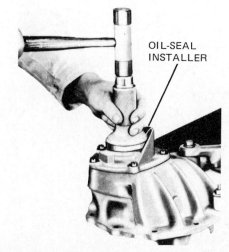

Fig. 20-55 Oil-seal installation. *(Ford Motor Company)*

257

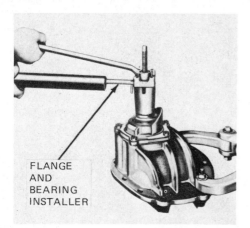

Fig. 20-56 Universal-joint-flange installation. *(Ford Motor Company)*

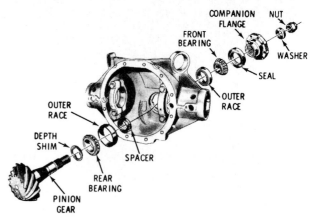

Fig. 20-58 Differential-pinion gear and companion flange. *(Oldsmobile Division of General Motors Corporation)*

14. Check the lubricant level with the vehicle and differential in running position. Add lubricant as necessary.

⚙ 20-15 Integral-carrier-differential disassembly
This differential (Fig. 20-7) is disassembled as follows. Support the car by the frame on safety stands and lower the axle housing to its lower limit. You may need to disconnect the shock absorbers to get additional clearance.

Careful: Do not allow the rear-brake hose to kink or stretch; this could damage the hose and then a new hose would be required. This would mean bleeding the rear brakes and refilling the master cylinder.

Figure 20-7 shows a completely disassembled rear axle and integral-carrier differential. Figures 20-57 and 20-58 show details of a disassembled integral-carrier differential.

NOTE: If a new ring-gear-and-drive-pinion set is installed, the driver should be told not to accelerate rapidly or drive over 50 mph [80 km/h] for the first 50 miles [80 km].

The removal procedure follows:

1. Back off the cover bolts and break the cover loose at the bottom to drain the lubricant into a container.
2. Remove the cover and gasket.
3. Check the ring-gear-to-pinion backlash with a dial indicator (Fig. 20-23). This will tell you whether gear or bearing wear or excessive backlash is causing noise.
4. Check the ring-gear runout with a dial indicator (Fig. 20-21).
5. Remove the bearing-cap bolts and caps. They should be marked *L* and *R* to make sure they are returned to their original positions (Fig. 20-57).
6. With care, use a pry bar to pry the differential case out of the housing (Fig. 20-59). The bearings have preload, and so the case will be tight until it is moved past a certain point. It will then suddenly come loose. Be prepared for this by having it properly supported so that it will not fall and be damaged.
7. With the case out, put the right and left bearing outer races and shims aside in sets so that they can be reinstalled in their original positions.

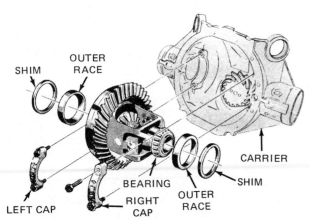

Fig. 20-57 Differential case and bearings. *(Oldsmobile Division of General Motors Corporation)*

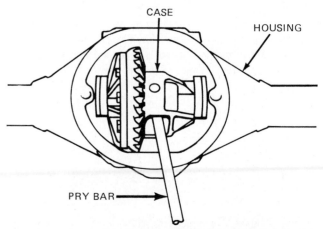

Fig. 20-59 Removing case assembly from the housing. *(Chevrolet Motor Division of General Motors Corporation)*

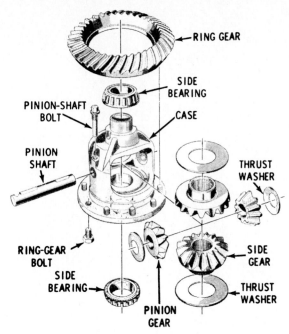

Fig. 20-60 Disassembled differential. *(Chevrolet Motor Division of General Motors Corporation)*

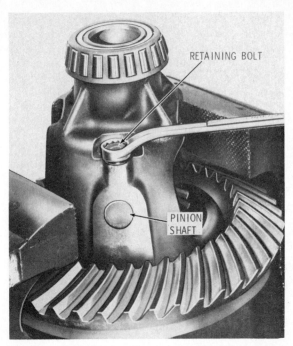

Fig. 20-62 Removing differential-pinion-shaft retainer bolt. *(Chevrolet Motor Division of General Motors Corporation)*

8. Disassemble the differential case as necessary (Fig. 20-60) as follows:

a. If the side bearings are to be replaced, pull them with the special puller as shown in Fig. 20-61.

b. Remove the pinion-shaft bolt (Fig. 20-62). Re-

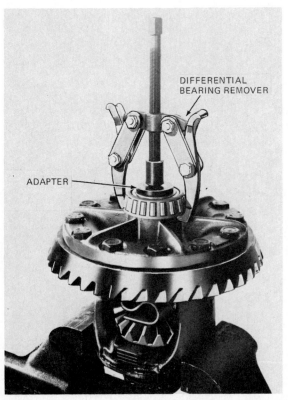

Fig. 20-61 Removing side bearing from differential case. *(Chevrolet Motor Division of General Motors Corporation)*

move the pinion shaft, pinions, side gears, and thrust washers.

NOTE: Mark the side gears and case so that they can be installed in their original positions.

c. Remove the ring gear if necessary. The bolts attaching the ring gear have left-hand threads. If the ring gear sticks on the case, tap it off using a brass drift punch and hammer. Do not try to pry it off. You could ruin both the gear and the case.

NOTE: Ring gears and drive pinions are supplied as matched sets. If one requires replacement, both must be replaced.

9. Remove the drive pinion, if necessary, as follows:

a. Check the drive-pinion-bearing preload as explained in ✹ 20-14. If you find no preload, the drive pinion is loose either because of defective bearings or because of a worn pinion flange. If the car has been operated for a long time with this condition, the ring gear and drive pinion are probably in bad condition and should be replaced. See ✹ 20-9 on gear-tooth-contact patterns.

b. Remove the flange nut and flange as explained in ✹ 20-14.

c. Reinstall the differential cover temporarily to prevent the drive pinion from falling to the floor when it is loosened. Put the flange nut partway back on the pinion shaft to protect the shaft threads. Tap the nut with a soft drift punch and hammer.

d. When the drive pinion breaks loose, remove the cover and the pinion from the housing. Examine the bearings. If they appear worn or damaged, replace them. A new pinion oil seal, nut, and collapsible spacer will be required on reassembly.

10. To replace the drive-pinion bearing, first remove the drive pinion as explained above. Then drive the pinion oil seal from the housing. Remove the front pinion-bearing outer race if this bearing requires replacement (Fig. 20-63). If the rear pinion bearing is to be replaced, drive the outer race out in the same way, working from the other side of the housing.

✿20-16 Cleaning and inspecting parts Clean all bearings in solvent. Do not spin-dry bearings with compressed air. Oil them and apply hand pressure as you rotate them. They should turn smoothly and without any roughness. See ✿ 7-6 for basic bearing-inspection procedures.

Inspect the other parts as follows:

1. Examine sealing surfaces for nicks or scratches that would cause leakage of lubricant.
2. Examine the ring-gear and drive-pinion teeth. If they are worn or damaged, a new ring-gear-and-drive-pinion set should be installed.
3. Check the other gears for wear or looseness in their bores.
4. If gears or bearings are worn or chipped, it could mean foreign material is present. The axle housing must be cleaned.

✿20-17 Setting the pinion depth on an integral-carrier differential In the integral-carrier differential, the shims are between the drive pinion and the inner race (Fig. 20-13). To change the shim thickness, the case must be removed. This process requires re-setting the backlash. An alternative method is offered by Chevrolet, as follows.

NOTE: If the drive pinion has not been removed and the original ring gear is to be reused, then the pinion depth should be okay. The procedure that follows is to be used if a new ring-gear-and-drive-pinion set is installed. Also, this procedure is to be used if, as a result of wear of the original set or bearings, the pinion depth is not

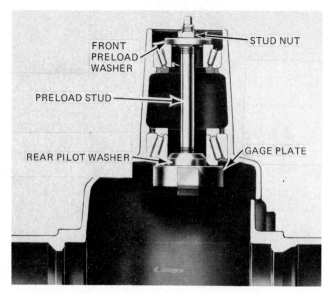

Fig. 20-64 Tool setup to set drive-pinion depth. *(Chevrolet Motor Division of General Motors Corporation)*

correct so that the proper gear-tooth pattern is not obtained (✿ 20-7 and 20-8).

1. Clean the gauge parts and housing (Figs. 20-64 and 20-65).
2. Lubricate and install the front and rear drive-pinion bearings in the housing.
3. Install the preload stud as shown in Fig. 20-64. Tighten the preload stud by holding it with a wrench and turning the stud nut until the bearings are tight. The bearings should require the specified torque to turn them.
4. Put the gauge disks on the gauge shaft and secure the assembly in the housing with the bearing caps

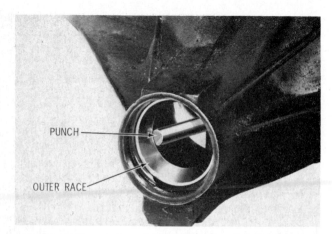

Fig. 20-63 Removing drive-pinion front-bearing outer race. *(Chevrolet Motor Division of General Motors Corporation)*

Fig. 20-65 Complete tool setup to set drive-pinion depth. *(Chevrolet Motor Division of General Motors Corporation)*

(Fig. 20-65). Turn the bearing-cap bolts up finger-tight.

5. Position the dial indicator on the mounting post of the shaft with the contact button resting on top of the gauge-plate plunger.

6. Preload the dial indicator one half revolution of the needle. Tighten it in this position.

7. Put the plunger on the gauge plate. Rotate the plate until the plunger rests on the contact button. Rock the plunger rod slowly back and forth to get the greatest deflection. Then, set the dial indicator to zero.

8. Swing the indicator probe off the gauge plate. The dial-indicator needle will now swing to a reading that indicates the basic shim thickness.

9. Pinions are marked (Fig. 20-66) to indicate if shim thickness should vary from the basic measurement. If the drive pinion has no marking, the basic shim thickness is correct. If the drive pinion is marked, the marking will indicate whether to add to or subtract from the basic shim thickness.

10. Install the correct shim thickness on the drive pinion and install the drive-pinion rear bearing on the pinion shaft. Then assemble the differential as described in ✹ 20-19.

✹ **20-18 Setting the pinion depth on a removable-carrier differential** This adjustment is made by adding or removing shims between the retainer and housing (Figs. 20-14 and 20-15). The procedure is described in ✹ 20-9.

✹ **20-19 Integral-carrier differential reassembly** After the pinion depth has been set, proceed as follows:

1. Put the side-gear thrust washers over the gear hubs and install the side gears in the case. If the gears are being reused, install them in the original positions.

2. Position one pinion, without a washer, between the side gears. Rotate the gears until the pinion is directly opposite the housing opening. Put the other pinion in place, making sure the pinion holes line up with the holes in the case.

3. Rotate the pinions back toward the opening just enough to allow you to slide the pinion thrust washers in place.

4. Secure the assembly with the pinion shaft and install the pinion-shaft retaining bolt. Torque the bolt to specifications.

5. If the ring gear has been removed, install it after making sure it fits snugly against the case. Mating surfaces must be clean and free of any burrs. Thread two bolts on opposite sides of the case to center the gear. Then install the rest of the bolts.

Careful: Always use *new* bolts. Never reuse old bolts.

6. Torque the bolts alternately (one side and then the other) in several stages to bring the ring gear down with even tension all around.

7. If the case side bearings were removed, install them (Fig. 20-67).

8. Side-bearing preload adjustment is made in either of two ways, according to the differential design. One design has adjustable bearing nuts on the outside of the case bearings (Figs. 20-7, 20-13, and 20-14). On these, turning the adjusting nuts inward adds preload to the bearings. At the same time, turning one adjusting nut inward and the other out-

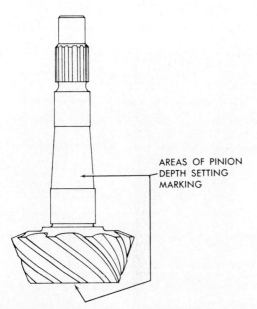

Fig. 20-66 Locations of drive-pinion markings. *(Chevrolet Motor Division of General Motors Corporation)*

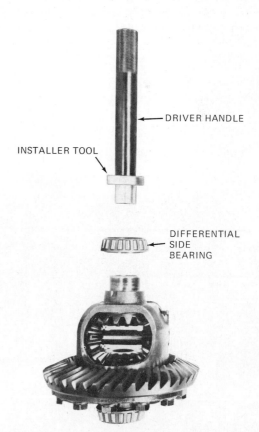

Fig. 20-67 Installing side bearing on differential case. *(Chevrolet Motor Division of General Motors Corporation)*

ward changes the backlash and gear-tooth-contact pattern (✿ 20-9).

The other design uses shims outside the side bearings (Fig. 20-57). With this design, changing the thickness of the shims adjusts the bearing preload. Also, with this design, adding a thicker shim on one side and a thinner shim on the other changes the backlash.

a. To determine the correct shim thickness for correct bearing preload (differential design, Fig. 20-57), install the differential case in the housing and secure it with one bearing cap.

NOTE: This procedure is done without the drive pinion installed. If the drive pinion is in place, remove the ring gear from the differential case.

Bearing-cap bolts should be left loose so that the case is loosely held in place. Then install the service spacer (Fig. 20-68). Next, install a thick shim as shown in Fig. 20-69. Start with a shim that goes in and use progressively thicker shims, at the same time checking the clearance with a thickness gauge (Fig. 20-70). When the thickness gauge drags, the clearance is zero, which is the beginning of preload.

b. Work the case in and out and to the left to insert the thickness gauge.

c. Once there is zero clearance, remove the left bearing cap and shim. Add the thickness of the left shim, plus the thickness of the right shim, plus the thickness of the thickness gauge. Divide by 2 to get the correct shim thickness to be installed on each side. On final assembly, use slightly thicker shims to produce the preload, driving them into position with light taps of a soft hammer (Fig. 20-69). Also, you may need to install a thicker shim on one side and a thinner shim on the other to shift the case and ring gear sideways and get the correct backlash (✿ 20-9).

9. After side-bearing preload is adjusted, tighten the bearing-cap bolts to the proper torque. Then re-

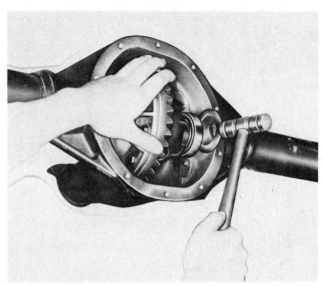

Fig. 20-69 Installing service shim. *(Chevrolet Motor Division of General Motors Corporation)*

check the gear-tooth pattern (✿ 20-7 to 20-9) and make further adjustments if necessary.

10. Install the cover and add the correct amount of the specified lubricant.

11. Road test the car. Check for differential noise to make sure the trouble has been corrected.

✿ 20-20 Removable-carrier-differential service This type of differential is shown in Figs. 20-8 and 20-15. The service procedure is as follows:

1. **Removal** First remove the driveshaft as described in the procedure given in ✿ 20-14 for replacing the oil seal on a removable-carrier differential. Then remove the carrier-housing attaching nuts and lift off the assembly. Catch the lubricant in a container. Install the assembly on the bench with the special holding tool (Fig. 20-54).

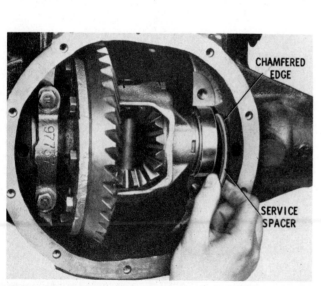

Fig. 20-68 Installing service spacer. *(Chevrolet Motor Division of General Motors Corporation)*

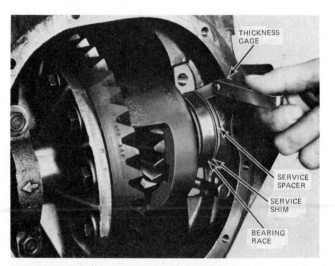

Fig. 20-70 Checking clearance between service spacer and housing with a thickness gauge. *(Chevrolet Motor Division of General Motors Corporation)*

2. **Disassembly** Use a punch to mark one bearing cap and mating bearing support. Then mark one of the bearing adjustment nuts and the carrier so that the parts can be reassembled in the proper relationship.

Remove the adjustment-nut locks, bearing caps, and adjustment nuts. Lift the differential assembly from the carrier. Pull the differential bearings with a bearing press (Fig. 20-71). Detach the ring gear from the differential case, using a soft-face hammer or a shop press to loosen it. Use a drift punch to drive out the differential-pinion-shaft lockpin. Separate the two-piece differential case. Drive out the pinion shaft with a brass drift punch and remove the gears and thrust washers.

Disassemble the drive pinion and bearing retainer (Fig. 20-15) by removing the U-joint flange (Figs. 20-52 and 20-53) and pinion seal (Fig. 20-54). Then, press out the bearings. Use a protective sleeve or rubber hose on the drive-pinion pilot-bearing surface and a fiber block on the end of the shaft to protect finished surfaces.

3. **Inspection** All parts should be carefully inspected. Worn or otherwise defective bearings, gears, or other parts should be discarded. The basic bearing-inspection procedure is given in ✿ 7-6.

If the ring-gear runout is excessive, the cause may be a warped gear or case or worn differential bearings. To determine the cause, assemble the two halves of the differential case without the ring gear, and press the two differential side bearings on the case hubs. Put the cups on the bearings, and set the differential case in the carrier. Install the bearing caps and adjusting nuts (see item 4 below) and adjust the bearings, as described in ✿ 20-9.

Check the runout of the differential-case flange. If the runout exceeds specifications, the case is defective or the bearings are worn. However, if the runout is

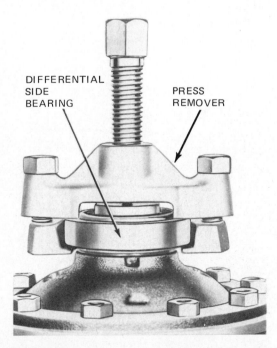

Fig. 20-71 Removing a differential side bearing (removable-carrier type). *(Ford Motor Company)*

within specifications, the trouble is with the ring gear and a new ring gear is required.

4. **Reassembly** First reassemble the drive pinion and bearing retainer (Fig. 20-15), and install the bearings. Install a new oil seal (Fig. 20-55) and the U-joint flange (Fig. 20-56).

Refer to Fig. 20-8 for assembly of the differential case. Lubricate all parts liberally during assembly. Place the side gear in the case half that has the pinion-shaft hole. Then drive the pinion shaft into the case only far enough to retain a pinion thrust washer and pinion gear. Put the second pinion gear and thrust washer into position, and drive the pinion shaft into place. Be sure to line up the pinion-shaft-lockpin holes.

Put the second side gear and thrust washer into position, and install the cover on the differential case. Insert an axle-shaft spline into a side gear to check for free rotation of the gears.

Insert two 7/16 N.F. bolts 2 inches [50.8 mm] long through the case flange. Thread them several turns into the ring gear as an aid to aligning it with the case. Press or tap the gear into place. Install and tighten the gear attachment bolts, using washers. Torque them alternately across the gear to ensure alignment. If the differential bearings have been removed, press them on. Install the bearing-retainer-and-drive-pinion assembly (✿ 20-9). Make sure that the gear-set timing marks are properly aligned, as shown in Fig. 20-27 (nonhunting type).

Adjust the backlash and drive-pinion location (✿ 20-9). Make a final tooth-pattern check before installing the carrier assembly in the axle housing.

✿ **20-21 Servicing limited-slip differentials** The following sections describe the servicing of three general types of limited-slip differentials. These are the two-clutch-pack type (Fig. 20-72), the spring-loaded single-clutch-pack type (Fig. 20-73), and the spring-loaded cone type (Fig. 20-74).

All operations described below are shown being performed on a differential that has been removed from the car. However, some of the operations can be performed on the car.

CAUTION: Never raise one wheel and run the engine with the transmission in gear. The driving force to the wheel on the floor will cause the car to move and jump off the car support. Do not use "on-the-car" wheel balancers on the rear wheels unless both wheels are off the floor.

✿ **20-22 Servicing the two-clutch-pack limited-slip differential** This differential is shown in Fig. 20-72. If the side bearings require replacement, remove them using standard bearing pullers as shown in Figs. 20-61 and 20-71.

1. Remove the ring gear if it is to be replaced.
2. Remove the pinion-shaft lock screw and shaft.
3. Drive the preload spring from the case (Fig. 20-75).
4. Rotate the side gears until the pinions are in the

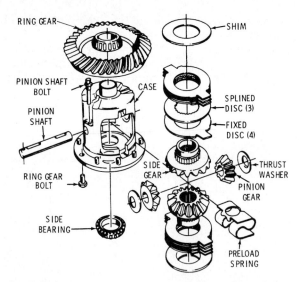

Fig. 20-72 Disassembled limited-slip differential, two-clutch-pack type. *(Chevrolet Motor Division of General Motors Corporation)*

openings of the case. Remove the pinion gears and thrust washers (Fig. 20-76).

5. Remove a side gear, clutch pack, and shims from the case, noting their locations in the case so that you can put them back in the same position.
6. Remove the side gear, clutch pack, and shims from the other side.
7. Remove the shims and clutch plates from the side gears. Keep the clutch plates in their original positions in the clutch pack.
8. Inspect and clean the components.
   a. Clean the side bearings in solvent and inspect them (❂ 7-6).
   b. Examine the ring-gear and drive-pinion teeth for nicks, burrs, or scoring. Replace the set (ring gear and drive pinion) if there is damage.

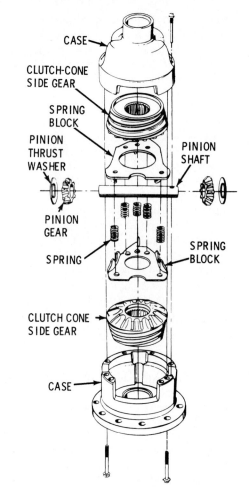

Fig. 20-74 Disassembled limited-slip differential, cone type. *(Chevrolet Motor Division of General Motors Corporation)*

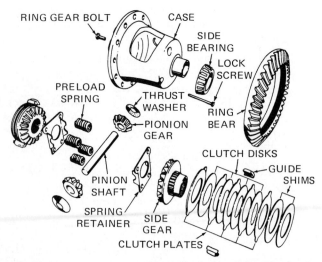

Fig. 20-73 Disassembled limited-slip differential, spring-loaded single-clutch-pack type. *(Chevrolet Motor Division of General Motors Corporation)*

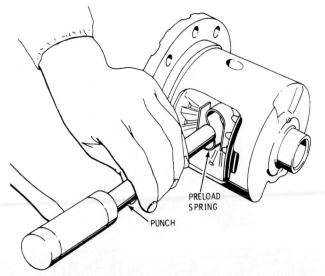

Fig. 20-75 Removing preload spring. *(Chevrolet Motor Division of General Motors Corporation)*

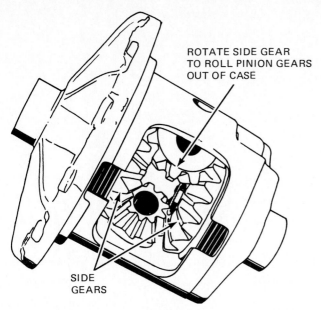

ROTATE SIDE GEAR
TO ROLL PINION GEARS
OUT OF CASE

SIDE
GEARS

Fig. 20-76 Removing or installing pinions and washers. *(Chevrolet Motor Division of General Motors Corporation)*

c. Inspect the pinion shaft, pinion gears, and side gears. Replace defective parts.

d. Check the fit of the side-bearing inner races on the case. They must be a tight fit.

e. Check the clutch plates for scores, wear, cracks, or distortion. Replace the pack if there are defects.

9. If side bearings were removed, install new ones.
10. Assemble a seven-plate clutch pack on a side gear, starting and ending with a clutch plate having external lugs. Line up the lugs (Fig. 20-72). Reinstall the spacer (if used) and shim.
11. Repeat the procedure for the other side gear.
12. Check the pinion-to-side-gear clearance as follows:

   a. Install one side gear with its clutch pack and shim in the case.

   b. Put the two pinion gears and thrust washers on

the side gear and install the pinion shaft to hold them in place (Fig. 20-77).

   c. Compress the clutch pack by inserting a screwdriver or wedge between the side gear and the pinion shaft. Figure 20-77 shows a screwdriver being used.

   d. Install a dial indicator with the contact button against a tooth of the pinion gear. Try to rotate the pinion gear. Clearance should be between 0.001 and 0.006 inch [0.03 and 0.15 mm].

   e. If clearance is less than 0.001 inch [0.03 mm], remove a shim. If it is excessive, add a shim. Repeat to get the specified clearance. Remove the side gear and clutch pack, with shims.

   f. Repeat the procedure for the other side gear and clutch pack.

13. Remove the pinion gears and shaft, with washers.
14. Install the second side gear and clutch pack, with shims.
15. Roll the pinion gears into place with their bores lined up with the holes in the case.
16. Install the pinion-gear thrust washers and insert the pinion shaft. Insert it through only one pinion gear.
17. Install the preload spring (Fig. 20-78).
18. Push the pinion shaft on into place and secure it with the pinion-shaft lock screw, torqued to specifications.
19. If a new ring gear is being installed, use new attaching bolts.

✿ 20-23 Servicing the spring-loaded single-clutch-pack limited-slip differential This differential is illustrated in Figs. 19-20 and 20-73. If it is necessary to remove the ring gear and side bearings, use the procedures for the standard differentials (✿ 20-19 and 20-20). Disassemble as follows:

1. Drive the preload spring retainer and springs through the observation hole in the case only far enough to secure them with a C clamp (Fig. 20-79). Install ¼-inch [6.4-mm] bolts and nuts through the retainers to hold everything in place.
2. If it is necessary to disassemble the spring pack,

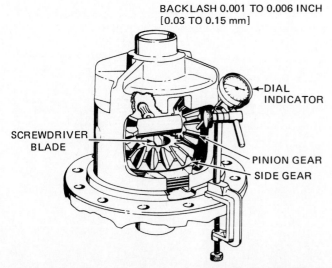

BACKLASH 0.001 TO 0.006 INCH
[0.03 TO 0.15 mm]

DIAL INDICATOR

SCREWDRIVER BLADE

PINION GEAR
SIDE GEAR

Fig. 20-77 Checking side-gear-to-pinion backlash. *(Chevrolet Motor Division of General Motors Corporation)*

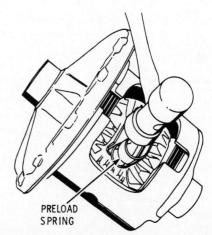

PRELOAD SPRING

Fig. 20-78 Installing preload spring. *(Chevrolet Motor Division of General Motors Corporation)*

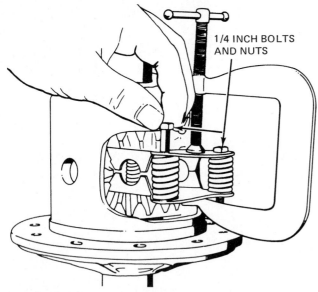

1/4 INCH BOLTS AND NUTS

Fig. 20-79 Removing preload springs and retainers. *(Chevrolet Motor Division of General Motors Corporation)*

position it in a vise to hold it temporarily. Remove the C clamp and bolts and then back off the vise slowly to relieve the spring force.

3. Remove the pinion thrust washers from behind the pinion gears.
4. Remove the pinion gears from the case by rotating them in one direction only (Fig. 20-80). Rotate the case clockwise to remove the first pinion gear. Then rotate the case in the opposite direction to remove the other pinion gear. You may need to pry the second pinion gear through the observation hole in the case.

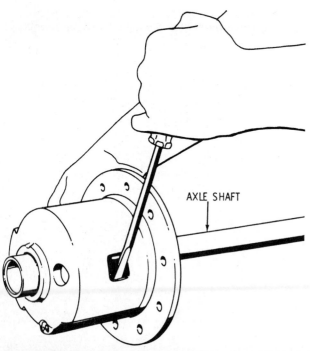

AXLE SHAFT

Fig. 20-80 Removing pinion gears. *(Chevrolet Motor Division of General Motors Corporation)*

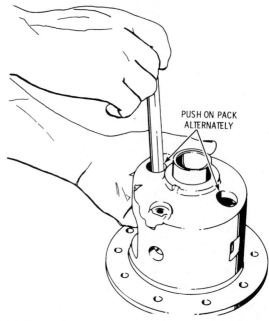

PUSH ON PACK ALTERNATELY

Fig. 20-81 Removing clutch pack. *(Chevrolet Motor Division of General Motors Corporation)*

5. Remove one side gear, the clutch pack, shims, and guides from the case. Tap the clutch pack out with a brass drift punch as shown in Fig. 20-81.
6. Remove the other side gear.
7. Inspect all parts as described in ⚙ 20-22, item 8.

Reassemble the spring-loaded single-clutch-pack differential as follows:

1. First reinstall the side gears and ring gear if they were removed. Then lubricate the clutch plates.
2. Position the clutch plates and disks alternately on the side gear. Figure 20-73 shows the positions. Disk splines should engage easily with side-gear splines. Add the shims.
3. Install guides on the clutch-plate lugs (Fig. 20-82).
4. Install the side gear with clutch pack and shim in the case.
5. Position the pinion gears and thrust washers on the side gear and secure with the pinion shaft.

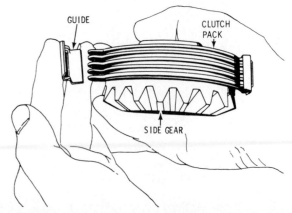

GUIDE   CLUTCH PACK

SIDE GEAR

Fig. 20-82 Installing clutch-pack guides. *(Chevrolet Motor Division of General Motors Corporation)*

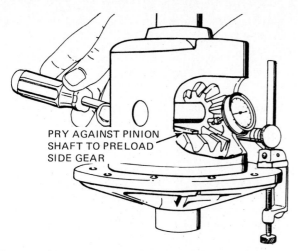

Fig. 20-83 Checking pinion-to-side-gear backlash. (Chevrolet Motor Division of General Motors Corporation)

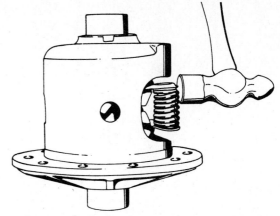

Fig. 20-85 Driving preload springs and retainers into place. (Chevrolet Motor Division of General Motors Corporation)

6. Install a dial indicator so that the contact button rests on a pinion-gear tooth (Fig. 20-83).

7. Use a screwdriver as shown in Fig. 20-83 to compress the clutch pack. Move the pinion gear while watching the dial indicator. The amount of movement is the clearance. It should be between 0.001 and 0.006 inch [0.03 to 0.15 mm]. Change shims if necessary to get this clearance.

8. Remove the side-gear assembly. Install the other side gear and check for clearance on that side. Make adjustments as necessary.

9. Remove the pinion shaft, pinion gears, and thrust washers.

10. Install the other side gear with clutch pack and shims in the case.

11. Install the pinion gears and thrust washers, aligning them with the holes in the case.

12. Install the springs in the spring retainers and clamp the assembly in a vise. Position a C clamp to hold them in place. Put ¼-inch [6.4-mm] bolts through the two springs closest to the clamp (Fig. 20-84) and secure them with nuts.

13. Put the assembly into the case, between the side gears (Fig. 20-85) and remove the bolts and C clamp. Drive the spring pack on into position.

14. Check the alignment of the spring retainers with the side gears. Adjust as necessary.

15. Secure the pinion gears, thrust washers, and spring retainers by installing the pinion shaft. Secure with the lock screw.

16. Install the side bearings and ring gear if they have been removed. Use new ring-gear bolts.

✿ 20-24 Servicing the cone-type limited-slip differential This differential is shown disassembled in Figs. 19-21 and 20-74. If the ring gear or side bearings require replacement, use the tools and procedures for

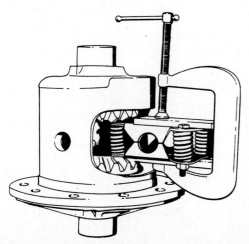

Fig. 20-84 Installing preload springs and retainers. (Chevrolet Motor Division of General Motors Corporation)

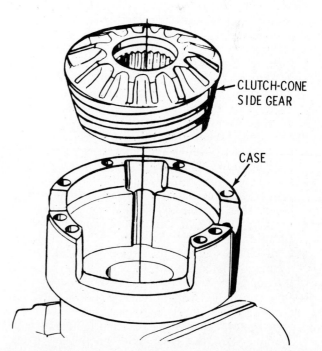

Fig. 20-86 Installing clutch-cone side gear. (Chevrolet Motor Division of General Motors Corporation)

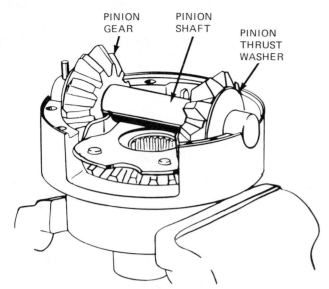

PINION GEAR   PINION SHAFT   PINION THRUST WASHER

Fig. 20-87 Installing parts in cap half of case. *(Chevrolet Motor Division of General Motors Corporation)*

standard differentials (✿ 20-19). Proceed with the disassembly as follows.

1. Before disassembling the differential, inspect the side bearings for visible damage of rollers and outer races. The basic bearing-inspection procedure is described in ✿ 7-6.
2. Place one outer race onto its matched inner-race-and-roller assembly and turn slowly, applying hand load.
3. If the bearing outer race turns smoothly and no visible damage is found, the bearing can be reused.
4. Repeat the above operation with the other race and matched bearing and check for smoothness.

   Both side bearings and their outer races are matched parts. If either bearing is to be replaced, its matching outer race must also be replaced.
5. Inspect the fit of the inner races on the case hubs by prying against the shoulders at puller recesses. The bearing inner races must be tight on the case hubs. If either bearing is loose on the case, the entire case must be replaced.
6. If bearing inspection indicates that bearings should be replaced, remove the side bearings by using tools as shown in Fig. 20-61.
7. If you remove the ring gear, clamp the case in a vise so that the jaws are 90° to the pinion shaft holes. Then remove 10 ring-gear retaining bolts.
8. Partially install two bolts on opposite sides of the ring gear.
9. Remove the ring gear from the case by alternately tapping on bolts. Do not pry between the case and the ring gear.
10. Remove the bolts that connect the two halves of the case.
11. Lift the cap half of the case from the flange half. Remove the clutch-cone side gears (Fig. 20-86), spring blocks, preload springs, pinion gears, and shaft. Be certain that each clutch-cone side gear

and each pinion gear is marked so that they can be installed in their original location.

Clean and inspect the parts of the disassembled differential:

1. Clean and dry all parts.
2. Inspect the pinion shaft, pinion and side gears, brake-cone surfaces, and corresponding cone seats in the case. The cone seats in the case should be smooth and free of any excessive scoring. Slight grooves or scratches, indicating the passage of foreign material, are permissible and normal. The land surface on the heavy spirals of the male cones will duplicate case surface conditions. If the case or clutch-cone side gear is damaged, it is necessary to replace the case assembly. All other parts are serviceable.

Reassemble the differential as follows:

1. Install the proper cone-gear assembly, seating it into position in the cap half of the case (Fig. 20-87).

NOTE: Be certain that each cone gear is installed in the proper half of the case. Tapers and surfaces become matched and their positions should not be changed.

2. Place one spring block in position over the gear face, in alignment with the pinion-gear-shaft grooves. Install the pinion shaft, pinion gears, and thrust washers in the cap half of the rear axle case so that the pinion-shaft retaining dowel can be inserted through the pinion-gear shaft into the rear-axle case. This prevents the pinion shaft from sliding out and causing damage to the carrier (Fig. 20-86). Pinion gears must be installed in their original locations.
3. Insert five springs into the spring block that is already installed in the case. Then place a second spring block over the springs (Fig. 20-88).
4. Install the second cone-gear assembly face down on the spring block so that the gear will mesh with the pinion gears.
5. Install the flange half of the rear axle case over the

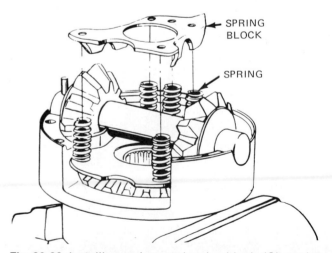

SPRING BLOCK

SPRING

Fig. 20-88 Installing springs and spring block. *(Chevrolet Motor Division of General Motors Corporation)*

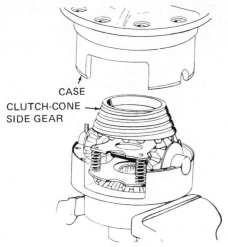

Fig. 20-89 Installing flange half of case. *(Chevrolet Motor Division of General Motors Corporation)*

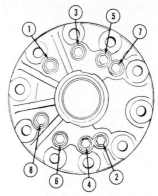

TIGHTEN BOLTS ONE TURN AT A TIME
THEN TORQUE TO 40 N·m [30 FT-LB]

Fig. 20-90 Sequence in which case bolts should be tightened. *(Chevrolet Motor Division of General Motors Corporation)*

cone. Then insert the case bolts finger-tight (Fig. 20-89).

6. Tighten the bolts one turn at a time in the sequence shown in Fig. 20-90. Then torque the case bolts to specifications.
7. If the side bearings were removed, lubricate outer bearing surfaces. Then press on bearings as shown in Fig. 20-67.
8. After making sure that mating surfaces are clean

and free of burrs, position the ring gear on the case so that the holes are in line.

9. Lubricate *new* attaching bolts with clean engine oil. Then install the bolts.
10. Pull the ring gear onto the case by tightening bolts alternately around the case. When all bolts are snug, tighten them evenly and alternately across the diameter to specifications. Do not use a hammer to force the ring gear onto the case.

Chapter 20 review questions

Select the *one* correct, best, or most probable answer to each question. Then check your answers against the correct answers given at the end of the book.

1. Most often, the condition that draws attention to trouble in the differential is:
   a. rough operation
   b. noise
   c. power loss
   d. lubricant discharge
2. A humming noise in the differential is often caused by improper tooth contact between the:
   a. drive pinion and ring gear
   b. axle and side gear
   c. pinion and side gears
   d. axle gear and drive pinion
3. If noise is present in the differential only when the car rounds a curve, the trouble probably is due to some condition in the:
   a. drive-pinion assembly
   b. differential-case assembly
   c. wheel bearing
   d. axle housing
4. To correct heavy face contact on the ring-gear teeth, move the:
   a. drive pinion in
   b. drive pinion out

   c. ring gear in
   d. none of the above
5. To correct heavy flank contact on the ring-gear teeth, move the:
   a. drive pinion in
   b. drive pinion out
   c. ring gear out
   d. drive gear out, and then adjust backlash as necessary
6. To correct heavy heel contact on the ring-gear teeth, move the:
   a. drive pinion in
   b. ring gear toward the pinion
   c. ring gear away from the pinion
   d. drive pinion out, and adjust backlash
7. To correct heavy toe contact on the ring-gear teeth, move the:
   a. ring gear toward the pinion
   b. ring gear away from the pinion
   c. drive pinion in
   d. none of the above
8. The drive pinion is adjusted with:
   a. adjustment nuts
   b. bearing adjusters
   c. snap rings
   d. shims

9. The ring gear is adjusted with:
   a. shims of proper thickness
   b. adjustment nuts
   c. adjustment screws
   d. both *a* and *b*
10. Bearing adjusters are held in place in the differential carrier by:
    a. bearing caps
    b. selective washers
    c. lockpins
    d. snap rings
11. The two basic rear-suspension systems are:
    a. integral- and removable-carrier
    b. coil spring and leaf spring
    c. direct drive and differential
    d. differential and transfer case
12. In the integral-carrier differential, the:
    a. differential is removed from the rear of the axle housing
    b. drive pinion is mounted in an extension of the axle housing
    c. differential side bearings mount between recesses in the axle housing and bearing caps
    d. all of the above
13. In the integral-carrier differential, the gear backlash is adjusted by:
    a. moving the drive pinion in or out
    b. turning the adjustment screws or nuts
    c. turning adjustment nuts or changing shims
    d. none of the above
14. In the removable-carrier differential, the:
    a. differential is removed from the front of the axle housing
    b. drive pinion is mounted in the carrier housing
    c. differential side bearings mount between the carrier housing and bearing caps
    d. all of the above
15. In the removable-carrier differential, the gear backlash is adjusted by:

    a. moving the drive pinion in or out
    b. turning adjustment screws or nuts
    c. turning adjustment nuts or changing shims
    d. all of the above
16. The first step in removing the rear-axle housing, after raising the car and placing safety stands under it, is to:
    a. disconnect the brake hose
    b. mark the rear U-joint and pinion flange
    c. disconnect the driveshaft
    d. none of the above
17. Rear-wheel axles are held in position in two ways, with:
    a. C washers at the inner ends or retainers at the outer ends
    b. C washers at the inner ends and retainers at the outer ends
    c. lock bolts at the inner ends or outer ends
    d. adjusting nuts or shims
18. To remove a C washer from the inner end of an axle, you must first:
    a. adjust gear backlash
    b. remove the differential-pinion shaft
    c. remove the differential side gears
    d. remove the differential drive pinion
19. On the retainer-type axle, the bearing, seal, and outer retainer come off with the:
    a. differential
    b. drive-pinion retainer
    c. axle
    d. differential-pinion shaft
20. The first step in disassembling an integral-carrier differential, after draining lubricant and removing the cover and gasket, is to:
    a. check the ring-gear-to-pinion backlash
    b. remove the differential case
    c. disassemble the differential
    d. remove the drive-pinion retainer

GLOSSARY

This glossary of automobile words and phrases provides a ready reference for the automotive technician. The definitions may differ somewhat from those given in a standard dictionary. They are not intended to be all-conclusive but to cover only what specifically applies to the automotive service field.

Acceleration   An increase in velocity or speed.

Accelerator   A foot-operated pedal linked to the throttle valve in the carburetor; used to control the flow of fuel to the engine.

Accessories   Devices not considered essential to the operation of a vehicle, such as the radio, car heater, and electric window lifts.

Additive   A substance added to gasoline or oil to improve some property of the gasoline or oil.

Adjust   To bring the parts of a component or system into a specified relationship, or to a specified dimension or pressure.

Adjustments   Necessary or desired changes in clearances, fit, or settings.

Alignment   The act of lining up; also, the state of being in a true line.

Antifriction bearing   Name given to almost any type of ball, roller, or tapered-roller bearing.

Arbor press   A small, hand-operated shop press used when only a light force is required against a bearing, shaft, or other part.

Assembly   A component part, itself made up of assembled pieces which form a self-contained, independently mounted unit. For example, in the automobile, the transmission is an assembly.

Automatic transmission   A transmission in which gear ratios are changed automatically, eliminating the necessity of hand-shifting gears.

Axis   The center line of a rotating part, a symmetrical part, or a circular bore.

Axle   A theoretical or actual crossbar supporting a vehicle and on which one or more wheels turn.

Axle ratio   The ratio between the rotational speed (rpm) of the driveshaft and that of the driven wheel; gear reduction through the differential, determined by dividing the number of teeth on the ring gear by the number of teeth on the drive pinion.

Axle shaft   In a vehicle drive axle, a shaft which transmits power from the differential side gear to the wheel hub.

Babbit   A metal consisting of tin, antimony, copper, and other metals; used to line bearings.

Backlash   In gearing, the clearance between the meshing teeth of two gears.

Ball-and-trunnion universal joint   A nonconstant-velocity universal joint which combines the universal joint and slip joint. It uses a drivable housing connected to a shaft with a ball head through two other balls mounted on trunnions.

Ball bearing   An antifriction bearing with an inner race and an outer race, and one or more rows of balls between them.

Bearing   A part that transmits a load to a support and in so doing absorbs the friction of moving parts.

Bearing caps   In the differential, caps held in place by bolts or nuts which, in turn, hold bearings in place.

Bearing oil clearance   The space purposely provided between a shaft and a bearing through which lubricating oil can flow.

Bevel gear   A gear shaped like the lower part of a cone; used to transmit motion through an angle.

bhp   Abbreviation for *brake horsepower*.

Body   On a vehicle, the assembly of sheet-metal and plastic panels and sections, together with windows, doors, seats, and other parts, that provides enclosures for the passengers, engine, and luggage compartments.

Boil tank   A large tank of boiling parts-cleaning solution, used for cleaning cylinder blocks, axle housings, and other large metal parts; also called a *hot tank*.

Bolt   A type of fastener having a head on one end and threads on the other; usually used with a nut.

Borderline lubrication   A type of poor lubrication resulting from greasy friction; moving parts are coated with a very thin film of lubricant.

Bore   An engine cylinder, or any cylindrical hole. Also, the process of enlarging or accurately refinishing a hole, as "to bore an engine cylinder." The bore size is the diameter of the hole.

Brake horsepower (bhp)   Power delivered by the engine and available for driving the vehicle; bhp = torque × rpm/5252.

Burr   A feather edge of metal left on a part being cut with a file or other cutting tool.

Bushing   A one-piece sleeve placed in a bore to serve as a bearing surface.

Calibrate   To check or correct the initial setting of a test instrument.

Caliper   A tool that can be set to measure the thickness of a block, the diameter of a shaft, or the bore of a hole (inside caliper).

Cam   A rotating lobe or eccentric which can be used with a cam-follower to change rotary motion to reciprocating motion.

Capacity The ability to perform or to hold.

Cardan universal joint A nonconstant-velocity universal joint which consists of two drivable yokes connected by a cross through four bearings.

Case hardening The carburizing method used on low-carbon steel or other alloys to make the case or outer layer of the metal harder than its core.

cc See *Cubic centimeter*.

Centimeter (cm) A unit of linear measure in the metric system; 1 centimeter equals approximately 0.390 inch.

Centrifugal See *Centrifugal force*.

Centrifugal clutch A clutch that uses centrifugal force to apply a higher force against the friction disk as the clutch spins faster.

Centrifugal force The force acting on a rotating body which tends to move it outward and away from the center of rotation. The force increases as rotational speed increases.

Chassis The assembly of mechanisms that make up the major operating systems of the vehicle; usually assumed to include everything except the car body.

Check To verify that a component, system, or measurement complies with specifications.

CID Abbreviation for *Cubic-inch displacement*.

Clearance The space between two moving parts or between a moving and a stationary part, such as a journal and a bearing. The bearing clearance is filled with lubricating oil when the mechanism is running.

Clutch The device used to transmit rotary motion from one shaft to another while permitting engagement or disengagement of the shafts during rotation of one or both. Normally, the two shafts are in line and rotate about the same axis.

Clutch disk See *Friction disk*.

Clutch fork In the clutch, a Y-shaped member into which the throwout bearing is assembled.

Clutch gear See *Clutch shaft*.

Clutch housing A metal housing that surrounds the flywheel-and-clutch assembly.

Clutch pedal A pedal in the driver's compartment that operates the clutch.

Clutch safety switch See *Neutral-start switch*.

Clutch shaft The shaft on which the clutch is assembled, with the gear that drives the countershaft in the transmission at one end. At the clutch-gear end, the shaft has external splines that can be used by a synchronizer drum to lock the clutch shaft to the main shaft for direct drive.

cm³ See *Cubic centimeter*.

Coil spring A spring made of an elastic metal, such as steel, formed into a wire and wound into a coil.

Coil-spring clutch A clutch using coil springs to hold the pressure plate against the friction disk.

Companion flange A mounting flange that fixedly attaches a driveshaft to another drive-train component.

Component A part of a whole assembly, system, or unit which may be identified and serviced separately. For example, a bulb is a component of the lighting system.

Constant-velocity universal joint Two closely coupled universal joints arranged so that their acceleration-deceleration phases cancel each other out. This results in an output speed that is always identical with the input speed, regardless of the angle of drive.

Corrosion Chemical action, usually by an acid, that eats away (decomposes) a metal.

Cotter pin A type of fastener, made from soft steel in the form of a split pin, that can be inserted in a drilled hole. The split ends are spread to lock the pin in position.

Countershaft In the transmission, a shaft which is driven by the clutch gear. Gears on the countershaft drive gears on the main shaft when they are engaged.

Crankcase The part of the engine that surrounds the crankshaft; usually the lower section of the cylinder block.

Cubic centimeter (cu cm, cm³, or cc) A unit of volume in the metric system; equal to approximately 0.03 fluid ounce.

Cubic-inch displacement The cylinder volume displaced or swept out by the pistons of an engine as they move from bottom dead center (BDC) to top dead center (TDC); measured in cubic inches.

cu cm See *Cubic centimeter*.

Dashpot A device on the carburetor that prevents the throttle valve from closing too suddenly, thereby causing the engine to stall. Sometimes used to prevent drive-line clunk on deceleration.

Dead axle An axle that supports weight and attached parts but does not turn or deliver power to a wheel or other rotating member.

Deceleration A decrease in velocity or speed. Also, allowing the car or engine to coast to idle speed from a higher speed with the accelerator at or near the idle position.

Degree Part of a circle; 1° is 1/360 of a complete circle.

Detent A small depression in a shaft, rail, or rod into which a pawl or ball drops when the shaft, rail, or rod is moved. This provides a locking effect.

Detergent A chemical added to engine oil; helps keep internal parts of the engine clean by preventing the accumulation of deposits.

Device A mechanism, tool, or other piece of equipment designed to serve a special purpose or perform a special function.

Diagnosis A procedure followed in locating the cause of a malfunction. Also, the act of specifically identifying and answering the question "What is wrong?"

Dial indicator A gauge that has a dial face and a needle to register movement; used to measure variations in dimensions and movements too small to be measured accurately by other means.

Diaphragm spring A spring shaped like a disk with tapering fingers pointed inward, or like a wavy disk (crown type).

Diaphragm-spring clutch A clutch in which a diaphragm spring, rather than a coil spring, applies pressure against the friction disk.

Differential A gear assembly between axle shafts that permits one wheel to turn at a different speed than the other, while transmitting power from the driveshaft to the wheel axles.

Differential case The metal unit that encases the differential side gears and pinion gears, and to which the ring gear is attached.

Differential side gears The gears inside the differential case which are internally splined to the axle shafts, and which are driven by the differential pinion gears.

Direct-bonded bearing A bearing formed by pouring liquid babbitt (bearing metal) directly into the bearing housing and machining the cooled metal to the desired bearing diameter.

Disassemble To take apart.

Displacement In an engine, the total volume of air-fuel mixture an engine is theoretically capable of drawing into all cylinders during one operating cycle. Also, the volume swept out by the piston in moving from one end of a stroke to the other.

Double-cardan universal joint A near-constant-velocity universal joint which consists of two Cardan universal joints connected by a coupling yoke.

Double-reduction differential A differential containing an extra set of gears to provide a second gear reduction.

Dowel A metal pin attached to one object which, when inserted into a hole in another object, ensures proper alignment.

Downshift To shift a transmission into a lower gear.

Driveability The general operation of a vehicle, usually rated from good to poor; based on characteristics of concern to the average driver, such as smoothness of idle, evenness of acceleration, ease of starting, rate of warmup, and tendency to overheat at idle.

Drive line An assembly of one or more driveshafts, usually with universal joints and some type of slip joint, used to transmit torque through varying angles from one shaft to another.

Driven disk The friction disk in a clutch.

Drive pinion A rotating shaft with a small gear on one end that transmits torque to another gear; used in the differential. Also called the *clutch shaft*, in the transmission.

Driveshaft An assembly of one or two universal joints connected to a shaft or tube; used to transmit power from the transmission to the differential. Also called the *propeller shaft*.

Drive train See *Power train*.

Dry-disk clutch A clutch in which the friction faces of the friction disk are dry, as opposed to a wet-disk clutch, which runs submerged in oil. The conventional type of automobile clutch.

Dry friction The friction between two dry solids.

Duration The length of time during which something exists or lasts.

Dynamic balance The balance of an object when it is in motion; for example, the dynamic balance of a rotating driveshaft.

Dynamometer A device for measuring the power output, or brake horsepower, of an engine. An engine dynamometer measures the power available at the flywheel; a chassis dynamometer measures the power available at the drive wheels.

Eccentric A disk or offset section (of a shaft, for example) used to convert rotary motion to reciprocating motion. Sometimes called a *cam*.

Efficiency The ratio between the power of an effect and the power expended to produce the effect; the ratio between an actual result and the theoretically possible result.

Electromechanical A term describing a device whose mechanical movement is dependent upon an electric current.

End play The distance that a shaft can move forward and backward in its housing or case.

Energy The capacity or ability to do work; usually measured in work units of foot-pounds (meters-newton) lb-ft [N–m] but also expressed in heat-energy units (Btu's [joules]).

Engine A machine that converts heat energy into mechanical energy. A device that burns fuel to produce mechanical power; sometimes referred to as a *power plant*.

Extreme-pressure lubricant A special lubricant for use in hypoid-gear differentials; needed because of the heavy wiping loads imposed on the gear teeth.

Fatigue failure A type of metal failure resulting from repeated stress, which finally alters the character of the metal so that it cracks. In engine bearings, frequently caused by excessive idling or by low engine idling speed.

Feeler gauge See *Thickness gauge*.

Fluid Any liquid or gas.

Fluid coupling A device in the power train consisting of two rotating members; transmits power from the engine, through a fluid, to the transmission.

Flywheel A heavy metal wheel that is attached to the crankshaft and rotates with it; helps smooth out the power surges from the engine power strokes; also serves as part of the clutch and engine-cranking system.

Flywheel ring gear A gear, fitted around the flywheel, that is engaged by teeth on the starting-motor drive to crank the engine.

Force Any push or pull exerted on an object; measured in pounds and ounces, or in newtons [N] in the metric system.

Four on the floor Slang for a four-speed transmission with the shift lever mounted on the floor of the driving compartment, frequently as part of a center console.

Four-speed A manual transmission having four forward gear ratios.

Four-wheel drive On a vehicle, driving axles at both front and rear, so that all four wheels can be driven.

Frame The assembly of metal structural parts and channel sections that supports the car engine and body and is supported by the wheels.

Friction The resistance to motion between two bodies in contact with each other.

Friction bearing A bearing in which there is sliding contact between the moving surfaces. Sleeve bearings, such as those used in connecting rods, are friction bearings.

Friction disk In the clutch, a flat disk, faced on both sides with friction material and splined to the clutch shaft. It is positioned between the clutch pressure plate and the engine flywheel. Also called the *clutch disk* or *driven disk*.

Friction horsepower The power used up by an engine in overcoming its own internal friction; usually increases as engine speed increases.

Front-wheel drive A vehicle having its drive wheels located on the front axle.

Full-coil suspension A vehicle suspension system in which each of the four wheels has its own coil spring.

Full-floating rear axle An axle which only transmits driving force to the rear wheels. The weight of the vehicle (including payload) is supported by the axle housing.

Full throttle Wide-open throttle position, with the accelerator pedal pressed all the way down to the floorboard.

Fusion Melting; conversion from the solid to the liquid state.

Gallery A passageway inside a wall or casting. The main oil gallery within the block supplies lubrication to all parts of the engine.

Gap The air space between two parts or electrodes.

Gasket A layer of material, usually made of cork, paper, plastic, composition, or metal, or a combination of these, placed between two parts to make a tight seal.

Gasket cement A liquid adhesive material, or sealer, used to install gaskets.

GCW Abbreviation for *gross combination weight;* the total weight of a tractor and semitrailer or trailer, including payload, fuel, and driver.

Geared speed The theoretical vehicle speed, based on engine rpm, transmission-gear ratio, drive-axle ratio, and tire size.

Gear lubricant A type of grease or oil blended especially to lubricate gears.

Gear ratio The number of revolutions of a driving gear required to turn a driven gear through one complete revolution. For a pair of gears, the ratio is found by dividing the number of teeth on the driven gear by the number of teeth on the driving gear.

Gears Toothed wheels that transmit power between shafts.

Gearshift A linkage-type mechanism by which the gears in an automobile transmission are engaged and disengaged.

Gradeability Ability of a truck to negotiate a given grade at a specified GCW or GVW.

Grease Lubricating oil to which thickening agents have been added.

Greasy friction The friction between two solids coated with a thin film of oil.

Grommet A device, usually made of hard rubber or a similar material, used to encircle or support a component.

GVW Abbreviation for gross vehicle weight, the total weight of a vehicle, including the body, payload, fuel, and driver.

Helical gear A gear in which the teeth are cut at an angle to the center line of the gear.

Heli-Coil See *Threaded insert.*

Hooke universal joint See *Cardan universal joint.*

Horsepower A measure of mechanical power, or the rate at which work is done. One horsepower equals 33,000 ft-lb (foot-pounds) of work per minute. It is the power necessary to raise 33,000 pounds a distance of 1 foot in 1 minute.

Hotchkiss drive A type of rear suspension in which leaf springs absorb the rear-axle-housing torque.

Hub The center part of a wheel, to which the wheel is attached.

Hydraulic clutch A clutch that is actuated by hydraulic pressure; used in cars and trucks when the engine is some distance from the driver's compartment so that it would be difficult to use mechanical linkages.

Hydraulic press A piece of shop equipment that develops a heavy force by use of a hydraulic piston-and-jack assembly.

Hydraulic pressure Pressure exerted through the medium of a liquid.

Hydraulics The use of a liquid under pressure to transfer force or motion, or to increase an applied force.

Hydraulic valve A valve in a hydraulic system that operates on, or controls, the hydraulic pressure in the system. Also, any valve that is operated or controlled by hydraulic pressure.

Hypoid gear A type of gear used in the differential (drive pinion and ring gear); cut in a spiral form to allow the pinion to be set below the center line of the ring gear so that the car floor can be lower.

Idle Engine speed when the accelerator pedal is fully released and there is no load on the engine.

Indicated horsepower The power produced within the engine cylinders before any friction loss is deducted.

Indicator A device used to make some condition known by use of a light or a dial and pointer.

Inertia A property of an object that causes it to resist any change in its speed or in the direction of its travel.

Inspect To examine a part or system for surface condition or function to answer the question "Is something wrong?"

Install To set up for use on a vehicle any part, accessory, option, or kit.

Integral Built into, as part of the whole.

Interaxle differential A two-position differential located between two driving axles. In the unlocked position it can divide the power unevenly, permitting one axle to turn faster than the other. In the locked position the power is divided evenly between the two axles.

Interchangeability The manufacture of similar parts to close tolerances so that any one of them can be substituted for another in a device, and the part will fit and operate properly; the basis of mass production.

Internal gear A gear with teeth pointing inward, toward the hollow center of the gear.

Journal The part of a rotating shaft which turns in a bearing.

Key A wedgelike metal piece, usually rectangular or semicircular, inserted in a groove to transmit torque while holding two parts in the same relative position. Also, the small strip of metal with coded peaks and grooves used to operate a lock, such as that for the ignition switch.

kg/cm² Abbreviation for *kilograms per square centimeter,* a metric engineering term for the measurement of pressure. One kilogram per square centimeter equals 4.22 pounds per square inch.

Kilogram (kg) In the metric system, a unit of weight and mass; approximately equal to 2.2 pounds.

Kilometer (km) In the metric system, a unit of linear measure; equal to 0.621 mile.

Kilowatt (kW) 1000 watts; a unit of power, equal to about 1.34 horsepower.

Kinetic energy The energy of motion; the energy stored in a moving body through its momentum; for example, the kinetic energy stored in a rotating flywheel.

Knock A heavy metallic sound which varies with engine speed; usually caused by a loose or worn bearing.

kPa Abbreviation for kilopascals, the metric unit of pressure. One kilopascal equals 0.145 pound per square inch.

kW Abbreviation for *kilowatt.*

Lash The amount of free motion in a gear train, between gears, or in a mechanical assembly, such as the lash in a valve train.

Leaf spring A spring made up of a single flat steel plate or of several plates of graduated lengths assembled one on top of another; used on vehicles to absorb road shocks by bending or flexing.

Light-duty vehicle Any motor vehicle manufactured primarily for transporting persons or property and having a gross vehicle weight of 6000 pounds (2727.6 kg) or less.

Limited-slip differential A differential designed so that when one wheel is slipping, a major portion of the drive torque is supplied to the wheel with the better traction; also called a *nonslip differential.*

Linear measurement A measurement taken in a straight line; for example, the measurement of shaft end play.

Linkage An assembly of rods, or links, used to transmit motion.

Liter (L) In the metric system, a measure of volume; approximately equal to 0.26 gallon (U.S.), or about 61 cubic inches (33.8 fluid ounces, or 1 quart 1.8 ounces); used as a metric measure of engine-cylinder displacement.

Live axle An axle that drives wheels which are rigidly attached to it.

Locknut A second nut turned down on a holding nut to prevent loosening.

Lockwasher A type of washer which, when placed under the head of a bolt or nut, prevents the bolt or nut from working loose.

Lubricant Any material, usually a petroleum product such as grease or oil, that is placed between two moving parts to reduce friction.

Lugging Low-speed, full-throttle engine operation in which the engine is heavily loaded and overworked; usually caused by failure of the driver to shift to a lower gear when necessary.

Machining The process of using a machine to remove metal from a metal part.

Magna-Flux A process in which an electromagnet and a special magnetic powder are used to detect surface and subsurface cracks in iron and steel, which otherwise might not be seen.

Make A distinctive name applied to a group of vehicles produced by one manufacturer; may be further subdivided into car lines, body types, etc.

Malfunction Improper or incorrect operation.

Manufacturer Any person, firm, or corporation engaged in the production or assembly of motor vehicles or other products.

Mass production The manufacture of interchangeable parts and similar products in large quantities.

Master cylinder The liquid-filled cylinder in the hydraulic brake system or clutch, where hydraulic pressure is developed when the driver depresses a foot pedal.

Matter Anything that has weight and occupies space.

Measuring The act of determining the size, capacity, or quantity of an object.

Mechanical advantage In a machine, the ratio of the output force to the input force applied to it.

Mechanical efficiency In an engine, the ratio between brake horsepower and indicated horsepower.

Mechanism A system of interrelated parts that make up a working assembly.

Member Any essential part of a machine or assembly.

Meshing The mating, or engaging, of the teeth of two gears.

Meter (m) A unit of linear measure in the metric system, equal to 39.37 inches. Also, the name given to any test instrument that measures a property of a substance passing through it, as an ammeter measures electric current. Also, any device that measures and controls the flow of a substance passing through it, as a carburetor jet meters fuel flow.

Micrometer A precision measuring device used to measure small bores, diameters, and thicknesses. Also called a *mike*.

Mike Slang term for *micrometer*.

Millimeter (mm) In the metric system, a unit of linear measure, approximately equal to 0.039 inch.

Mode Term used to designate a particular set of operating characteristics.

Model year The production period for new motor vehicles or new engines, designated by the calendar year in which the period ends.

Modification An alteration; a change from the original.

Motor vehicle A vehicle propelled by a means other than muscle power, usually mounted on rubber tires, which does not run on rails or tracks.

mph Abbreviation for *miles per hour,* a measure of speed.

Multiple-disk clutch A clutch with more than one friction disk; usually there are several driving disks and several driven disks, alternately placed.

Neck A portion of a shaft that has a smaller diameter than the rest of the shaft.

Needle bearing An antifriction bearing of the roller type, in which the rollers are very small in diameter (needle-size).

Neoprene A synthetic rubber that is not affected by the various chemicals that are harmful to natural rubber.

Neutral In a transmission, the setting in which all gears are disengaged and the output shaft is disconnected from the drive wheels.

Neutral-start switch A switch wired into the ignition switch to prevent engine cranking unless the transmission shift lever is in neutral or the clutch pedal is depressed.

NHTSA Abbreviation for *National Highway Traffic Safety Administration*.

Nonslip differential See *Limited-slip differential*.

Nut A removable fastener used with a bolt to lock pieces together; made by threading a hole through the center of a piece of metal which has been shaped to a standard size.

Odometer The meter that indicates the total distance a vehicle has traveled, in miles or kilometers; usually located in the speedometer.

OEM Abbreviation for *original-equipment manufacturer*.

Oil A liquid lubricant made from crude oil and used to provide lubrication between moving parts.

Oil cooler A small radiator that lowers the temperature of oil flowing through it.

Oil seal A seal placed around a rotating shaft or other moving part to prevent leakage of oil.

One-way clutch See *Sprag clutch*.

Orifice A small calibrated hole in a line carrying a liquid or gas.

O ring A type of sealing ring, usually made of rubber or a rubberlike material. In use, the O ring is compressed into a groove to provide the sealing action.

Oscillating Moving back and forth, as a swinging pendulum.

Output shaft The main shaft of the transmission; the shaft that delivers torque from the transmission to the driveshaft.

Overcenter spring A spring used in some clutch linkages to reduce the force required to depress the clutch pedal.

Overdrive A device or transmission-gear arrangement which causes the output shaft to overdrive, or rotate faster than the input shaft.

Overflow Spilling of the excess of a substance; also, to run or spill over the sides of a container, usually because of overfilling.

Overhaul To completely disassemble a unit, clean and inspect all parts, reassemble it with the original or new parts, and make all adjustments necessary for proper operation.

Overheat To heat excessively; also, to become excessively hot.

Oxidation Burning or combustion; the combining of a material with oxygen. Rusting is slow oxidation, and combustion is rapid oxidation.

Parallel The quality of two items being the same distance from each other at all points; usually applied to lines and, in automotive work, to machined surfaces.

Part A basic mechanical element or piece—which normally cannot be further disassembled—of an assembly, component, system, or unit. Also applied to any separate entry in a parts catalog, or one that has a "part number."

Particle A very small piece of metal, dirt, or other impurity which may be contained in the air, fuel, or lubricating oil used in an engine.

Passage A small hole or gallery in an assembly or casting, through which air, coolant, fuel, or oil flows.

Passenger car Any four-wheeled motor vehicle manufactured primarily for use on streets and highways and carrying 10 or fewer passengers.

Pawl An arm pivoted so that its free end can fit into a detent, slot, or groove at certain times to hold a part stationary.

Payload The weight of the cargo carried by a truck, not including the weight of the truck body.

Peen To mushroom, or spread, the end of a pin or rivet.

Percent of grade The quotient obtained by dividing the height of a hill by its length; used in computing the power requirements of trucks.

Petroleum The crude oil from which gasoline, lubricating oil, and other such products are refined.

Pilot bearing A small bearing, such as in the center of the flywheel end of the crankshaft, which carries the forward end of the clutch shaft.

Pilot shaft A shaft which is used to align parts and which is removed before final installation of the parts; a dummy shaft.

Pinion gear The smaller of two meshing gears.

Piston A movable part, fitted in a cylinder, which can receive or transmit motion as a result of pressure changes in a fluid.

Piston displacement The cylinder volume displaced by the piston as it moves from the bottom to the top of the cylinder during one complete stroke.

Pitch The number of threads per inch on any threaded part.

Pivot A pin or shaft upon which another part rests or turns.

Planetary-gear system A gear set consisting of a central sun gear surrounded by two or more planet pinions which are, in turn, meshed with a ring (or internal) gear; used in overdrives and automatic transmissions.

Planet carrier In a planetary-gear system, the carrier or bracket that contains the shafts upon which the planet pinions turn.

Planet pinions In a planetary-gear system, the gears that mesh with, and revolve about, the sun gear; they also mesh with the ring (or internal) gear.

Plastic gasket compound A plastic paste which can be squeezed out of a tube to make a gasket in any shape.

Plunger A sliding reciprocating piece driven by an auxiliary power source and having the motion of a ram or piston.

Pounds per horsepower A measure of vehicle performance; the weight of the vehicle divided by the engine horsepower.

Pour point The lowest temperature at which an oil will flow.

Power The rate at which work is done. A common power unit is the horsepower, which is the power necessary to raise 33,000 pounds a distance of 1 foot in 1 minute. In the metric system, the Watt (W) is the unit of power. 1 kilowatt equals 1.34 horsepower (hp).

Power hop The loss of traction at the drive wheels caused by an excessive transfer of power and resulting in wheel bounce.

Power plant The engine or power source of a vehicle.

Power takeoff An attachment for connecting the engine to devices or other machinery when its use is required.

Power team The combination of an engine, transmission, and specific axle ratio.

Power tool A tool having a power source other than muscle power; a tool powered by air or electricity.

Power train The mechanisms that carry the power from the engine crankshaft to the drive wheels; these include the clutch, transmission, drive line, differential, and axles.

Preload In bearings, the amount of load placed on a bearing before actual operating loads are imposed. Proper preloading requires bearing adjustment and ensures alignment and minimum looseness in the system.

Press fit   A fit between two parts so tight that one part has to be pressed into the other, usually with a shop press.

Pressure   Force per unit area, or force divided by area. Usually measured in pounds per square inch (psi) or in kilopascals (kPa) in the metric system.

Pressure plate   That part of the clutch which exerts force against the friction disk; it is mounted on and rotates with the flywheel.

Pressurize   To apply more-than-atmospheric pressure to a gas or liquid.

Preventive maintenance   The systematic inspection of a vehicle to detect and correct failures, either before they occur or before they develop into major defects. A procedure for maintaining a vehicle in a satisfactory and dependable operating condition.

Propeller shaft   See *Driveshaft*.

Prussian blue   A blue pigment; in solution, useful in determining the area of contact between two surfaces.

psi   Abbreviation for *pounds per square inch*, a measurement of pressure.

psig   Abbreviation for *pounds per square inch gauge*, a measurement of pressure.

Pull   The result of an unbalanced condition. For example, uneven braking at the front brakes or unequal front-wheel alignment will cause a car to swerve (pull) to one side when the brakes are applied.

Puller   Generally, a shop tool used to separate two closely fitted parts without damage. Often contains a screw, or several screws, which can be turned to apply a gradual force.

Pulley   A metal wheel with a V-shaped groove around the rim; drives or is driven by a belt.

Pump   A device that transfers gas or liquid from one place to another.

Quadrant   A term sometimes used to identify the shift-lever selector mounted on the steering column.

Races   The metal rings on which ball or roller bearings rotate.

Ratio   The relationship in size or quantity of two or more objects. A gear ratio is derived by dividing the number of teeth on the driven gear by the number of teeth on the driving gear.

Reamer   A round metal-cutting tool with a series of sharp cutting edges; enlarges a hole when turned inside it.

Rear-end torque   The reaction torque that acts on the rear-axle housing when torque is applied to the wheels; tends to turn the axle housing in the direction opposite to wheel rotation.

Reassembly   Putting back together the parts of a device.

Reciprocating motion   Motion of an object between two limiting positions; motion in a straight line either back and forth or up and down.

Release bearing   See *Throwout bearing*.

Release fingers   See *Release levers*.

Release levers   In the clutch, levers that are moved by throwout-bearing movement, causing clutch-spring force to be relieved so that the clutch is disengaged, or uncoupled from the flywheel.

Remove and reinstall (R and R)   To perform a series of servicing procedures on an original part or assembly; includes removal, inspection, lubrication, all necessary adjustments, and reinstallation.

Replace   To remove a used part or assembly and install a new part or assembly in its place; includes cleaning, lubricating, and adjusting as required.

Retaining ring   A removable fastener used as a shoulder to retain and position a round bearing in a hole.

Return spring   A "pull-back" spring, often used in brake systems.

Reverse idler gear   In a transmission, an additional gear that must be meshed to obtain reverse gear; a gear used only in reverse that does not transmit power when the transmission is in any other position.

Ring gear   A large gear carried by the differential case; meshes with and is driven by the drive pinion.

Roadability   The steering and handling qualities of a vehicle while it is being driven on the road.

Road load   The power required to hold a constant vehicle speed on a level road.

Room temperature   68 to 72°F [20 to 22°C].

Rotary   Term describing the motion of a part that rotates, or turns.

rpm   Abbreviation for *revolutions per minute*, a measure of rotational speed.

RTV sealer   Room-temperature vulcanizing gasket material, which cures at room temperature; a plastic paste squeezed from a tube to form a gasket of any shape.

Runout   Wobble; said of an assembly that does not run true about its centerline.

SAE   Abbreviation for *Society of Automotive Engineers;* used to indicate a grade or weight of oil measured according to Society of Automotive Engineers standards.

Safety   Freedom from injury or danger.

Safety stand   A pinned or locked type of stand placed under a car to support its weight after the car has been raised with a lift or floor jack. Also called a *car stand* or *jack stand*.

Schematic   A pictorial representation, most often in the form of a line drawing. A systematic positioning of components, showing their relationship to each other or to an overall function.

Scored   Scratched or grooved, as a cylinder wall may be scored by abrasive particles moved up and down by the piston rings.

Scraper   A device used in engine service to scrape carbon from the engine block, pistons, or other parts.

Screens   Pieces of fine-mesh metal fabric; used to prevent solid particles from circulating through any liquid or vapor system and damaging vital moving parts.

Screw   A metal fastener with threads that can be turned into a threaded hole. There are many different types and sizes of screws.

Scuffing   A type of wear in which there is a transfer of material between parts moving against each other; shows up as pits or grooves in the mating surfaces.

Seal   A material, shaped around a shaft, used to close off the operating compartment of the shaft, preventing oil leakage.

Sealer   A thick, tacky compound, usually spread with a brush, which may be used as a gasket or sealant to seal small openings or surface irregularities.

Seat   The surface upon which another part rests, as a valve seat. Also, to wear into a good fit; for example, new piston rings seat after a few miles of driving.

Sediment   The accumulation of matter which settles to the bottom of a liquid.

Self-locking screw   A screw that locks itself in place without the use of a separate nut or lockwasher.

Self-tapping screw   A screw that cuts its own threads as it is turned into an unthreaded hole.

Semifloating rear axle   An axle that supports the weight of the vehicle on the axle shaft in addition to transmitting driving forces to the rear wheels.

Serviceable   A term describing parts or systems that can be repaired and maintained to continue in operation.

Service manual   A book published annually by each vehicle manufacturer, listing the specifications and service procedures for each make and model of vehicle. Also called *shop manual*.

Servo   A device in a hydraulic system that converts hydraulic pressure to mechanical movement; consists of a piston which moves in a cylinder as hydraulic pressure acts on it.

Setscrew   A type of metal fastener that holds a collar or gear on a shaft when its point is turned down into the shaft.

Shackle   The swinging support by which one end of a leaf spring is attached to the car frame.

Shift lever   The lever used to change gears in a transmission. Also, the lever on the starting motor which moves the drive pinion into or out of mesh with the flywheel teeth.

Shim   A slotted strip of metal used as a spacer to adjust the front-end alignment on many cars; also used to make small corrections in the position of body sheet metal and other parts.

Shimmy   Rapid oscillation. In wheel shimmy, for example, the front wheel turns in and out alternately and rapidly; this causes the front end of the car to oscillate, or shimmy.

Shim stock   Sheets of metal of accurately known thickness which can be cut into strips and used to measure or correct clearances.

Shock absorber   A device placed at each vehicle wheel to regulate spring rebound and compression.

Shop layout   The location of aisles, work areas, machine tools, etc., in a shop.

Shrink fit   A tight fit of one part into another, achieved by heating or cooling one part and then assembling it to the other part. A heated part will shrink on cooling to provide the tight fit; a cooled part will expand on warming to provide the tight fit.

Side clearance   The clearance between the sides of moving parts when the sides do not serve as load-carrying surfaces.

Slip joint   In the power train, a variable-length connection that permits the driveshaft to change its effective length.

Snap ring   A metal fastener, available in two types: the external snap ring fits into a groove in a shaft; the internal snap ring fits into a groove in a housing. Snap rings often must be installed and removed with special snap-ring pliers.

Solvent   A petroleum product of low volatility used in the cleaning of engine and vehicle parts.

Solvent tank   In the shop, a tank of cleaning fluid in which most parts are brushed and washed clean.

Specifications   Information provided by the manufacturer that describes each automotive system and its components, operation, and clearances. The service procedures that must be followed for a system and its components, operation, and clearances. Also, the service procedures that must be followed for a system to operate properly.

Specs   Short for *specifications*.

Speed   The rate of motion; for vehicles, measured in miles per hour or kilometers per hour.

Speedometer   An instrument that indicates vehicle speed; usually driven from the transmission.

Speed shift Performance term for shifting gears in a manual transmission without releasing the accelerator.

Splines Slots or grooves cut in a shaft or bore. Splines on a shaft are matched to splines in a bore to ensure that two parts turn together.

Sprag clutch A one-way clutch. It can transmit power in one direction, but not in the other.

Spring A device that changes shape when it is stretched or compressed, but returns to its original shape when the force is removed; the component of the automotive suspension system that absorbs road shocks by flexing and twisting.

Spring shackle See *Shackle*.

Sprung weight That part of the car which is supported on springs (includes the engine, frame, and body).

Spur gear A gear in which the teeth are parallel to the center line of the gear.

Squeak A high-pitched noise of short duration.

Squeal A continuous high-pitched noise.

Stabilizer bar An interconnecting shaft between the two lower suspension arms; reduces body roll on turns.

Stalls The condition in which an engine quits running, at idle or while driving.

Static balance The balance of an object while it is not moving.

Static friction The friction between two bodies at rest.

Steam cleaner A machine used for cleaning large parts with a spray of steam, often mixed with soap.

Steering-and-ignition lock A device that locks the ignition switch in the OFF position and locks the steering wheel so that it cannot be turned.

Steering-column shift An arrangement in which the transmission shift lever is mounted on the steering column.

Stud A headless bolt that is threaded on both ends.

Stud extractor A special tool used to remove a broken stud or bolt.

Substance Any matter or material; may be a solid, a liquid, or a gas.

Sun gear In a planetary-gear system, the center gear that meshes with the planet pinions.

Surface grinder A grinder used to resurface flat surfaces, such as cylinder heads.

Suspension system The springs and other parts which support the upper part of a vehicle on its axles and wheels.

Sway bar See *Stabilizer bar*.

Synchromesh transmission A transmission with a built-in device that automatically matches the rotating speeds of transmission gears as they are about to mesh, thereby eliminating the need for "double-clutching."

Synchronize To make two or more events or operations occur at the same time or at the same speed.

Synchronizer A device in the transmission that synchronizes gears about to be meshed so that there will not be any gear clash.

Synthetic oil An artificial oil that is manufactured; not a natural mineral oil made from petroleum.

System A combination or grouping of two or more parts or components into a whole which in operation performs some function that cannot be done by the separate parts.

Tachometer A device for measuring the speed of an engine in revolutions per minute (rpm).

Tap A tool used for cutting threads in a hole.

Taper A gradual reduction in the width of a shaft or hole. Also, in a cylinder, uneven wear, more pronounced at one end that at the other.

Technology The applications of science.

Temperature The measure of heat intensity in degrees. Temperature is not a measure of heat quantity.

Thickness guage Strips of metal made to an exact thickness, used to measure clearances between parts.

Thread chaser A device, similar to a die, that is used to clean threads.

Thread class A designation indicating the closeness of fit between a pair of threaded parts, such as a nut and bolt.

Threaded insert A threaded coil that is used to restore the original thread size to a hole with damaged threads. The hole is drilled oversize and tapped, and the insert is threaded into the tapped hole.

Thread series A designation indicating the pitch, or number of threads per inch, on a threaded part.

Throwout bearing In the clutch, the bearing that can be moved inward to the release levers by clutch-pedal action to cause declutching, which disengages the engine crankshaft from the transmission.

Thrust bearing In the engine, the main bearing that has thrust faces to prevent excessive end play, or forward and backward movement of the crankshaft.

Torque Turning or twisting effort; usually measured in pound-feet or newton-meters. Also, a turning force such as that required to tighten a connection.

Torque wrench A wrench that indicates the amount of torque, or turning force, being applied with the wrench.

Torsional vibration Rotary vibration that causes a twist-untwist action on a rotating shaft, so that a part of the shaft repeatedly moves ahead of, or lags behind, the remainder of the shaft; for example, the action of a crankshaft responding to the cylinder firing impulses.

Torsion-bar spring A long, straight bar that is fastened to the vehicle frame at one end and to a suspension part at the other. Spring action is produced by a twisting of the bar.

Tracking Rear wheels following the front wheels in a parallel path when the vehicle is moving straight ahead.

Tractive effort The force available at the road surface in contact with the driving wheels of a truck. It is determined by engine torque, transmission ratio, axle ratio, tire size, and frictional losses in the drive line. Rim pull is also known as tractive effort.

Tramp Up-and-down motion (hopping) of the front wheels at higher speeds, due to unbalanced wheels or excessive wheel runout. Also called *high-speed shimmy*.

Transaxle A power transmission device that combines the functions of the transmission and the drive axle (differential) into a single assembly; used in front-wheel-drive cars with front-mounted engines and in rear-wheel-drive cars with rear-mounted engines.

Transducer Any device which converts an input signal of one form into an output signal of a different form. For example, the automobile horn converts an electric current into sound.

Transfer case An auxiliary transmission mounted behind the main transmission. Used to divide engine power and transfer it to both front and rear differentials, either full-time or part-time.

Transmission An assembly that transmits power from the engine to the driving axle. It provides the different forward-gear ratios, neutral, and reverse through which engine power is transmitted to the differential.

Transmission-oil cooler A small, sometimes finned, tube or tank, either mounted separately or included as part of the engine radiator, which cools the transmission fluid.

Trouble diagnosis The detective work necessary to find the cause of a trouble.

Truck Any motor vehicle primarily designed for the transportation of property, which carries the load on its own wheels.

Truck tractor Any motor vehicle designed primarily for pulling truck trailers and constructed so as to carry part of the weight and load of a semi-trailer.

Two-disk clutch A clutch with two friction disks for additional holding power; used in heavy-duty equipment.

Two-speed rear axle See *Double-reduction differential.*

U-bolt An iron rod with threads on both ends, bent into the shape of a U and fitted with a nut at each end.

Unit An assembly or device that can perform its function only if it is not further divided into its component parts.

Universal joint In the drive line, a connecting joint that can transmit torque between shafts while the drive angle varies.

Unsprung weight The weight of that part of the car which is not supported on springs; for example, the wheels and tires.

Upshift To shift a transmission into a higher gear.

Vacuum A pressure less than atmospheric pressure; a negative pressure. Vacuum can be measured in pounds per square inch, but is usually measured in inches or millimeters of mercury (Hg); a reading of 30 inches [762 mm] Hg would indicate a perfect vacuum.

Vacuum gauge In automotive-engine service, a device that measures intake-manifold vacuum and thereby indicates actions of engine components.

Valve Any device that can be opened or closed to allow or stop the flow of a liquid or gas. There are many different types.

Vaporization A change of state from liquid to vapor or gas by evaporation or boiling; a general term including both evaporation and boiling. In the carburetor, vaporization refers to breaking gasoline into fine particles and mixing it with incoming air.

V block A metal block with an accurately machined V-shaped groove; used to support an armature or shaft while it is checked for roundness.

Vehicle See *Motor vehicle.*

Vehicle identification number (VIN) The number assigned to each vehicle by its manufacturer, primarily for registration and identification purposes.

Ventilation The circulating of fresh air through any space, to replace impure air; the basis of crankcase ventilation systems.

VIN Abbreviation for *vehicle identification number.*

Viscosity The resistance to flow exhibited by a liquid. A thick oil has greater viscosity than a thin oil.

Viscous Thick; tending to resist flowing.

Viscous friction The friction between layers of a liquid.

Volatility A measure of the ease with which a liquid vaporizes; has a direct relationship to the flammability of a fuel.

Vulcanizing A process of treating raw rubber with heat and pressure; the treatment forms the rubber and gives it toughness and flexibility.

Weight distribution The percentage of a vehicle's total weight that rests on each axle.

Weight, sprung See *Sprung weight.*

Weight, unsprung See *Unsprung weight.*

Welding The process of joining pieces of metal by fusing them together with heat.

Wet-disk clutch A clutch in which the friction disk (or disks) is operated in a bath of oil.

Wheel A disk or spokes with a hub at the center which revolves around an axle, and a rim around the outside for mounting the tire on.

Wheel balancer A device that checks a wheel-and-tire assembly (statically, dynamically, or both) for balance.

Wheelbase The distance between the center lines of the front and rear axles. For trucks with tandem rear axles, the rear center line is considered to be midway between the two rear axles.

Wheel tramp The tendency of a wheel to move up and down so that it repeatedly bears down hard, or "tramps," on the road. Sometimes called *high-speed shimmy.*

Wire thickness gauge A set of round wires of known diameters; often used to check clearances between electric contacts, such as distributor points and spark-plug electrodes.

Work The changing of the position of an object against an opposing force; measured in foot-pounds or meters-newton. The product of a force and the distance through which the force acts.

Worm A type of gear in which the teeth resemble threads; used on the lower end of the steering shaft.

WOT Abbreviation for *wide-open throttle.*

Yoke In a universal joint, the drivable torque-and-motion input and output member, attached to a shaft or tube.

INDEX

A

Asbestos, health cautions about, 36
Automatic transmission, 1–2
Automatic-transmission fluid, 78
Automobile, components of, 1
Axle housing, 238–239, 248
  intallation of, 248
  removal of, 238–239
Axle shafts, rear wheel, 250–251
Axles:
  drive (*see* Drive axle)
  rear (*see* Rear axle)

B

Backlash, adjusting of, 248
Backup lights, transmission control of, 80
Balancing, driveshaft, 206–209
Ball-and-trunnion universal joint, 194
Ball stud, clutch-fork pivot, 20
Bearings, 19–21, 96–97, 240, 251–253
  inspection of, 96–97
  noise in, 240
  pilot, 19
  servicing rear-axle, 251–253
  throwout, 20–21
Belleville spring clutch, 22–23
  (*See also* Clutch)
Bevel gears, 5
Birfield universal joint, 202–203
  servicing of, 217–219
Borg and Beck clutch, 20–22
  (*See also* Clutch)
Bushing:
  clutch pilot, 86
  extension housing, 98

C

Cardan universal joint, 195–196
  (*See also* Universal joint)
Cleveland type universal joint, servicing,
  212–213
Cluster gear, 6
  (*See also* Transmission)
Clutch, 1–2, 15–30, 32–46, 86
  asbestos in, 36
  ball stud in, 20
  Belleville spring, 22–23
  Borg and Beck, 20–22
  coil-spring, 20–22
  construction of, 17–19
  diaphragm-spring, 22–23
  double-disk, 23
  fork in, 20
  friction disk in, 19–20
  functions of, 2
  housing alignment, 43, 46
  hydraulic linkage for, 27
    troubles in, 36

Clutch (*Cont.*)
  inspection and repair of, 43–46
  linkage for, 20, 27, 36–41
  Long, 22
  noise in, 35
  pilot bearing in, 19
  pilot bushing in, 86
  pressure plate in, 19–27
  purpose of, 15–16
  removal and replacement of, 41–43
  safety switch, 27–30
  self-adjusting, 30
  semicentrifugal, 22
  servicing of, 32–46
  torsional-vibration damping in, 20
  for transaxle, 23–27
  trouble diagnosis of, 32–36
  types of, 16–17
  wet, 17
Column-shift linkage, 51–54
Constant-velocity universal joints, 194–197,
  202–203, 213–214
  servicing of, 213–214
Crankshaft, pilot bearing in, 19
Cross-and-two yoke universal joint, 194

D

Diaphragm-spring clutch, 22–23
  (*See also* Clutch)
Differential, 11–12, 223–231, 238–269
  adjusting backlash in, 248
  construction of, 224
  double-reduction, 227, 230
  function of, 223
  gear-tooth-contact patterns in, 243–248
  gearing in, 227–228
    double-reduction, 230
    (*See also* Gear ratio; Gears)
  integral-carrier, 241–246, 248–249,
    253–256, 258–262
    setting pinion depth for, 250–251,
      260–261
  limited-slip, 230–231, 240–241,
    263–269
    servicing clutch-pack type of,
      263–267
    servicing cone type of, 267–269
    types of, 241
  noise in, 239–240
  operation of, 224–227
  overhauling case for, 148–150
  pre-repair diagnosis of, 241
  removable-carrier, 241, 244–245, 250,
    253, 256–258, 262–263
    servicing of, 262–263
  servicing of, 238–269
  trouble diagnosis of, 238–240
Double-Cardan universal joint, 195–196
Double-disk clutch, 23
Double-offset universal joint, 203

Double-reduction differential, 227, 230
Downshift, 49
Drive axle, 223–233, 239–269
  differential in, 223
  four-wheel drive, 231–233
  locking hubs for, 233
  servicing of, 238–269
  (*See also* Front-wheel drive; Rear axle)
Drive line, 189–190, 199–204, 206–222
  servicing of, 206–222
  (*See also* Driveshaft; Slip joint; Universal joint)
Driveshaft, 9–10, 12, 190–191, 206–222
  balancing of, 206–209
  checking runout of, 209–210
  function of, 190–191
  installation of, 214
  servicing of, 211–215
  servicing front-drive, 217–222
  trouble diagnosis of, 216
Dual-range transaxle, 69, 137, 152–172
  servicing of, 152–172
  servicing of range-selector control for, 137
Dynamometer, power train, 12–13

E

Extension housing, transmission, 98

F

Final drive (*see* Differential)
Five-speed transmission, 67, 116–125
  servicing of, 116–125
Floor shift, 6, 51–54
Fluid coupling, 15
Fork, clutch, 20
  (*See also* Clutch)
Four-speed transmission, 62–67, 102–106, 108–115
  overdrive type of, 65–67
    servicing of, 108–115
  servicing of, 102–106
Four-wheel drive, 11, 49–50, 190, 231–233
  drive axles for, 223–233
  locking hubs for, 233
Friction disk, 19–20
  (*See also* Clutch)
Front-end torque, 203
Front-engine:
  four-wheel drive, 190
  front-wheel drive, 189
  rear-wheel drive, 189
Front-wheel drive, 9, 11, 189–202, 215–222
  driveshafts for, 201–202
    servicing, 215–222
  front-engine, 189
Full-time transfer case, 72

G

Gear ratio, 3–5
Gear sets:
  hunting and nonhunting, 228, 230
  planetary, 72–77
Gear-tooth-contact patterns, 241–248
  adjusting, 243–248
  reading of, 243
Gearing, differential, 227–228
  double-reduction, 227, 230
  (*See also* Gear ratio; Gears)
Gears, 2–9, 58, 77–78, 227, 241
  cluster, 6
  lubricants for, 77–78
    transaxle leakage of, 130
  pinion, 3
  radius of, 4
  reverse idler, 9, 58
  tooth nomenclature, 228, 241
  torque in, 4
  types of, 5
Gearshift linkage, 6, 51–54, 64
  floor shift type of, 6, 51–54

H

Helical gears, 5, 9
Hubs, locking, 233
Hunting gear sets, 228, 230
Hydraulic clutch linkage, 27
  troubles in, 36

I

Integral-carrier differential, 241–246, 253–256, 258–262
  setting pinion depth for, 245–246, 260–261
Interlock, transmission, 58

L

Limited-slip differential, 230–231, 240–241, 263–269
  servicing clutch-pack type of, 263–267
  servicing cone type of, 267–269
  types of, 231
Linkage:
  clutch, 20, 27, 36–41
    hydraulic types of, 27
      troubles in, 36
  gearshift, 6, 51–54, 64
    floor shift type of, 6, 51–54
Locking hubs, 233
  (*See also* Four-wheel drive)
Long clutch, 22
  (*See also* Clutch)
Lubricants, gear, 77–78, 130
  transaxle leakage of, 130

M

Manual transaxle (*see* Transaxle)
Manual transmission (*see* Transmission)
Mounts, transaxle, 130

N

Noise, bearing, 240
  clutch, 35
  differential, 238–240
  rear-axle, 238–240
  transaxle, 126–129
  transmission, 85–87
Nonhunting gear sets, 223, 228, 230

O

Oil leaks:
  transaxle, 130
  transmission, 87
Oil seal:
  extension housing, 98
  replacing differential pinion, 253–258
  transmission drive-gear-bearing, 98
Overdrive transmission, 5, 64–67
  (*See also* Five-speed transmission; Four-speed transmission)

P

Part-time transfer case, 72–74
Pilot bearing, clutch, 19
Pilot bushing, clutch, 86
Pinion gear, 3
  (*See also* Differential)
Planetary gears:
  operation of, 74–77
  in transfer case, 72–74
Power flow, 6
Power train, components of, 1–13
Pressure plate, clutch, 19–27
Propeller shaft (*see* Drive line; Driveshaft)

R

Ratio:
  gear, 3
  torque, 4
Rear axle, 9–10, 238–241, 248–253
  installation of, 248–249
  noise in, 238–240
  removal of, 250–251
  servicing of, 240–241, 251–253
  trouble diagnosis of, 238–240
  wheel axles in, 250–251
  (*See also* Differential)
Rear-end torque, 197–201
Rear-engine:
  four-wheel drive, 190
  rear-wheel drive, 11, 189–190

Rear-suspension system, 197–201
Rear-wheel drive, 9–11, 189–190
  front-engine, 189
  rear-engine, 11, 189–190
Removable-carrier differential, 241, 244–245, 248, 253, 256–258, 262–263
  servicing of, 262–263
Reverse idler gear, 9, 58
Runout, driveshaft, 209–210
Rzeppa universal joint, 196, 202, 203

S

Safety switch, clutch, 27–30
Self-adjusting clutch, 30
Semicentrifugal clutch, 22
Sliding gear transmission, 9
Slip joint, 10, 197
Speedometer drive, 80
Spur gear, 5
Steering-column shift, 51–54
Synchromesh transmission, 9
  (*See also* Five-speed transmission; Four-speed transmission)
Synchronizers, transmission, 9, 58–62, 86, 97, 142–143
  servicing of, 86, 97, 142–143
  trouble diagnosis of, 86

T

Three-ball-and-trunnion universal joint, 196–197
Three-speed transmission, 56–58, 93–99
  servicing of, 93–99
Throwout bearing, 20–21
Torque, 2–5, 197–201, 203
  calculation of, 3–5
  front-end, 203
  ratio in gears, 4
  rear-end, 197–201
Torque converter, 15
Torsional vibration in clutch, 20
Transaxle, 11, 23–27, 49, 67–69, 126–137, 152–172
  adjusting shift cables for, 133
  checking lubricant level in, 130
  clutch for, 23–27
  dual-range, 69
    servicing of, 152–172
    servicing of range-selector control for, 137
  lubricant leaks from, 130
  mounts for, 130
  noises in, 126–129
  removal and installation of, 131–137
  servicing of, 135–137, 139–150
  trouble diagnosis of, 126–130
Transfer case, 11, 49–50, 69–74, 173–188
  adjusting linkage for, 175–177
  full-time, 72

Transfer case (*Cont.*)
  installation of, 173
  lubrication of, 173–175
  part-time, 72–74
  removal of, 173
  servicing of, 178–188
  servicing lockout-clutch in, 183–184
  servicing range box in, 184–185
  trouble diagnosis of, 173
  types of, 69–72
Transmission, 2–9, 48–67, 79–80, 82–91,
    93–104, 106–123
  automatic, 1–2
  backup-light switch on, 80
  countershaft in, 6
  designations of, 54–56
  five-speed, 67, 116–125
    servicing of, 117–125
  four-speed, 62–67, 102–106, 108–115
    overdrive type of, 65–67
      servicing of, 108–115
    servicing of, 102–106
  function of, 48–49
  gears in, 56–58
  installation of, 90
  interlock in, 58
  oil leaks from, 87
  operation of, 5
  overdrive, 5, 64–67
    (*See also* Five-speed transmission;
      Four-speed transmission)
  power flow through, 6
  purpose of, 48
  removal of, 88–90
  shift-linkage adjustment, 91
  sliding gear, 9
  speedometer drive in, 80
  synchromesh, 9
  synchronizers in, 9, 58–62, 86, 97,
    142–143

Transmission (*Cont.*)
  servicing of, 86, 97, 142–143
  trouble diagnosis of, 86
  TCS switch on, 79–80
  three-speed, 56–58, 93–99
    servicing of, 93–99
  trouble diagnosis of, 82–87
  types of, 49
Transmission-controlled-spark switch, 79–80
Tripot universal joint, 196–197
Two-wheel drive, 11

U

Universal joint, 9–10, 191–197, 201–203,
    210–214
  ball-and-trunnion, 196–197
  Birfield, 201–202
    servicing of, 217–219
  checking angles of, 210–211
  Cleveland, servicing of, 212–213
  constant-velocity, 194–196, 202–203
    servicing of, 213–214
  cross-and-two yoke, 194
  double-Cardan, 195–196
  double-offset, 202–203
  Rzeppa, 196, 202, 203
  servicing of, 206, 212–214
  three-ball-and-trunnion, 196–197
  tripot, 196–197

V

Vibration, torsional, in clutch, 20

Y

Yoke, clutch, 20

ANSWERS TO REVIEW QUESTIONS

The answers to the chapter review questions are given here. If you want to figure your grade on any quiz, divide the number of questions in the quiz into 100. This gives you the value of each question. For instance, suppose there are 10 questions: 10 goes into a hundred 10 times. Each correct answer, therefore, gives you 10 points. If you answered 8 correct out of the 10, then your grade would be 80 (8 × 10).

If you are not satisfied with the grade you make on a test, restudy the chapter and retake the test. This review will help you remember the important facts.

Remember, when you take a course in school, you can pass and graduate even though you make a grade of less than 100. But in the automotive shop, you must score 100 percent all the time. If you make 1 error out of 100 service jobs, for example, your average would be 99. In school that is a fine average. But in the automotive shop that one job you erred on could cause such serious trouble (a ruined engine or a wrecked car) that it would outweigh all the good jobs you performed. Therefore, always proceed carefully in performing any service job and make sure you know exactly what you are supposed to do and how you are to do it.

Chapter 1

1. (b) 2. (a) 3. (c) 4. (d) 5. (b)
6. (c) 7. (a) 8. (c) 9. (b) 10. (d)
11. (a) 12. (a) 13. (d) 14. (c) 15. (c)

Chapter 2

1. (a) 2. (c) 3. (a) 4. (c) 5. (b)
6. (c) 7. (d) 8. (b) 9. (a) 10. (c)

Chapter 3

1. (d) 2. (c) 3. (a) 4. (c) 5. (b)
6. (b) 7. (a) 8. (c) 9. (c) 10. (b)

Chapter 4

1. (a) 2. (d) 3. (b) 4. (b) 5. (b)
6. (c) 7. (c) 8. (b) 9. (d) 10. (b)
11. (d) 12. (b) 13. (c) 14. (d) 15. (d)
16. (b) 17. (a) 18. (d) 19. (b) 20. (a)

Chapter 5

1. (d) 2. (d) 3. (b) 4. (d) 5. (c)
6. (b) 7. (d) 8. (a) 9. (c) 10. (b)

Chapter 6

1. (c) 2. (b) 3. (c) 4. (a) 5. (d)

Chapter 7

1. (c) 2. (d) 3. (a) 4. (c) 5. (c)
6. (a) 7. (c) 8. (b)

Chapter 8

1. (b) 2. (d) 3. (a) 4. (c) 5. (b)

Chapter 9

1. (d) 2. (b) 3. (c) 4. (b) 5. (a)

Chapter 10

1. (b) 2. (d) 3. (c) 4. (d) 5. (c)

Chapter 11

1. (d) 2. (d) 3. (a) 4. (d) 5. (b)
6. (c) 7. (d) 8. (c)

Chapter 12

1. (b) 2. (c) 3. (b) 4. (d) 5. (c)

Chapter 13

1. (c) 2. (b) 3. (b) 4. (c) 5. (a)
6. (a) 7. (d) 8. (c) 9. (b) 10. (c)

Chapter 14

1. (c) 2. (d) 3. (a) 4. (b) 5. (c)
6. (b) 7. (c) 8. (b)

Chapter 15

1. (d) 2. (b) 3. (d) 4. (b) 5. (d)

Chapter 16

1. (b) 2. (a) 3. (d) 4. (c) 5. (b)

Chapter 17

1. (b) 2. (d) 3. (c) 4. (a) 5. (d)
6. (c) 7. (a) 8. (b) 9. (a) 10. (d)

Chapter 18

1. (c) 2. (d) 3. (b) 4. (a) 5. (b)
6. (c) 7. (d) 8. (a) 9. (b) 10. (a)

Chapter 19

1. (b) 2. (a) 3. (c) 4. (c) 5. (a)
6. (b) 7. (a) 8. (d) 9. (b) 10. (c)

Chapter 20

1. (b) 2. (a) 3. (b) 4. (a) 5. (b)
6. (b) 7. (a) 8. (d) 9. (d) 10. (c)
11. (b) 12. (d) 13. (c) 14. (d) 15. (c)
16. (b) 17. (a) 18. (b) 19. (c) 20. (a)